Sustainable Development Goals Series

The **Sustainable Development Goals Series** is Springer Nature's inaugural cross-imprint book series that addresses and supports the United Nations' seventeen Sustainable Development Goals. The series fosters comprehensive research focused on these global targets and endeavours to address some of society's greatest grand challenges. The SDGs are inherently multidisciplinary, and they bring people working across different fields together and working towards a common goal. In this spirit, the Sustainable Development Goals series is the first at Springer Nature to publish books under both the Springer and Palgrave Macmillan imprints, bringing the strengths of our imprints together.

The Sustainable Development Goals Series is organized into eighteen subseries: one subseries based around each of the seventeen respective Sustainable Development Goals, and an eighteenth subseries, "Connecting the Goals," which serves as a home for volumes addressing multiple goals or studying the SDGs as a whole. Each subseries is guided by an expert Subseries Advisor with years or decades of experience studying and addressing core components of their respective Goal.

The SDG Series has a remit as broad as the SDGs themselves, and contributions are welcome from scientists, academics, policymakers, and researchers working in fields related to any of the seventeen goals. If you are interested in contributing a monograph or curated volume to the series, please contact the Publishers: Zachary Romano [Springer; zachary.romano@springer.com] and Rachael Ballard [Palgrave Macmillan; rachael.ballard@palgrave.com].

Venkatesh Dutta • Priyanka Ghosh
Editors

Sustainability: Science, Policy, and Practice in India

Challenges and Opportunities

Springer

Editors
Venkatesh Dutta
School of Earth and Environmental Sciences (SEES)
Babasaheb Bhimrao Ambedkar University
Lucknow, India

Priyanka Ghosh
VIT-AP School of Social Sciences and Humanities (VISH)
VIT-AP University
Amaravati, Andhra Pradesh, India

ISSN 2523-3084 ISSN 2523-3092 (electronic)
Sustainable Development Goals Series
ISBN 978-3-031-50131-9 ISBN 978-3-031-50132-6 (eBook)
https://doi.org/10.1007/978-3-031-50132-6

This Springer imprint is published by the registered company Springer Nature Switzerland AG
The registered company address is: Gewerbestrasse 11, 6330 Cham, Switzerland

Paper in this product is recyclable.

Preface

Sustainability in practice creates solutions that could balance *economic*, *social*, and *environmental* priorities. The "triple bottom line" of sustainability refers to these three pillars. Economic sustainability refers to the ability of an economy to support growth and development without depleting natural resources or causing long-term damage to the environment. Social sustainability refers to the ability of a society to meet the needs of its citizens and ensure that everyone has access to basic human rights and opportunities. Environmental sustainability refers to the ability of the planet to support life, and it includes efforts to reduce pollution, protect biodiversity, and mitigate the effects of climate change.

The Sustainable Development Goals (SDGs) are a set of 17 goals adopted by the United Nations in 2015, with the aim of ending poverty, protecting the planet, improving the quality of life, and ensuring that all people enjoy peace and prosperity by 2030. There has been progress made in achieving some of the SDGs, but progress has been uneven across different goals and regions. Overall, while some significant improvement has been made toward achieving some of the SDGs, there is still much work to be done.

There are numerous important gaps that must be filled in order to mainstream sustainable development goals in our policies, plans, and programs. This requires tackling a complex set of interrelated social, economic, and environmental problems. Firstly, there is a paucity of solid scientific information and data to guide efforts toward developing standard benchmarks. Lack of data on critical variables, such as biodiversity loss, climate change impacts, and sustainable consumption, etc., might be a contributing factor.

Secondly, there are policy gaps, meaning there aren't enough rules or policies in place to properly promote sustainable growth. Among these concerns is the absence of political will to enact required laws and regulations, as well as a lack of coordination and integration across various sectors and levels of the government. Thirdly, sustainable development policies and practices are not being effectively implemented or enforced. Lack of capacity and resources to undertake sustainable projects and a lack of involvement and engagement from important players, including the corporate sectors and civil societies, are examples.

Deficiencies in institutional arrangements and governance systems prevent sustainable development from being tackled head-on. Lack of efficient monitoring and evaluation methods to track progress and identify areas for improvement can be a contributing factor, as can a lack of coordination and collaboration across different levels of government, a lack of involvement and engagement from important stakeholders, and so on.

Finally, the absence of sufficient funding and resources to promote large sustainable development projects is a major hurdle. This might be due to a lack of adequate financial instruments to promote sustainable growth, a lack of investment in sustainable infrastructure and technologies, or both.

India has made significant progress in achieving the SDGs, but there is still a long way to go. The country needs to continue to focus on addressing the issue of inequality and promoting sustainable development. The government and private sectors should work together to mobilize resources, build capacity, and create an enabling environment to achieve the SDGs. The civil society should also be engaged to ensure that the SDGs are inclusive and responsive to the needs of all citizens. In order to address these critical gaps, a multidisciplinary and holistic approach is required. This approach must involve collaboration between different levels of government, and the private sectors, with cooperation from civil societies, and individuals. Additionally, it must incorporate scientific, economic, and social perspectives.

This book reveals the contemporary sustainability narratives in India within the context of 17 SDGs, including the eradication of poverty and hunger, the promotion of sustained economic growth, the empowerment of women and children, the reduction of inequalities between regions and states, the promotion of gender equality, and the promotion of the protection, conservation, and restoration of natural resources.

The book analyzes the existing difficulties and potential benefits of achieving SDGs in India against the backdrop of the UN global sustainability framework. Each chapter of this book delves further into a particular component of SDGs, evaluating the current state and progress achieved by India and giving empirical case studies from various parts of the country. Several case studies are presented, each of which sheds light on the challenges and constraints that are impeding the process of achieving SDGs and investigates potential long-term answers to these issues.

The following six thematic areas are covered in this book: Poverty, Gender, and Health; Quality Education; Clean Water and Energy; Economic Growth, Sustainable Cities, and Communities; Land and Water; Global Partnership and Sustainable Development: A Way Forward.

The book consolidates the research work of international and national scholars, academicians, and practitioners in the fields of sustainability, environmental studies, environmental science, development studies, gender, women studies, public health, and epidemiology.

The volume will immensely benefit scientists, researchers, policymakers, and practitioners as it offers a thematic and comprehensive understanding of challenges associated with mainstreaming SDGs at national, subnational, and microscales in India.

Lucknow, India — Venkatesh Dutta

Amaravati, India — Priyanka Ghosh

Contents

List of Figures

List of Tables

Sustainability as the Development Paradigm: Evolving Frames and Interpretations

1

Venkatesh Dutta and Priyanka Ghosh

Abstract

The concept of sustainability has evolved over time to encompass a broader range of issues and perspectives, and this evolution continues as new challenges and opportunities arise. There are several gaps and challenges in mainstreaming Sustainable Development Goals (SDGs) into India's development policies, plans, and programs. These challenges encompass various aspects, including lack of scientific data, policy implementation, enabling institutions and governance, and funding. Insufficient data on critical variables like biodiversity loss, climate change impacts, and sustainable consumption hampers progress. There is a deficiency in policies and regulations aimed at promoting holistic and sustainable growth. This is partly due to a lack of political will to enact enabling framework and a lack of coordination across different government sectors. This can be attributed to a lack of capacity and resources for sustainable projects, as well as limited engagement from corporate sectors and civil societies. Insufficient monitoring and evaluation methods, lack of coordination, and limited involvement of key stakeholders may further hinder desired progress. A major obstacle is the absence of adequate funding and resources for large-scale sustainable development projects. This can be due to a lack of financial instruments, insufficient investment in sustainable infrastructure and technologies, or both. Achieving some SDGs may require greater amount of international cooperation and partnerships, which can be challenging due to complex geopolitical environment. Despite these challenges, India has made significant progress towards achieving the SDGs in many sectors, particularly renewable energy, education, and housing. To continue this progress, it is crucial to address issues of inequality and promote inclusive development. Collaboration between the government, private sector, civil society, and individuals is essential to mobilize resources, build capacity, and create an enabling environment for meeting SDG targets. Success in reaching these goals is contingent on collective actions that promote a harmonious relationship with nature. A multidisciplinary and holistic approach, encompassing scientific, economic, and social perspectives, is necessary to bridge these critical gaps and drive sustainable development forward.

V. Dutta
Department of Environment Science,
School of Earth and Environmental Sciences (SEES),
Babasaheb Bhimrao Ambedkar University
(A Central University), Lucknow, India
e-mail: dvenks@gmail.com

P. Ghosh (✉)
VIT-AP School of Social Sciences and Humanities
(VISH), VIT-AP University, Amaravati,
Andhra Pradesh, India
e-mail: priyanka.ghosh@vitap.ac.in

V. Dutta, P. Ghosh (eds.), *Sustainability: Science, Policy, and Practice in India*,
Sustainable Development Goals Series, https://doi.org/10.1007/978-3-031-50132-6_1

Keywords

Equity · Inclusive growth · Policy · Governance · Ecological resilience

1.1 Introduction

The idea of sustainability has developed over the course of time to incorporate a wider spectrum of issues and points of view, and this evolving process is ongoing as new possibilities and problems come into play (Horne and Fichter 2022; Truffer et al. 2022). The word 'sustain' has been in use for many centuries, but the term 'sustainable' was originally used in the context of natural resources, namely, forests (Hahn and Knoke 2010), to mean 'able to be sustained' or 'able to be maintained at a given pace or condition'. The following is a brief overview of how the concept of sustainability has evolved over time.

In the latter half of the eighteenth century, the idea of management and conservation of natural resources, in particular forests, was where the notion of sustainability was first introduced (Gutierrez Garzon et al. 2022; Adamowicz 2003). The concept of a 'sustained yield' was established, which centred on the concept of maintaining forests in a way that would provide a steady supply of wood without diminishing the forest resources. This conception was the impetus for the development of the concept of a 'sustained yield'. In order to guarantee the continued and effective management of the resource, the optimal method for cutting down trees in a forest has been determined. An optimal rotation rule was first developed by economist Harold Hotelling in the year 1929 (Livernois et al. 2006; Gutierrez Garzon et al. 2022; Hotelling 1931). According to Hotelling's rule, a resource ought to be exploited at a pace that, over the course of time, will bring the marginal revenue from the resource into harmony with the marginal cost of extraction. To put it another way, the optimal opportunity to harvest a tree is when both the rate at which the tree is growing and the amount of money that can be made from harvesting the tree are at their peaks (Reynolds 2013).

Later, the optimal rate of harvesting a fish stock was thought of in order to ensure sustainable management of the fisheries. It was first proposed by economist William F. Lloyd in the early 1950s and later developed by mathematicians, biologist, and fisheries scientist George Clark (1973, 1990). Optimum harvesting was based on the idea that there was an optimal level of fishing that maximized the long-term yield of a fish stock, known as the maximum sustainable yield (MSY). By harvesting fish at the optimal rate, fishery managers can ensure that the fish stock is managed in a sustainable way while also maximizing the economic benefits from the resource. This became a crucial principle for fisheries management to ensure sustainable use of fish stock while maximizing the economic benefits from the resource. Subsequently, there has been a lot of work done in the field of optimal control theory to figure out the best way to exploit renewable resources like fish populations (Duncan et al. 2011).

The concept of sustainability began to further expand in the 1960s and 1970s, as people became increasing concerned about environmental degradation and depletion of natural resources. In the context of economic development, the word 'sustainable' was first used by the World Conservation Union in 1980 (IUCN 1980). The concept of sustainability gained further momentum in the late 1980s with the publication of the Brundtland Report in 1987 (Burton 1987), which defined sustainable development as 'development that meets the needs of the present without compromising the ability of future generations to meet their own needs' (Brundtland 1987). This report helped to define the concept of sustainability and brought it to the mainstream.

The concept of sustainability has continued to evolve in the twenty-first century, with increasing emphasis on its social, economic, and environmental dimensions. The United Nations adopted the Sustainable Development Goals (SDGs) in 2015, which aims to achieve a sustainable future for all by 2030 (Costanza et al. 2016). This framework goes beyond the environmental dimension of sustainability and incorporates economic and social dimensions as well.

The idea of sustainable development has undergone significant change over a considerable amount of time, transitioning from a 'buzzword' to the 'new paradigm of development' while maintaining a lack of uniformity and consistency in terms of interpretation (Feng et al. 2022; Lélé 1991). In the beginning, their definitions were not entirely clear, and the notion did not take into account issues of poverty, community engagement, and social well-being. This finally led to inconsistencies in the formulation of policies (Daly 1990; Costanza and Daly 1992; Vogel 2016). In addition, the implementation of it was not clear or rigorous enough. However, because of this deficit, many studies and research outcomes, as well as the political openness of action initiatives by international agencies and development banks, started to fill the gaps between policies and practices (Harris 2000; Pearce 2014).

It is obvious that our environment is always changing, and it is also indisputable that people all over the world are always confronted with new problems and difficulties that are associated with evolving socio-economic and geopolitical situations. However, development practitioners knew that the advantages of progress did not immediately 'trickle-down' to the base of the pyramid, to those who needed them the most. This is what Harris claims (2000) – powerful frames of thinking are typically simplistic and direct, making them simple to comprehend. When it comes to the field of social science, concepts that influence the lives of millions of people and shape the actions of nations need to be open and accessible to everyone and not just a select few. Only in this way will they be able to get into agencies on a scale ranging from the local to the global and become the part of human landscape, a component of the framework under which we define our collective existence.

The notion of the Human Development Index (HDI), which was first developed in the 1990s (Lind 1992), is acknowledged as a more logical measure of development despite the fact that the majority of nations throughout the world define growth primarily by the GDP. The Human Progress Index (HDI) incorporates social and sustainable goals in a far more relevant way than other measures of economic success, despite the fact that it is not recognized as a standard measure of economic development. However, even the HDI does not take into account more widespread environmental difficulties and challenges, such as climate change and the destruction of natural systems (Hickel 2020). When one focuses on income growth, it is possible to overlook the consequences that growth has on environmental disasters and pollution. This naturally defies the ideals of sustainability, as is evident from the fact that nations with high HDI scores also account for a larger per capita contribution to carbon emissions and pollution. Therefore, it is possible that the Sustainable Development Index (SDI) will be required as an integrative principle that includes social and ecological resilience and better gauges the environmental efficiency of nations in attaining social and human development goals (Barrera-Roldán and Saldıvar-Valdés 2002).

1.2 The 'Triple Bottom Line' of Sustainability

The meaning of 'sustainability' has changed over time as new interpretations and perspectives have been applied to the concept. It is the capacity to achieve a balance between economic development and protecting natural resources and ecosystems for future generations. This term is used in discussions of social and economic progress to describe a state in which current demands may be met without jeopardizing those of future generations. It incorporates economic, social, and environmental aspects. The 'triple bottom line' is a term used to describe these three aspects of sustainability (Slaper and Hall 2011). The concept of triple bottom line is that in order to be truly sustainable, an organization or society must consider and balance the economic, social, and environmental impacts of its actions (Hacking and Guthrie 2008) (Fig. 1.1).

How economic development may be achieved without depleting natural resources or causing long-term environmental damage is the primary emphasis of economic sustainability (Khan et al. 2022). It explores how to build a sustainable

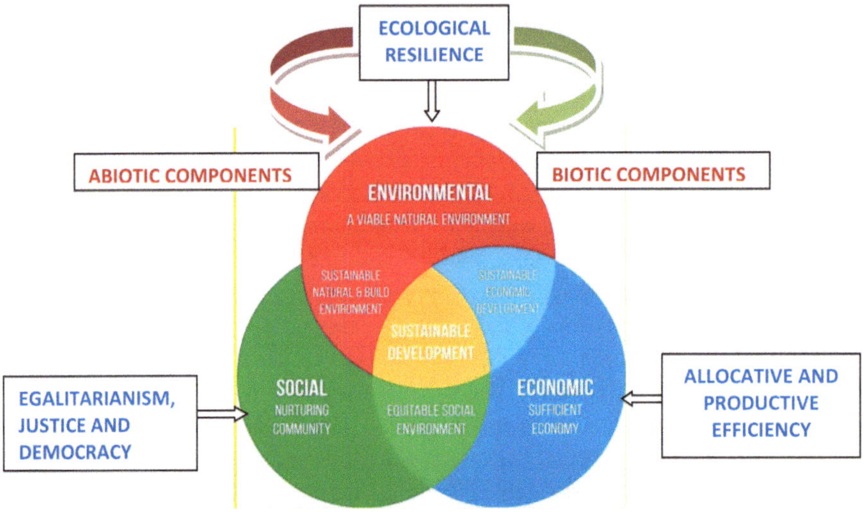

Fig. 1.1 Three pillars of sustainable development and their interdependencies. (Verma and Dutta 2021)

economy that grows over time and gives people a reasonable wage while simultaneously safeguarding the planet's dwindling resources for future generations. Social sustainability examines the social dimensions of sustainability, such as how to build a just and equitable society in which everyone is granted their basic human rights and opportunities. Poverty, inequality, education, and healthcare are all examples of such problems. In order to safeguard the planet's resources and biodiversity while simultaneously reducing the impact of climate change, environmental sustainability places a strong emphasis on environmental issues. Cleaner energy, more efficient transportation, and more environmentally friendly land use are all part of this movement. Together, these three pillars aim to promote a more balanced and sustainable future for all.

1.3 The Notion of Sustainability in India: A Brief History

Both the idea of sustainability and the practice of sustainable development can be traced back to India's rich cultural history and long-standing traditions (Chauhan et al. 2022). The *Vedic* literature, which dates back to around 1500 BCE, incorporates concepts of *sustainable living*. These ideas include the utilization of natural resources in a manner that is both balanced and responsible. There are several references made throughout the *Vedas* to the need of defending natural resources, retaining forest cover, and maintaining biological richness. At several places, the values of balance, harmony, and intergenerational equality were brought to the forefront (Mukherjee 2022).

There is a great focus, throughout the canon of Vedic literature, on the interdependence of all living species as well as the significance of preserving balance and harmony in the natural world. The *Vedas*, the *Upanishads*, and other ancient texts contain principles and practices that are pertinent to sustainable development. Some examples of these include the principle of 'non-violence' (*Ahimsa*) towards all living beings, the idea of intergenerational equity, and the concept of *minimalist* living (Mulia et al. 2022; Tiwari et al. 2022).

The post-Vedic period saw the development of the Indian tradition known as *Vrikshayurveda*, which translates to 'science of tree life'. This tradition places a strong emphasis on the necessity of conserving and maintaining natural resources (Muralivallabhan 2022). The Indian tradition is responsible for the development of a philosophy that is referred to as 'eco-spirituality'. This philosophy is based on environmental consciousness or the understanding that the spiritual and material aspects of life are interconnected and that the health of the earth and the people who live on it are interdependent (Abhijat 2022; Allen 2022).

The path, on the other hand, has not been linear, and there have been moments when the principles of sustainability have been subordinated to the objectives of economic progress. The British created a system of resource exploitation throughout the colonial period, which resulted in the deterioration of natural resources, most notably forests. On the other hand, British officials and scientists also understood the need of preserving natural resources, and as a result, they enacted policies and legislation to save the nation's forests, woodlands, and wild animals. During the 1970s and 1980s, India started to come to terms with the necessity of sustainable development. As a result, the country passed a number of policies and regulations that are designed to preserve natural resources and safeguard the environment.

India has implemented several policy reforms and programs in the area of sustainable development in recent years. Overall, India's sustainable development policies and programs have emphasized the importance of reduction of greenhouse gases, use of renewable energy sources, and managing natural resources such as water and land. These policy reforms demonstrate India's commitment to sustainable development and its efforts to address key environmental challenges. However, the effectiveness of these policies is still under evaluation and need more stringent implementation to meet the global and national SDGs.

1.4 From Development-Centric 'Environmentalism' to Targeted 'Sustainability Goals'

It wasn't until the early 1960s that the idea of 'environmentalism' became firmly established in the broader framework of the development. The concept was not fixed and continued to change its meaning over time. It shifted in response to the shifting landscape of the global economy and was further shaped by the larger context of both social and ecological resilience. This global scenario also applies to the Indian landscape, which is where environmentalism evolved and changed over the course of several decades, shifting from an arbitrary approach in the 1970s to a more responsive framework in the late 1980s. This global scenario applies to both the Indian and global landscapes. Only in the early 1990s did predictive and integrative methods begin to be understood and addressed in the policy and planning literature. Following the meeting that took place in Stockholm in 1972, modern environmental legislation in India started getting drafted. The National Council for Environmental Policy and Planning was created in 1972. In 1985, this organization was elevated to the status of a full-fledged Ministry of Environment and Forests (MoEF). In later years, the Indian judiciary issued a number of rulings that demonstrated their interest for the preservation and safeguarding of the natural environment. It wasn't until after the adoption of Environmental Impact Assessment notification in 1994 that environmental issues were substantially incorporated into the evaluation of projects. In social and economic development debates, the idea of an allocative and productive efficiency in natural resource use was still not a chosen theme to examine. As a result of this, it can be observed that too many methods have been chosen at different points in time throughout the course of India's economic policy's evolution as a long-term path of conflicts and adaptation (Fig. 1.2). Initially, the focus was placed on accelerating industrialization and urbanization, as well as fostering more intensive agriculture through the application of technology developed during the green revolution.

An integrative and holistic framework to sustainability science is largely missing, and there is a pressing need for further progress in this regard. The policies regarding environmental protection in India have adjusted very rapidly as a consequence of the international negotiations and diplomacy, as well as numerous different pieces of legislation, the creation of new organizations, the setting up of the National Green Tribunal, and the crucial judicial inferences. Despite these changes, there is still a lack of a holistic and integrative approach to sustainability science. Scholars argue that the world is going through a

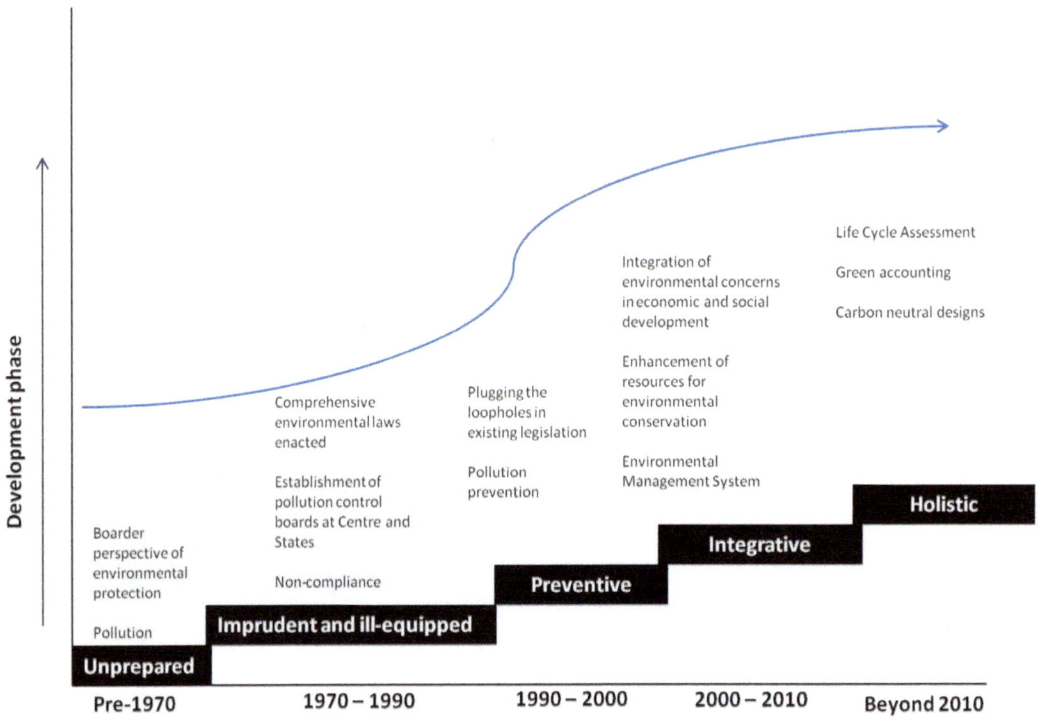

Fig. 1.2 Evolution of environmental framework in development policy. (Verma and Dutta 2021)

desperate time, in which socio-economic problems are man-made as a result of the disruptions brought on by modern civilization (Kaul and Adhikary 2021).

The decade of the 1980s saw the development of state and central pollution control boards, which signalled a change in emphasis towards the formulation of broad environmental laws and regulations. In this phase, relatively major structural changes were made, for example, the commencement of economic reform with the liberalization of trade, the elimination of budget deficits through cess and taxation, and the demolition of inefficient public sector and state-owned institutions. The precautionary phase of policy-making, in which pollution prevention and the polluter pays principle were given emphasis, happened during the decade that lasted from 1990 to 2000. Several pieces of legislation were analysed, and changes were made to them as well as new ones being drafted. Even though there was an increase in both productive and distributive efficiencies as a result of the economic reform, market-oriented policies were not successful in addressing the more serious problem of income disparity and the plight of the economically disadvantaged. In light of this, one school of thought contends that social inclusion is the most important factor in determining sustainable development, which in turn encourages the effective use of resources and inclusive economic expansion (Verma and Dutta 2021).

The world started framing laws and policies to safeguard the environment after the discussion that began during the 1972 Stockholm conference. Within the subsequent half-century, the number of international environmental regulations increased by a factor of 38. Worsening conditions on Earth persist, though. The annual amount of carbon dioxide released into the atmosphere has reached 36.7 billion tonnes, which is more than twice as much as was released in 1972. Consequences will result from this impact globally, mainly on health and biodiversity within the next half-century unless urgent actions to reverse the trend are taken.

1.5 Why Millennium Development Goals (MDGs) Could Not Achieve the Ambitious Targets Set for 2015?

The United Nations introduced eight Millennium Development Goals (MDGs) in the year 2000, with the deadline for their completion in 2015. Poverty, hunger, disease, a lack of sufficient housing, and gender discrimination were only some of the social and economic development issues that the goals aimed to address. To be more explicit, the MDGs aimed to end world hunger and extreme poverty; provide all children with access to quality primary education; advance gender parity and the empowerment of women; decrease infant and maternal mortality; enhance child health and maternal well-being; combat HIV/AIDS, malaria, and other diseases; safeguard future generations' ability to live in a healthy environment; and create a global partnership for development.

Even while poverty and inequality reduction were seen as a primary driver of the MDGs, there was no such specific target in the goals explicitly. Reducing gender inequality and expanding access to high-quality education and healthcare were some areas where the MDGs fell short. Some work was made towards these targets, although they were not all completed by the 2015 target. This occurred for many reasons, the most prominent of which were insufficient resources, a lack of political will, and inefficient programme execution. The goals were also criticized for being overly general and failing to address the unique difficulties faced by each region.

An important gap that prevented the MDGs from achieving sustainable development was their failure to incorporate environmental considerations. Although the eighth target of the MDGs was to promote sustainable development, this goal did not account for environmental and ecological considerations. As a result, the MDGs did not fully accomplish the grandiose objectives set for 2015; however, they did serve as a useful framework for focusing international attention on a set of key development challenges, and the progress made in some areas provides a foundation for the post-2015 development agenda.

1.6 Evolution of SDGs as Transformative Goals

The United Nations adopted the Sustainable Development Goals (SDGs) in 2015 as a successor to the Millennium Development Goals (MDGs) to address a range of social and economic development challenges such as inclusive growth, social equity, and ecological sustainability. The SDGs include 17 goals and 169 targets, and they cover a wide range of issues including poverty, hunger, health, education, gender equality, clean water and sanitation, affordable and clean energy, decent work and economic growth, industry, innovation and infrastructure, reduced inequalities, sustainable cities and communities, responsible consumption and production, climate action, life below water, life on land, peace, justice and strong institutions, and partnerships for the goals.

Because of the SDGs, there is now a significant amount of pressure on national and regional economies as well as corporate entities to accomplish these goals by the year 2030. While progress has been made in some areas, overall progress on the SDGs has been mixed. However, the SDGs have served as a useful tool for governments, organizations, and businesses to set and work towards common goals for sustainable development. Additionally, the SDGs have helped to raise awareness and mobilize action on these issues at national and global levels. Overall, the SDGs are an important mechanism for driving progress towards a more sustainable future, but it is still an ongoing effort to reach the target.

SDGs are not the first or only attempt to establish global development objectives. While they have been challenged in the past, the SDGs represent the UN's most extensive and thorough effort to date towards advancing sustainable development. The question arises, after 7 years of implementing, if these 17 SDGs have had any political

Fig. 1.3 SDGs transformative impacts through conceptual, regulatory, institutional, and ESG framework

influence across national and global governments to solve some of the urgent issues facing humanity like inclusive growth, social equity, and ecological sustainability. In order to do this, government, civic society, and scientific and business organizations will need to work together to align growth with social equality and the resilience of ecosystems (Sachs et al. 2019; Silvestre and Ţîrcă 2019). All 17 SDGs are intricately connected to one another, but it is unclear how these 17 goals might be implemented across the various thematic areas.

Based on the review of policy papers and SDGs dashboard of many countries, four types of impacts are visible: The first is *conceptual* impact. They may be seen in things like increased use of SDG-related terminology in international and domestic debates. The second is *regulatory* impact which is reflected by changes in legal and regulatory structures and policies that are made to reflect and support the SDGs. The third visible impact is *institutional* impact seen through evidence of new institutions being formed or old ones being restructured as a result of work towards the SDGs. And the fourth impact is on business and industries with new *ESG framework* being adopted by them in line with SDGs. Transformative impact is the ultimate aim of the 2030 Agenda, which is possible through synergies of all four types of impacts throughout a whole socio-economic and political system (Fig. 1.3). Both the SDGs and the Paris Agreement on Climate Change are significant contemporary milestones in the process of formulating global environmental policy. Both of these documents call for more deeper and transformative actions to be taken across every nation.

1.7 Operationalizing Sustainable Development Goals (SDGs)

However, it is still unclear whether or not regulating by such overarching global targets is effective. There have been no thorough meta-studies conducted to evaluate the political influence of the goals on a global or cross-national scale. Biermann et al. (2022) gathered data from an assessment of more than 3000 scientific papers on the SDGs that were published during 2016 and April 2021. According to their results, the goals appear to have exerted some political

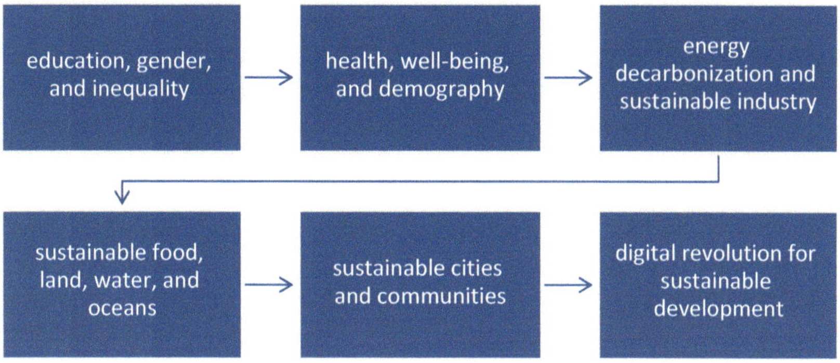

Fig. 1.4 Six building blocks for operationalizing SDGs as proposed by Sachs et al. (2019)

influence on the policies and institutions of governance at both local and global levels. The majority of this impact has been conceptual, having an effect on how people perceive and convey about sustainable development. It is unlikely to have a deeper normative and systemic outcome, such as influencing the legislative activity or the distribution of resources. The authors conclude that SDGs have had only a modest influence in terms of political transformation so far.

Sachs et al. (2019) have proposed six building blocks for the purpose of accomplishing these goals. These building blocks are as shown in Fig. 1.4.

Each of these foundational elements defines the primary aspect of the capital and regulatory environments that need substantial revolutionary changes in the institutional and governance architecture of the various levels, including local, regional, and national.

According to Fukuda-Parr and Muchhala (2020), the South was vital in placing the sustainable development agenda on the global map, which brought into question the unanimity of the developing economies of the North. These authors suggest that the South played a significant role in this quest. An alternative vision of Sustainable Development Goals (SDGs) was suggested by development activists and entrepreneurs from the Global South. It was primarily a result of challenging the dominant beliefs on development and pushing forward ideas that were on the margins.

1.8 India's Progress in Achieving SDGs

Considering India's massive population of 1.4 billion, the achievement of fulfilling 17 sustainable development objectives would also make an impact on a global scale. The progress India has made towards fulfilling some of the SDGs could be termed satisfactory, while large development deficits remain given the country's status as a developing nation. Several academicians, researchers, policymakers, and practitioners have contributed to this work. The challenges and opportunities of attaining the SDGs in India are examined via the lens of in-depth empirical case studies from various regions of the country. For this reason, a book that focuses on India and includes actual case studies would be a welcome addition to and expansion of the literature on SDGs. Moreover, it can be difficult for the government, policymakers, and common people in a multicultural country like India to work together to achieve the SDGs. The various chapters aim to analyse the obstacles and potential solutions to reaching the SDGs in India by 2030 and to speculate on what this could mean for the country in the long run.

Today, India is making large strides towards achieving a sustainable future by actively focusing on sustainable development and adopting policies and initiatives. One major initiative is the *National Action Plan on Climate Change* (NAPCC), which was launched in 2008 to address the challenges of climate change. The plan includes eight national missions, such as the National Solar Mission and the National Mission for Enhanced Energy Efficiency to reduce greenhouse gas emissions and promote the use of renewable energy. Another significant reform is the National Clean Energy Fund, which was established in 2010 to promote research and development in clean energy technologies. However, in recent years, many large projects are being undertaken in sensible ecosystems. Development experts have been warning against constructing large roads and hydroelectric dams in the Himalayas for decades. Nonetheless, the magnitude of these developments has been beyond the ability of the mountains to withstand them. The 1976 Mishra Committee's advice to avoid forest clearing was ignored. The planners did not provide the care that mountain communities need. The panel warned against hasty urbanization in hill towns. In particular, it warned against cutting down trees and emphasized the need of avoiding disturbing the foot of the unstable slopes. It is imperative that we develop a fresh perspective on how to construct urban areas in mountainous regions, one that is informed by studies of carrying capacity. As it is, the failure to take action on climate change is the greatest global threat. However, India's progress towards SDG 13's climate action targets has slowed between 2019 and 2020. When it comes to SDG 13, the performance of 28 states/UTs is still below par.

As of April 2022, there were five states with unemployment rates that were greater than 15%. The unemployment rates in the states of Haryana, Rajasthan, and Bihar remained at alarmingly high levels, at 34%, 28%, and 21%, respectively. In 2021–2022, about 60% of rural families requested employment opportunities under the *Mahatma Gandhi National Rural Employment Guarantee Act.*

The *Atal Bhujal Yojana* (ABY) is a scheme launched in 2018, which focuses on improving groundwater management and conservation in priority areas in the country. The scheme's main objective is to improve the groundwater situation in priority areas in the country by involving community and other stakeholders in the management of the resource. It will also focus on capacity building, institutional strengthening, and awareness campaigns. The scheme is being implemented in 8 states (Gujarat, Haryana, Karnataka, Madhya Pradesh, Maharashtra, Rajasthan, Uttar Pradesh, and Uttarakhand) across 78 districts, covering about 9650-gram *panchayats* and 6937 villages. As per the plan, ABY is expected to have a positive impact on groundwater management in India. However, it is not clear how it will help in addressing the issue of over-extraction of groundwater and recharge the depleting groundwater table. It is worth mentioning that the scheme is still in its early stages and its impact will be known in the long run, but the government's efforts to address the groundwater management and depletion issues are commendable.

The *Swachh Bharat Abhiyan* (Clean India Campaign) is a nationwide campaign launched in 2014 to clean up the streets, roads, and infrastructure of India's cities, towns, and rural areas. The program has been successful in changing people's attitudes towards cleanliness in certain regions of India, increasing awareness of the importance of cleanliness and hygiene, and improving the infrastructure and facilities for waste management. Despite this, there are still many obstacles to overcome in order to completely execute the programme, for example, improved urban waste management and reduction in open defecation, particularly in rural areas. One may argue that the campaign as a whole has advanced, but this does not negate the fact that there is an opportunity for more development.

The *National Green Highways Policy* was launched in 2015 to promote the development of green corridors along national highways in India, with the goal of reducing air pollution and promoting sustainable transportation. The policy has led to the creation of green belts along highways, use of alternative energy sources, and adoption of sustainable construction practices. However, its implementation has faced challenges such as limited financial resources, lack of coordination between various agencies, and difficulties in land acquisition. Despite these challenges, the policy has been largely successful in promoting sustainable development and preserving the environment along the highways.

The percentage of the population living below the poverty line has decreased dramatically over the past several years, from 21.9% in 2011–2012 to 13.4% in 2016–2017. The Mahatma Gandhi National Rural Employment Guarantee Act, which ensures that all rural residents receive at least 100 days of paid work per year, and the *Pradhan Mantri Jan Dhan Yojana*, which seeks to ensure that all households have access to banking services, have both contributed to this success.

The country must prioritize the elimination of all types of poverty and deprivation and bring its poverty rate down from 21.92% to 10.96% by the year 2030. (SDG India Index & Dashboard 2020–2021). In spite of the fact that India has achieved significant strides towards the eradication of poverty over the course of the last two decades, an anti-poverty strategy that is more comprehensive is required for the long-term elimination of poverty. To accomplish this long-term goal, we need to focus on employment, equality, and empowerment (Bhukta 2020). The National Institution for Transforming India (NITI Aayog) believes that eradicating poverty should focus on two aspects: continuous economic growth that leads to employment generation and the successful implementation of anti-poverty programmes (NITI Aayog 2021).

Providing citizens with easy access to potable water and sanitary facilities is another area where India has made some progress. The government has spent a lot of money upgrading the country's water and sewage systems, so many people can use them now. Access to toilets and sanitation have been greatly improved in rural areas because of the government's *Swachh Bharat Abhiyan* (Clean India Mission). However, heavy toxic metals are present at worrying rates at 74% of river monitoring stations in India. Insufficient treatment of wastewater from industries, agriculture, and residences is evidenced by high amounts of total coliform found in 40% of the monitoring stations with 15% of the stations having biochemical oxygen demand over allowable levels (CSE 2022).

Renewable energy is another area where India has achieved breakthroughs. With non-fossil fuel sources accounting for 40% of its installed electrical capacity, the country is now ranked as the third largest generator of renewable energy in the world. Several initiatives, such as the *Jawaharlal Nehru National Solar Mission* and the *National Wind Energy Mission*, have been launched by the government to promote the growth of renewable energy sources. India has demonstrated its leadership in the fight against climate change, showcasing its firm commitment. With the goal of achieving net zero emissions by 2070, the country has also set short-term targets to be met by 2030, including increasing its renewable energy capacity to 500 GW, achieving 50% of its energy needs from renewable sources, reducing its cumulative emissions by 1 billion tonnes, and decreasing the emission intensity of its GDP by 45% by 2030. As a developing nation, India's efforts and experiences will serve as a valuable example for other countries as they take concrete steps towards fulfilling their climate commitments and transitioning towards a sustainable energy future.

The primary school attendance rate in India rose to 96.7% in 2016–2017, a sign that the country is making strides towards its goal of universal primary and secondary education. Government programmes like the *Sarva Shiksha Abhiyan* and the *Rashtriya Madhyamik Shiksha Abhiyan* aim to raise standards in classrooms across the country.

The advancement of women in India is another area where the country has achieved progress. The National Policy for the Empowerment of Women and the *Beti Bachao, Beti Padhao* initiative have been adopted to advance women and girls' rights and reverse the alarming trend of a dropping child sex ratio in India.

In a nation where there is a substantial divide between those who have wealth and those who do not, it is essential that inequalities be reduced. It has been seen that in the past, the government of India did not pay a lot of attention to reducing inequality due to several misconceptions. These misconceptions included the ideas that inequality cannot be prevented with rapid economic growth and that inequality in India is relatively lower than in other developing countries. Both of these ideas were incorrect. This problem should be addressed by the government through mainstreaming inclusive growth in all the sectors. In addition, the empowerment of the poor is essential for the eradication of poverty. The Indian government should place a primary emphasis on involving the poor and other marginalized groups in the formulation and execution of poverty eradication programs (Bhukta 2020). This requires an approach that is both collective and well-balanced.

When evaluating the performance of the SDGs, it is important to do so in the context of the global geopolitical circumstances, which are sometimes difficult and complicated, as well as the effects these conditions have on the climate, energy, food, and socio-economic conditions. Even if India is the fifth largest economy in the world when measured by GDP, the country's economic expansion will not be sustainable if the country does not prioritize the fight against climate change and the damage it causes (SDG-13). In addition, action on climate change has to be linked with other SDGs, such as the eradication of extreme poverty and hunger, gender equality, education, and the provision of clean water and affordable energy.

Many states are not quite on track to accomplish the SDGs by 2030, despite having just 7 years to do so. There are 295 key indicators that India has identified. However, only 115 of them are being tracked. While India has made significant progress towards its goal of eliminating poverty, it still confronts substantial obstacles on the path to reaching the SDGs. Under SDG 8, which focuses on economic growth and productive employment for everyone, we have, for instance, defined 26 indicators. Only nine are under India's close watch. There is no mention of key metrics like the expansion of micro-, small-, and medium-sized businesses that have registered.

Uneven economic development is a major obstacle, since many people still struggle to make ends meet and are denied access to opportunities because of their socio-economic status. Climate change and adaptation to extreme weather are another big obstacle India must overcome, since it threatens the country's progress and the lives of its vast vulnerable population. The nation must also work to improve healthcare infrastructure and lower the rate of maternal and newborn mortality.

Data collection for tracking the SDGs presents another major challenge. Given the ever-evolving nature of the aims, governments would benefit from regular and thorough data collection. Some indicators have a 10-year interval between data collecting points. This is why various indicators have shown no performance change. These metrics are out of sync with reality at the moment.

1.9 Structure of the Book

This book discusses the contemporary sustainability narrative in India from different parts of the country, as well as from different thematic areas illustrating the prospects and challenges in operationalizing the SDGs. The chapters are grouped into six thematic sections:

Part I: Poverty, Gender, and Health
Part II: Quality Education
Part III: Clean Water and Energy
Part IV: Economic Growth, Sustainable Cities, and Communities
Part V: Land and Water
Part VI: Global Partnership and Sustainable Development: A Way Forward

The first part of this book explores the imperfections and limitations of poverty measurement in India, which has a vast and diverse population. It suggests recalibrating the poverty line to account for changes in income, consumption patterns, and prices and highlights the need for strengthening food security and farmers' economy. One way of achieving this is by fortifying the beneficial microbial population in soil through soil-water management, which would combat desertification and loss of biodiversity.

The first part also focuses on gender and health issues in India, emphasizing the importance of using Summary Measures of Population Health (SMPH) to assess population health. This indicator combines morbidity and mortality data, and the papers in this section examine the significance of measuring women's health and well-being. The second part of the book explores SDG 4, which relates to quality education. It emphasizes the crucial role that inclusive and quality education plays in achieving other SDGs and examines the national educational policies in India. The third part of the book examines the dynamics surrounding clean water and sanitation (SDG 6) and affordable and clean energy (SDG 7). This section delves into the challenges and opportunities surrounding these important sustainability goals.

Sya Kedzior's paper sheds light on the challenges of providing sustainable access to water and sanitation in India and offers potential solutions for achieving SDG 6. Additionally, this section also examines India's current and future energy mix and energy efficiency paths.

The fourth part of the book delves into the right to decent working conditions, which is closely linked to economic development and sustainability. This section also evaluates the limitations of public-private partnerships (PPP) in promoting sustainable urban development. The fifth part explore the themes of life below water and life on land (SDG 14 and SDG 15). The chapters in this section critically examine the sustainability of Indian seas and Blue Economy discourse and analyse forest conservation initiatives in Odisha.

The last section of the book discusses India's diplomatic efforts in the areas of environment, development, and climate. A chapter on green diplomacy recommends that changing India's diplomatic communication approach would lead to better results in forming global partnerships for sustainable development.

The various chapters in this book illustrate that the SDGs are not a mere set of directives that should only be used by the governments and policymakers. India's success in achieving these targets is only possible if civil societies and citizens take up the collective actions to live a life that is in harmony with the nature. Pro-planet activities are not a complicated set of practices. Through small changes in lifestyles, for example, using environmentally friendly materials, sun-drying clothes, conserving electricity and water, banning single-use plastics, etc. can help preserve and protect the environment in a big way. Besides small changes at the individual level, conscious government's efforts at national and subnational scales are required to fulfil the gaps between specific SDG targets and outcomes. For instance, careful planning is needed in case of water resource management to reduce the gap between supply and demand especially in the urban areas. In other words, the proper management of hydrological cycle and its components such as surface water, groundwater, storm water, and wastewater is a necessary step to bridge the gap between demand and supply. In the recent past, the idea of water-sensitive urban design (WSUD) is being put into practices in the city's master plans – it conceptualizes cities as water supply catchments and important sites for provision of various ecosystem services.

1.10 Conclusion: Why Sustainability Goals May Be Difficult to Realize in India by 2030?

India is a very diverse country, and the sustainability challenges and solutions may vary based on the location, the regulatory and governance systems, and the level of development. Degradation has occurred through almost 30% of India's land area due to climate change and other anthropogenic factors. The soil still lacks the vital macronutrients and micronutrients necessary for long-term farming. The percentage of an agricultural family's income that comes from farming has decreased from 48% in 2012–2013 to 37% in 2018–2019, even though the cost of cultivation has climbed by roughly 35% over the same time period. Around 50% of farm families have credit card debt, and nearly 29 farmers and farm workers commit suicide every single day in the country, according to official statistics.

Addressing health disparities and improving healthcare infrastructure are crucial challenges. Several factors contribute to these disparities, such as uneven distribution of resources, varying levels of healthcare accessibility, and regional economic disparities. Only seven states, five of which are major (Telangana, Maharashtra, Kerala, Tamil Nadu, and Andhra Pradesh) and two of which are small (Tripura and Mizoram), demonstrate high overall performance on the Health Index, which evaluates important health indicators and infrastructure facilities (CSE 2022).

Implementing sustainability goals in India by 2030 may face several challenges, including:

Ineffective policy implementation: Despite the government's commitment, the lack of effective policy implementation and enforcement has hindered progress towards sustainability goals.

Conflicting economic growth: India's focus on economic growth can sometimes clash with sustainability goals. The country's fast-paced economic development often prioritizes growth over sustainability, leading to increased pollution and resource depletion.

Prevalent poverty and inequality: India has a significant population living in poverty and high levels of income inequality, making it challenging to implement sustainability measures that may be perceived as too expensive or onerous for those in poverty.

Lack of government coordination: India is a federal country, and often there is a lack of coordination between central and state governments when it comes to sustainability issues.

Limited resources and funding: India has limited resources to devote to sustainability initiatives, particularly in rural areas, where poverty is most prevalent. Many of the sustainability initiatives in India are underfunded, which limits their effectiveness.

Lack of public awareness and education: There is often a lack of public awareness about sustainability and the importance of environmental conservation. Many people in India may not be aware of the potential benefits of sustainable practices or may be resistant to change. The lack of public awareness and understanding about sustainability issues can make it difficult to achieve sustainability goals.

Lack of public accountability and corruption: Corruption may impede India's progress towards sustainability. It results in the mismanagement and misallocation of resources and weakens enforcement of environmental regulations. Both public and private sector corruptions could hinder the implementation of sustainable practices and the attainment of sustainability goals.

Limited data, measurement, and monitoring: There is often a lack of data and monitoring systems in place to track the progress of sustainability initiatives, which makes it difficult to identify and address problems, assess their effectiveness, and make adjustments as needed.

Lack of technology readiness: India lacks the necessary technology to effectively implement sustainable practices, such as renewable energy and efficient waste management systems.

Resistance to change: Many industries and individuals in India may be resistant to changes that could negatively impact their livelihoods or established ways of life. The resistance may stem from concerns about the economic, social, or personal costs of such transitions, making it difficult to implement sustainability initiatives that require significant shifts in behaviour or investment. Additionally, deeply ingrained cultural norms and beliefs may also contribute to resistance to change.

Since governments perceive increased economic growth as a means to reduce poverty and raise living standards, this growth is often given higher priority than protecting the environment. For countries like India with high populations and high unemployment rates, economic expansion is typically considered as a means to attract investment and generate jobs. The government may also put infrastructure projects and the engagement of foreign investors ahead of environmental protection initiatives as part of its efforts to stimulate economic growth. In addition, both the general public and political leaders are woefully uninformed about the devastating effects of environmental deterioration. For the sake of present and future generations, however, sustainable development that strikes a balance between economic growth and environmental conservation is crucial. More and more people are becoming aware of the detrimental effects that economic expansion may have on the environment, and steps are being taken to strike a balance between the two goals.

There are many sustainability targets where notable progress has been made. For example, extreme poverty has been reduced, access to primary education has improved, the maternal mortality rate has decreased, and there has been an increase in the use of renewable energy sources. However, many challenges still remain, particularly in areas such as reducing inequality, ensuring access to quality education and healthcare, and addressing the impacts of climate change. Additionally, the pandemic has had a significant impact on the progress of SDGs, mainly in areas such as poverty, hunger and health, and economic development.

Achieving the SDGs will require a concerted effort from governments, private sector, civil society, and individuals, and it will require a focus on addressing the underlying structural and systemic issues that are preventing progress in many areas.

References

Abhijat SR (2022) Environmental consciousness and Sri Aurobindo: learnings for the present. In: Reading Sri Aurobindo: metaphysics, ethics and spirituality. Singapore, Springer, pp 261–269

Adamowicz W (2003) Economic indicators of sustainable forest management: theory versus practice. J For Econ 9(1):27–40

Allen D (2022) The moral, philosophical, and spiritual basis of Gandhi's transformative nonviolence. In: Gandhi's global legacy: moral methods and modern challenges, p 65

Barrera-Roldán A, Saldıvar-Valdés A (2002) Proposal and application of a sustainable development index. Ecol Indic 2(3):251–256

Bhukta A (2020) No poverty: how much, how far. In: Sustainable development goals: an Indian perspective, Springer, pp 1–11

Biermann F, Hickmann T, Sénit CA et al (2022) Scientific evidence on the political impact of the sustainable development goals. Nat Sustain 5:795–800. https://doi.org/10.1038/s41893-022-00909-5

Brundtland GH (1987) What is sustainable development. Our common future 8(9)

Burton I (1987) Report on reports: our common future: the world commission on environment and development. Environ Sci Policy Sustain Dev 29(5):25–29

Chauhan R, Kaul V, Maheshwari N (2022) Indian indigenous knowledge system: sustainable approach toward waste management. In: Emerging trends to approaching zero waste. Elsevier, pp 37–57

Clark CW (1973) The economics of overexploitation. Science 181:630–634

Clark CW (1990) Mathematical bioeconomics: the optimal management of renewable resources. Willy, New York, p 775

CSE (2022) State of India's environment 2022. Centre for Science and Environment, New Delhi

Costanza R, Daly HE (1992) Natural capital and sustainable development. Conserv Biol 6(1):37–46

Costanza R, Daly L, Fioramonti L, Giovannini E, Kubiszewski I, Mortensen LF et al (2016) Modelling and measuring sustainable wellbeing in connection with the UN sustainable development goals. Ecol Econ 130:350–355

Daly HE (1990) Toward some operational principles of sustainable development. Ecol Econ 2(1):1–6

Duncan S, Hepburn C, Papachristodoulou A (2011) Optimal harvesting of fish stocks under a time-varying discount rate. J Theor Biol 269(1):166–173

Feng Y, Zeng Z, Searchinger TD, Ziegler AD, Wu J, Wang D et al (2022) Doubling of annual forest carbon loss over the tropics during the early twenty-first century. Nat Sustain 5(5):444–451

Fukuda-Parr S, Muchhala B (2020) The southern origins of sustainable development goals: ideas, actors, aspirations. World Dev 126:104706

Gutierrez Garzon AR, Bettinger P, Abrams J, Siry JP, Mei B (2022) Forest sustainability in state forest management plans: a content analysis. J Sustain For 41(1):92–113

Hacking T, Guthrie P (2008) A framework for clarifying the meaning of triple bottom-line, integrated, and sustainability assessment. Environ Impact Assess Rev 28(2–3):73–89

Hahn WA, Knoke T (2010) Sustainable development and sustainable forestry: analogies, differences, and the role of flexibility. Eur J For Res 129:787–801

Harris JM (2000) Basic principles of sustainable development. Dimensions of Sustainable Development, pp 21–41

Hickel J (2020) The sustainable development index: measuring the ecological efficiency of human development in the Anthropocene. Ecol Econ 167:106331

Horne J, Fichter K (2022) Growing for sustainability: enablers for the growth of impact startups–a conceptual framework, taxonomy, and systematic literature review. J Clean Prod 349, 131163

Hotelling H (1931) The economics of exhaustible resources. J Polit Econ 39(2):137–175

International Union for Conservation of Nature, & World Wildlife Fund (1980) World conservation strategy:

living resource conservation for sustainable development, vol 1. IUCN, Gland
Kaul A, Adhikary CD (2021) Sustainability: revisiting enduring modern and identified tradition. In: Verma MK (ed) Environment, development and sustainability in India: perspectives, issues and alternatives. Springer, Singapore. https://doi.org/10.1007/978-981-33-6248-2_4
Khan SAR, Yu Z, Farooq K (2022) Green capabilities, green purchasing, and triple bottom line performance: leading toward environmental sustainability. Bus Strateg Environ 32:2022
Lind NC (1992) Some thoughts on the human development index. Soc Indic Res 27:89–101
Livernois J, Thille H, Zhang X (2006) A test of the hotelling rule using old-growth timber data. Can J Econ 39(1):163–186
Lélé SM (1991) Sustainable development: a critical review. World Dev 19(6):607–621
Mukherjee P (2022) Ethics, consumerism and sustainability in the backdrop of the Kathopanishad. In: Global perspectives on Indian spirituality and management: the legacy of SK Chakraborty. Singapore, Springer, pp 393–400
Mulia P, Bartoszewski J, Kumari J, Behura AK, Jusko P (2022) Gandhi's concept of Sarvodaya for peace and sustainability in a technologically challenged world. Pedagogical Almanac 30:226
Muralivallabhan TV (2022) Protection of environment and promotion of development: the Hindu wisdom of sustainable development. Natl Secur 5(2):146–180
NITI Aayog (2021) India national multidimensional poverty index: baseline report 4, national institution for transforming India, pp 1–414
Pearce D (2014) Blueprint 3: measuring sustainable development. Routledge
Reynolds DB (2013) Uncertainty in exhaustible natural resource economics: the irreversible sunk costs of Hotelling. Resour Policy 38(4):532–541
Sachs JD, Schmidt-Traub G, Mazzucato M, Messner D, Nakicenovic N, Rockström J (2019) Six transformations to achieve the sustainable development goals. Nat Sustain 2(9):805–814
Silvestre BS, Ţîrcă DM (2019) Innovations for sustainable development: moving toward a sustainable future. J Clean Prod 208:325–332
Slaper TF, Hall TJ (2011) The triple bottom line: what is it and how does it work. Indiana Bus Rev 86(1):4–8
Tiwari MM, Goel V, Ahamad F (2022) State of the science of environment, spirituality and health: an overview. Environ Conserv J 23(3):471–478
Truffer B, Rohracher H, Kivimaa P, Raven R, Alkemade F, Carvalho L, Feola G (2022) A perspective on the future of sustainability transitions research. Environ Innov Soc Trans 42:331–339
Verma MK, Dutta V (2021) Introduction. In: Verma MK (ed) Environment, development and sustainability in India: perspectives, issues and alternatives. Springer, Singapore, pp 1–19. https://doi.org/10.1007/978-981-33-6248-2_1
Vogel N (2016) Municipalities' ambitions and practices: at risk of hypocritical sustainability transitions? J Environ Policy Plan 18(3):361–378

Venkatesh Dutta is a professor at the School of Earth and Environmental Sciences at Babasaheb Bhimrao Ambedkar Central University, Lucknow, India. His research interests include environmental management with a special emphasis on water policy, catchment planning, river restoration, and eco-hydrology. He has about twenty years of professional experience in areas of water quality management, land-use planning, and environmental impact assessment. He has contributed to significant changes in legislation and policy for sustainable water resource management in India. As part of his public policy work, he has served as a member of the expert group for the formulation of the State Water Policy of Government of Uttar Pradesh in 2020. He is a Fulbright-Nehru Fellow and a British Chevening Scholar.

Priyanka Ghosh is a Sr. Assistant Professor of Geography at VIT-AP University, Amaravati, Andhra Pradesh, India. Her areas of interest are political ecology, biodiversity conservation, protected area management, human-animal conflicts, traditional ecological knowledge, tourism, and sustainable development. Her research on the Indian Sundarbans has been published in leading international journals such as *GeoJournal*, *Geographical Review*, and *Environmental Management*. She has also published several chapters on environmental conservation, water resources, and tourism governance in peer-reviewed international books.

2 Poverty Alleviation with Sustainable Development in India

Ravi Patni

Abstract

Poverty is a multidimensional function of income and nonincome variables. The income aspect of poverty is currently measured as per the common international poverty line of \$1.25 (Rs. 91) per person per day based on the 2005 purchasing power parity, whereas the nonincome aspect focuses on food, drinking water, sanitation, housing, education, access to quality medical facilities, and electricity. The fundamental approach to measure poverty is to specify a minimum income that is required to purchase essential goods and services to fulfill basic human needs. The last official estimate of poverty was released by the erstwhile Planning Commission at 21.92% in 2011–2012 based on the recommendations of the Tendulkar Committee. The United Nations' Sustainable Development Agenda acknowledged the need for multidimensional approaches to poverty alleviation that go beyond economic deprivation. The Sustainable Development Goal "No Poverty" (SDG 1) envisions ending all forms of poverty (income and deprivations) in both rural and urban areas and among all groups (children, women, disabled, elderly) by 2030 through effective and sustainable state interventions. To achieve the vision of SDG 1, we should start by acknowledging the multidimensionality of poverty and the importance of capacity building in human development aimed at reducing all aspects of poverty. The Government of India has introduced various schemes such as Mahatma Gandhi National Rural Employment Guarantee Act 2005 (MGNREGA), National Social Assistance Programme (NSAP), National Rural Livelihood Mission (NRLM), PM Awas Yojana, PM Gram Sadak Yojana, Direct Benefit Transfer of LPG or PAHAL, and PM Jan Dhan Yojana along with social protection measures such as PM Jeevan Jyoti Bima Yojana and Suraksha Bima Yojana and Atal Pension Yojana for the inclusive social and economic development of the marginalized and vulnerable section. The unprecedented and unexpected advent of the COVID-19 pandemic could have significantly increased the gap between an influential and a marginalized section of the population. Several relief packages in the form of free rations, cash payments, loan moratorium, increased free healthcare facilities, and many more gave a sigh of relief to a majority of the population. The aim of this chapter is to provide an overview of various methods of poverty estimation along with the challenges of the poverty line approach in India. It also describes multiple dimensions of poverty and the performance of India based on the Global Multidimensional Poverty Index (MPI).

R. Patni (✉)
Research Officer (Engineering), State Planning Institute (New Division), Yojana Bhawan, Lucknow, Uttar Pradesh, India

V. Dutta, P. Ghosh (eds.), *Sustainability: Science, Policy, and Practice in India*,
Sustainable Development Goals Series, https://doi.org/10.1007/978-3-031-50132-6_2

Keywords

Capacity building · Multidimensional poverty · Poverty estimation · State intervention · Sustainable development

2.1 Introduction

Poverty is the lack of money to meet basic human needs—food, clothing, and shelter (The World Bank 2000–2001). As per the World Bank, poverty can be defined as deprivation in the well-being of an individual comprising many dimensions. This definition also includes low incomes and the inability to obtain or buy goods and services that are essential for humans to survive with dignity and safety. For an individual human, poverty can be measured based on the monthly (or annual) expenditure. The threshold on which this expenditure is compared is called the poverty line. However, the concept of poverty goes beyond the mere lack of money and material goods.

Poverty is a multidimensional function of income and nonincome variables. The income aspect of poverty is currently measured according to the World Bank's international poverty line of $1.90 (Rs. 141) per person per day at 2011 purchasing power parity (The World Bank 2022), whereas the nonincome aspect focuses on food, drinking water, sanitation, housing, education, access to quality medical facilities, and electricity.

The most fundamental approach to measure poverty is to determine a minimum expenditure (or income) required to purchase those goods and services that are necessary to satisfy basic human needs. This expenditure is termed the poverty line, and the poverty line can be defined in terms of the number of people living below the poverty line. The incidence of poverty can be expressed as the percentage of poor to the total population, which is termed the head count ratio (HCR) or the poverty ratio. Before trying to eliminate or reduce poverty, the first thing we need is to have reliable data that can reflect the correct situation of poverty in a country, or we can say that having an efficient system of poverty estimation is the first and the most crucial step to eradicate poverty.

The chapter aims to discuss poverty alleviation with sustainable development (SDG 1) with a prime focus on poverty and its estimation and multiple dimensions. It also describes the performance of India based on the Global Multidimensional Poverty Index (MPI) and the current situation of poverty in Indian states. The chapter also provides various measures undertaken by the government to achieve SDG 1.

2.2 Methodology of Estimation

In India, several expert committees, e.g., the Alagh Committee (1979), Lakdawala Committee (1993), and many others, have proposed various methods to calculate poverty. The erstwhile Planning Commission (now NITI Aayog) was the nodal agency in India for the estimation and calculation of poverty. Based on the methodology and techniques put forward by the experts of the Planning Commission, India has undertaken periodic assessments for the measurement of poverty.

The poverty ratio in India is measured through the poverty line, which is quantified in terms of the per capita consumption expenditure over a month. The consumer expenditure data are collected from the National Sample Survey (NSS). Households having consumption expenditures below the poverty line are said to be "below the poverty line (BPL) and are deemed poor" (Gaur and Rao 2020, 3).

2.3 History of Estimation of Poverty in India

The first step of poverty estimation includes providing a definition and quantification of a poverty line. Table 2.1 describes the various committees or working groups that estimated poverty based on different methodologies (Gour and Rao 2020):

Table 2.1 Poverty estimation in India—an overview

Poverty estimation	Methodology	Poverty line
Poverty and Unbritish Rule in India by Dadabhai Naoroji (1901)	Estimated based on the cost of a subsistence diet at the prices of 1867–1868 for the emigrant coolies while their voyage living in a state of quietude	Rs. 16 to Rs. 35 per capita per year
National Planning Commission under the chairmanship of Sh. Jawaharlal Nehru (1938)	Based on the minimum standard of living required by individuals for healthy living	Rs. 15 to Rs. 20 per capita per year
Working Group, Planning Commission (1962)	In 1962, India quantified the poverty line for the first time, and a minimum requirement (food and nonfood) of individuals for healthy living was determined	Rural areas, Rs. 20 per capita per month; urban areas, Rs. 25 per capita per month based on 1960–1961 prices
Study by VM Dandekar and N Rath (1971)	First systematic assessment of poverty in India based on National Sample Survey (NSS) data	For the first time, the minimum average calorie norm of 2250 calories per capita per day is established
Task Force on "Projections of Minimum Needs and Effective Consumption Demand" headed by Dr. Y. K. Alagh (1979)	The first official poverty counts began in India based on the approach of this Task Force To define the poverty line, the average daily calorie requirement in rural areas was set at 2400 kcal per capita per day. In urban areas, the requirement was 2100 kcal per capita per day	Rural areas, Rs. 49.09 per capita per month; urban areas, Rs. 56.64 per capita per month based on 1973–1974 prices
Lakdawala Expert Group, Planning Commission (1993)	Based on the minimum nutritional requirements as per Alagh Committee	Retained the separate rural and urban poverty lines recommended by the Alagh Committee at the national level. However, it disaggregated them into state-specific poverty lines in order to reflect the interstate price differentials
Tendulkar Expert Group (2009)	According to Lakdawala Expert Group the officially measured urban poverty line of 2004–2005 (25.7%) did not construct a poverty line. Per person per day, consumption expenditure resulted in the figure of Rs 32 and Rs 26 a day for urban and rural areas, respectively	For 2011–2012, the national poverty line was estimated per month as Rs. 816 per capita per month for rural areas and Rs. 1000 per capita per month for urban areas
Rangarajan Committee (2014)	Based on the consumption expenditure	Monthly per capita consumption expenditure was recommended as of Rs. 972 in rural areas and Rs. 1407 in urban areas for poverty line at the all-India level

2.4 Release of Poverty Line Estimates

The percentage of people living below the poverty line was released by the Planning Commission of India for the years 1973–1974, 1977–1978, 1983, 1987–1988, 1993–1994, 1999–2000, 2004–2005, 2009–2010, and 2011–2012.

In July 2013, the Planning Commission released poverty data for 2011–2012 based on the Tendulkar Committee report (Planning Commission 2013). For 2011–2012, using the Tendulkar methodology for rural areas, the national poverty line is estimated at Rs. 816 per capita per month and for urban area Rs. 1000 per capita per month. So, for a family of five,

Table 2.2 Percentage and number of poor estimated by the Tendulkar method

	Poverty ratio (%)			Number of poor (million)		
Year	Rural	Urban	Total	Rural	Urban	Total
1993–1994	50.1	31.8	45.3	328.6	74.5	403.7
2004–2005	41.8	25.7	37.2	326.3	80.8	407.1
2011–2012	25.7	13.7	21.9	216.5	52.8	269.3

Source: Press Information Bureau, Government of India website https://pib.gov.in/newsite/erelcontent.aspx?relid=97365

Table 2.3 Percentage distribution of people below poverty line

Five states with highest percentage of person below poverty line			Five states with lowest percentage of person below poverty line		
S. no.	State	Percentage of persons below poverty line	S. no.	State	Percentage of persons below poverty line
1	Chhattisgarh	39.93	1	Punjab	8.26
2	Jharkhand	36.96	2	Sikkim	8.19
3	Manipur	36.89	3	Himachal Pradesh	8.06
4	Arunachal Pradesh	34.67	4	Kerala	7.05
5	Bihar	33.74	5	Goa	5.09

Source: Press Information Bureau, Government of India website https://pib.gov.in/newsite/erelcontent.aspx?relid=97365

Rs. 4080 per month in rural areas and Rs. 5000 per month in urban areas are estimated in terms of consumption expenditure for the all-India poverty line. Because of interstate price differentials, these poverty lines would vary among different states (Table 2.2).

In rural areas, the percentage of people below the poverty line in 2011–2012 was 25.7%. In urban areas, the percentage of people below the poverty line due to the same period was 13.7%, while the percentage of people for the entire country was 21.9%. For 2004–2005, the respective ratios for rural areas were 41.8%, and for urban areas, they were 25.7% and 37.2% for the country as a whole. For 1993–1994, it was 50.1% in rural areas, 31.8% in urban areas, and 45.3% for the country. In 2011–2012, India had 269.3 million persons below the Tendulkar poverty line compared to 407.1 million in 2004–2005, which is a reduction of 137.8 million persons over the 7-year period. From 2011 to 2012, the percentage of poor people in India was 21.9%. There were no official poverty estimates in India after 2004–2005.

According to the report, Chhattisgarh has the highest 39.93% of persons below the poverty line, and Goa has the lowest 5.09% of persons below the poverty line. The top five and bottom five states with the highest and lowest percentages of persons below the poverty line are given in Table 2.3.

Among the union territories, Dadra and Nagar Haveli have the highest 39.31% of persons below the poverty line, and Andaman and Nicobar Islands have the lowest 1% of persons below the poverty line.

2.5 International Indices for the Estimation of Poverty

Estimation of poverty is performed by various international organizations. Some of the efforts to measure poverty are briefly discussed below:

2.5.1 World Bank Poverty Line

The World Bank defines extreme poverty as living on less than $1.90 or Rs 142.10 a day, measured in 2011 purchasing power parity prices. However, the intensity of poverty, i.e., individuals with consumption levels marginally below the poverty line and individuals with consumption levels much further below the poverty line, cannot be differentiated; both the categories are considered poor.

To measure the intensity of poverty, the World Bank has developed the "poverty gap index" as an alternative way of measuring poverty. The "poverty gap index" is defined as the mean shortfall in income or consumption from the International Poverty Line ($1.90 a day in 2011 international dollars), counting the nonpoor as having zero shortfall, expressed as a percentage of the poverty line.

The poverty gap for India is reported at 4.3% in 2011, a decline from 20% in 1977, according to the World Bank collection of development indicators compiled from officially recognized sources.

2.5.2 World Poverty Clock (WPC)

WPC is a systematically formulated analytical framework to measure the progress of Sustainable Development Goals (SDGs) by the World Data Lab. The World Poverty Clock (WPC) provides real-time tracking for the majority of countries in the world (Ramachandran 2019). It includes data for income distribution, production, and consumption. According to the WPC, the percentage of people living below the extreme poverty line in the world decreased from 36% to 10% in 2015. This represents a decrease from approximately 1.9 billion people to approximately 736 million in 2015 (ibid.). Based on USD 1.9 $ poverty line, the current poverty level for India is shown as 4%.

2.5.3 Global Multidimensional Poverty Index (MPI)

In 2010, the United Nations Development Programme (UNDP) and the Oxford Poverty and Human Development Initiative (OPHI) launched the Global MPI. Compared to the conventional methodology that measures poverty only in income or monetary terms, the MPI measures multidimensional poverty covering more than 100 developing countries. It goes beyond income as the sole indicator for poverty and tracks deprivation across the three dimensions of education, health, and standard of living. The MPI ranges from 0 to 1 on a scale, with higher values implying higher poverty. India's National Family Health Survey (NFHS) is used for MPI data collection. In India, the Multidimensional Poverty Index Coordination Committee (MPICC) under NITI Aayog is the nodal agency for MPI.

Global Multidimensional Poverty Index (MPI) Report 2018 frames India as a success story devoting a separate chapter to India. As per the report, India has a reduction of poverty rate from 55% to 28% in 10 years, i.e., from 2005–2006 to 2015–2016. It is in parallel with the phenomenal level of poverty reduction achieved in China a decade or so earlier (MPI 2018).

It is fascinating to know that with reference to the World Bank's 1.90$ poverty line for extreme poverty, Global MPI Reports 2019 and 2020 show India's poverty line for 2011–2012 as 21.2% (for the year 2011–2012). It is quite close to the Tendulkar Committee-based poverty line. The Global MPI 2020 Report points out that India is ranked 62nd among 107 countries, with an MPI score of 0.123 and 27.9% of the population is recognized as multidimensionally poor; the number was 9.2% for urban and 36.8% for rural India (MPI 2020).

The Global MPI 2021 Report, which was released in September 2021, indicates that India is 66th among 109 countries, with an MPI score of 0.123. As per this report, 27.9% of India's population has been identified as multidimensionally poor; the number was 36.85% for rural and 9.19% for urban India (MPI 2021). Table 2.4 provides a comparison of the rank of India and its neighbors in the Global MPI Reports 2019, 2020, and 2021.

As per Table 2.4, we can see that the ranks of India, China, Bhutan, Myanmar, and Pakistan are declining, as per the Global Multidimensional Poverty Index (MPI) Reports published in 2019, 2020, and 2021. Only Nepal has shown an improvement of rank from MPI 2020 to MPI 2021, whereas Bangladesh showed improvement of rank from MPI 2019 to MPI 2020 with a slight decline in MPI 2021. Additionally, we can see that among all the countries, China (32) has been ranked the best, and Pakistan (75) has been ranked the lowest as per the MPI 2021 Report.

Table 2.4 Rank of India and its neighbors in Global Multidimensional Poverty Index (MPI) Reports

Country	MPI 2019	MPI 2020	MPI 2021
India	53	62	66
China	24	30	32
Bhutan	62	68	70
Myanmar	63	69	71
Pakistan	67	73	75
Nepal	58	65	55
Bangladesh	66	58	61

Source: Global Multidimensional Poverty Index (MPI) Reports of 2019, 2020, and 2021

2.6 Current Level of Poverty in India

In India, information from the Socio Economic and Caste Census 2011 (SECC 2011), the latest of below poverty line (BPL) censuses, are used for the identification of poor by the state governments. Instruments other than the poverty ratio are used for the allocation of expenditures on anti-poverty programs. Various universal programs under the Mahatma Gandhi National Rural Employment Guarantee Act (MGNREGA) and Sarva Shiksha Abhiyan (SSA) are available to all making the question of allocation debatable.

In 2011–2012, the last official estimate of poverty was released by the Planning Commission at 21.92%, which was estimated using the Tendulkar Committee approach. The Tendulkar poverty line of 21.92% adopted in 2011 is the official poverty line in the SDG 2019 Report by NITI Aayog. After that, no estimates have been officially released.

2.7 Multidimensional Poverty Index by NITI Aayog

NITI Aayog is the nodal Ministry for the Multidimensional Poverty Index (MPI). The National MPI Project is focused on deconstructing the Global MPI and creating a globally aligned and yet customized India MPI for drawing up comprehensive reform action plans. The larger goal is to improve India's position in the ranking of Global MPI. As the nodal Ministry for MPI, NITI Aayog is also responsible for engaging with the publishing agencies of the index and ranking state and union territories based on their performance (NITI Aayog).

Including members from relevant line ministries and departments, NITI Aayog has established a Multidimensional Poverty Index Coordination Committee (MPICC). The exercise is targeted to compel states to take aggressive measures for poverty reductions by competing with each other. The results are expected to be added to the UNDP's Multidimensional Poverty Index (MPI).

NITI Aayog released the National Multidimensional Poverty Index 2021 (NITI MPI) Baseline Report based on the National Family Health Survey (NFHS)-4 (2015–2016) in September 2021. Under the Ministry of Health and Family Welfare, the International Institute for Population Sciences (IIPS), Government of India, conducts NFHS, and NFHS-4 was conducted in 2015–2016.

As per the NITI MPI 2021 Baseline Report, the percentage of the population who is multidimensionally poor has been estimated to be 32.75% in rural areas, 8.81% in urban areas, and 25.01% for the country as a whole. Kerala has the least 0.71%, and Bihar has the maximum 51.91% of the population, which is multidimensionally poor. There are wide variations in the head count ratio across states, which are shown in Fig. 2.1.

2.8 Sustainable Development Goals (SDGs)

The Sustainable Development Goals (SDGs) include 17 global goals ratified by the United Nations and its 193 members as a "blueprint to achieve a better and more sustainable future for all." The SDGs were set up by the United Nations General Assembly in 2015 and intended to be achieved by 2030. They are part of United Nations Resolution 70/1, the 2030 Agenda. Our experience has shown us that economic growth alone does not alleviate poverty or generate employment opportunities for youth. It is important to ensure that the benefits of economic development reach the lowermost strata of the

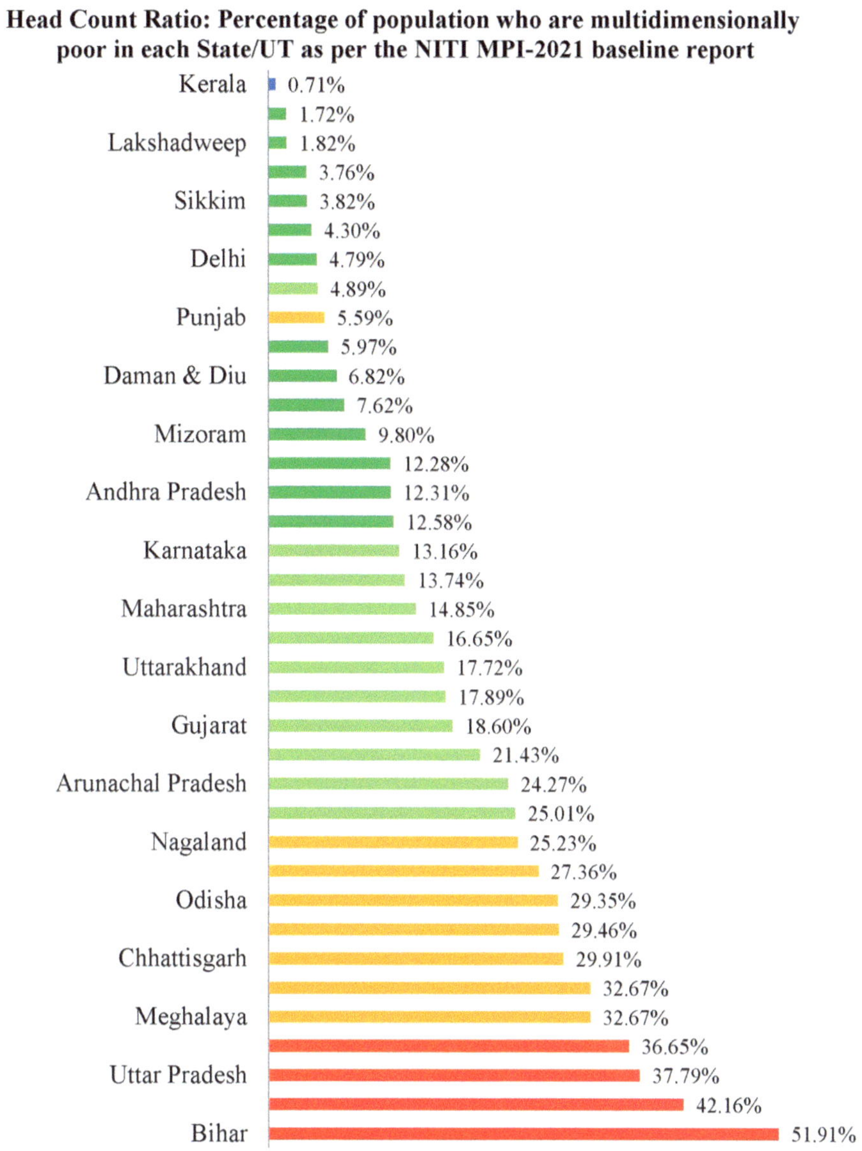

Fig. 2.1 Head count ratio as per the NITI MPI 2021 Baseline Report

population, especially the poor and marginalized families, farmers, laborers, women, and the differently abled (*divyang*), who have thus far not been able to enjoy the fruits of development in terms of improvement in human development indicators and access to resources.

The strength and potential of a country with reference to its natural resources require sustainable and systematic planning and acknowledging the fact that social, economic, and environmental factors are interlinked along the SDGs.

India is also a leading proponent of achieving the first goal, "No Poverty," which commits the signatories to eliminating poverty according to the common international poverty line and cutting it in half "according to national definitions." The importance of multidimensional approaches to poverty eradication is reaffirmed by the UN's 2030 Agenda for Sustainable Development, which goes beyond economic deprivation. India is committed to the 17 Sustainable Development Goals (SDGs), in which one of the most crucial goals is "ending poverty in all forms and hunger." SDG 1 aims to reduce the proportion of men, women, and children of all ages living in poverty in all its dimensions by at least half according to national definitions by 2030.

NITI Aayog is the nodal department for SDG in India and plays a pivotal role in coordinating with the state/UT governments to implement SDG at the grassroots level. NITI Aayog has published three SDG reports to date, i.e., SDG India Index Baseline Report 2018 and SDG India Index and Dashboard 2019–2020 and 2020–2021 (SDG Reports 2018). The latest report of SDG India Index and Dashboard 2020–2021 covers 17 goals, 70 targets, and 115 indicators.

2.9 Vision of SDG Goal 1

One of the visions of SDH Goal 1 is to eliminate all forms of poverty (income and deprivations) in both rural and urban areas and among all age groups, such as children, women, disabled individuals, and elderly individuals, by 2030 through effective and sustainable state interventions. These include investments in human capacities, quality basic services, optimizing employment potential of sectors, augmenting access to resources (physical and financial), and provision of social security to reduce economic shocks and other vulnerabilities.

Under Goal 1, five targets have been identified to monitor the progress in achieving the above-desired vision. India's performance against each target is described in Table 2.5.

As per Table 2.5, India is steadily progressing toward achieving the SDG target for Goal 1. The maximum gap can be seen in Target 2, as there is less awareness regarding the health insurance benefits, especially in rural areas. However, the Indian government's Ayushman Bharat Pradhan Mantri Jan Arogya Yojana (PM-JAY) is the largest health assurance scheme in the world aimed at providing a health coverage of Rs. 5 lakhs per family per year for secondary and tertiary care hospitalization to over 10.74 crores of poor and vulnerable families (approximately 50 crore beneficiaries) comprising the bottom 40% of the Indian population (https://pmjay.gov.in); progress in this target will be seen gradually.

From Table 2.6, it can be seen that compared to 11 states/union territories in 2018, only 6 states/union territories scored less than 50 in 2020 in the Goal 1 score. Most of the states/union territories have improved in the Goal 1 score. The importance of Goal 1 in overall SDGs can also be reflected from the fact that out of 37 states/union territories, 29 states have managed to show progress in the score of Goal 1 from 2018 to 2020 and their composite SDG scores have also improved. The country has overall improved the

Table 2.5 SDG Goal 1 targets and achievements

S. no.	Target	SDG target	SDG India 2018–2019	SDG India 2019–2020	SDG India 2020–2021
1	Percentage of population living below national poverty line	10.95%	21.92	21.92%	21.92%
2	Percentage of households with any usual member covered by any health scheme or health insurance	100%	28.70%	28.70%	28.70%
3	Persons provided employment as a percentage of persons who demanded employment under Mahatma Gandhi National Rural Employment Guarantee Act (MGNREGA)	100%	84.75%	85.26%	84.44%
4	Proportion of the population (out of total eligible population) receiving social protection benefit under maternity benefit	100%	36.40%	36.40%	91.38%
5	Percentage of households living in kutcha houses	0%	10.39%	4.20%	4.20%

Source: SDG India Reports 2018, 2019–2020, and 2020–2021 by NITI Aayog, Government of India

Table 2.6 Comparison of scores of states/UTs in Goal 1 and composite SDG

States/ Union Territories	SDG-2018 Goal 1 score	Composite SDG-2018 score	SDG-2019 Goal-1 score	Composit e SDG-2019 score	SDG-2021 Goal 1 score	Composit e SDG-2021 score
Andaman & Nicobar Islands	57	58	48	61	71	67
Andhra Pradesh	67	64	69	67	81	72
Arunachal Pradesh	52	51	34	53	54	60
Assam	53	49	48	55	51	57
Bihar	45	48	33	50	32	52
Chandigarh	39	68	48	70	75	79
Chhattisgarh	50	58	49	56	49	61
Dadra & Nagar Haveli	21	57	33	63	65	62
Daman & Diu	58	63	58	61	65	62
Delhi	30	62	54	61	81	68
Goa	62	64	53	65	83	72
Gujarat	48	64	47	64	66	69
Haryana	50	55	47	57	69	67
Himachal Pradesh	60	69	60	69	80	74
Jammu & Kashmir	61	53	58	59	69	66
Jharkhand	37	50	28	53	36	56
Karnataka	52	64	49	66	68	72
Kerala	66	69	64	70	83	75
Ladakh	NA	NA	NA	NA	79	66
Lakshadweep	43	62	56	63	61	68
Madhya Pradesh	44	52	40	58	44	62
Maharashtra	47	64	47	64	66	70
Manipur	44	59	42	60	60	64
Meghalaya	68	52	68	54	77	60
Mizoram	71	59	67	56	80	68
Nagaland	59	51	56	57	73	61
Odisha	59	51	47	58	41	61
Puducherry	61	65	56	66	75	68
Punjab	56	60	48	62	69	68
Rajasthan	59	59	56	57	63	60
Sikkim	64	58	65	65	80	71
Tamil Nadu	76	66	72	67	86	74
Telangana	52	61	52	67	68	69
Tripura	71	55	70	58	82	65
Uttar Pradesh	48	42	40	55	44	60
Uttarakhand	65	60	64	64	74	72
West Bengal	57	56	52	60	59	62
India	54	57	50	60	60	66

Aspirant (0-49) | Performer (50-64) | Front Runner (65-99)

Source: SDG India Reports 2018, 2019–2020, and 2020–2021 by NITI Aayog, Government of India

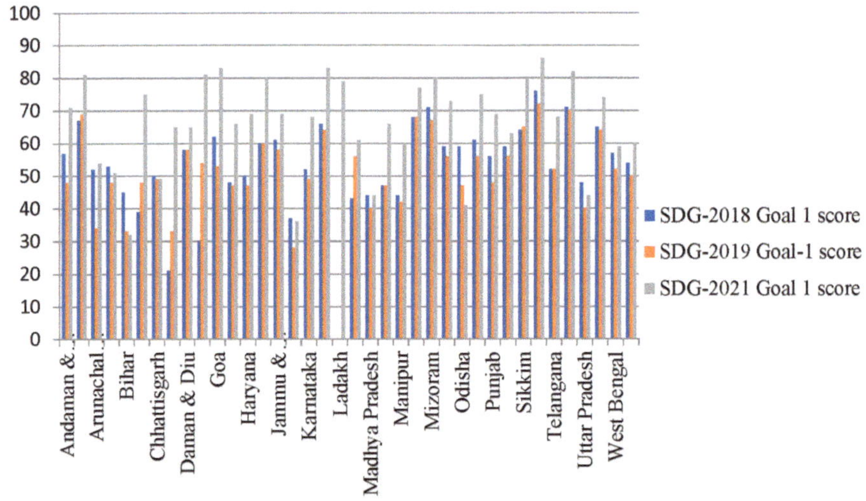

Fig. 2.2 Progress of SDG Goal 1 in states/UTs in SDG 2018, 2019, and 2021

Goal 1 score, which also enabled it to increase the composite SDG score from 2018 to 2020 (Fig. 2.2).

2.10 Measures Taken to Fight Poverty and Achieve SDG 1

Global evidence clearly indicates that India is on track for the rapid pace of poverty reduction and meeting its poverty elimination goals by 2030. A two-pronged strategy is in action to eliminate poverty, which holds a core position in India's national development agenda. Maintaining an average annual GDP growth rate of 8% is a critical and important element of the strategy for the formation of remunerative jobs for new entrants to the labor market as well as those facing redundancy in agriculture. Second, various targeted programs and schemes aim to directly attack and reduce various facets of poverty and help the poor. They facilitate income growth and increment for the economically disadvantaged by developing agriculture infrastructure and support services, creating productive assets, and developing skills and entrepreneurship. Social protection measures and mitigation of risks from natural, man-made, and other disasters aim to ensure that unforeseen exigencies do not disrupt poverty reduction efforts.

Over the last few years, the emphasis on poor households has been reflected across a range of interventions covering food security, nutrition support, housing for all with basic amenities, universal health coverage, education for all, road connectivity, social security, employment, livelihood diversification, financial inclusion, skill development, etc. Using the Socio Economic and Caste Census (SECC), 2011 data for the selection of targeted beneficiaries include the use of IT/direct benefit transfer (DBT), Aadhaar, geotagging, and other financial and governance reforms. It has transformed the delivery of benefits to the poor section of society. Various schemes (India Year Book 2021) launched by the Indian government that will directly and indirectly facilitate the achievement of SDG 1 are described below:

2.10.1 Pradhan Mantri Jan Dhan Yojana (PMJDY)

PMJDY provides universal access to banking facilities for all households in India to cover all households with at least one basic savings bank account. These savings accounts also provide the option of using the RuPay debit card with an in-built accident insurance cover of Rs. 1 lakh.

Its main objective is to cover all households across the country so that each household has at least one bank account per, with a view to increasing banking penetration and promoting financial inclusion. The long-term vision behind launching PMJDY and linking it with individual Aadhaar is to provide targeted benefits in the form of cash or kind transfers to the real beneficiaries.

2.10.2 Direct Benefit Transfer (DBT)

DBT aims to reform the government financial delivery system in welfare schemes by directly providing benefits (cash or kind) to the beneficiaries' Aadhaar-linked authenticated bank account. This ensures accurate targeting and end-to-end digitization of beneficiaries, deduplication, and reduction of fraud. DBT works on the pillars of JAM trinity, i.e., Jan Dhan, Aadhaar, and Mobile. They are the enablers of DBT, and more than 22 crore Jan Dhan account, more than 100 crore Aadhaar, and approximately 100 crore Mobile connections provide a unique opportunity to implement DBT in each welfare scheme across country, including states and UTs. The DBT mission under the cabinet secretariat of the Government of India monitors the progress of DBT across various schemes and has ranked states/UTs on their performance in DBT implementation (India Year Book 2021).

DBT along with other technical interventions has led to the removal of duplicate/ghost beneficiaries and plugging of leakages, as the result of which the government has been able to target the genuine and deserving beneficiaries. After the implementation of DBT, the estimated cumulative savings/benefits in various schemes up to March 2021 are Rs. 2,22,968.43 crores as per the DBT Bharat Portal, as of January 21, 2022.

As of January 2022, Haryana is ranked at the top, followed by Uttar Pradesh in the DBT performance rankings in states/UTs, and West Bengal is ranked at the bottom (DBT Bharat Portal).

2.10.3 Mahatma Gandhi National Rural Employment Guarantee Act (MGNREGA)

MGNREGA aims to provide a minimum of 100 days of guaranteed work in a fiscal year to each rural household whose adult members are interested in doing unskilled manual work, thereby increasing livelihood security in rural areas of the country. The four basic tenants of the MGNREGA are social security, gender parity, social inclusion, and equitable growth. The government has been working toward bringing positive changes in the program with new measures such as the public financial management system (PFMS), Aadhaar seeding of beneficiaries, geotagging of assets created under MGNREGA, and strengthening of social audit system to bring in more accountability and transparency in the program implementation. Currently, more than 84% of persons are provided employment as a percentage of persons who demanded employment under MGNREGA. It is one of the targets of SDG 1. An interesting case study undertaken in the state of Uttar Pradesh under MGNREGA is placed in Annexure.

2.10.4 Deendayal Antyodaya Yojana—National Rural Livelihoods Mission

The National Rural Livelihoods Mission (launched in 2011) has been renamed the Deendayal Antyodaya Yojana—National Rural Livelihoods Mission. It aims at providing livelihood opportunities to the rural population and seeks to reach out to 8–9 crore rural poor households through participatory processes. The identification of poor households is approved by Gram Sabha. It organizes one woman member from each household into affinity-based women SHGs and federations at the village level and at higher levels. DAY-NRLM ensures social inclusion of every section of society so that 50% of the beneficiaries belong to scheduled castes and scheduled tribes, 15% belong to minority groups, and 3% belong to persons with disability.

Providing livelihood indirectly helps reduce the percentage of the population below the national poverty line and thereby focuses on achieving Target 1 of SDG 1.

2.10.5 Ayushman Bharat Pradhan Mantri Jan Arogya Yojana (PM-JAY)

PM-JAY is considered the largest health assurance scheme of the world providing a health cover of Rs. 5 lakhs per family per year for secondary and tertiary care hospitalization to over 10.74 crores of poor and vulnerable families (approximately 50 crore beneficiaries) that form the bottom 40% of the Indian population (https://pmjay.gov.in). Providing health insurance directly helps increase the percentage of households with any usual member covered by any health scheme and thereby focuses on achieving Target 2 of SDG 1.

2.10.6 Pradhan Mantri Matru Vandana Yojana (PMMVY)

PMMVY is a maternity benefit program implemented in the country from January 2017. Under this program, pregnant women and lactating mothers are the beneficiaries of Rs. 5000, i.e., deposited directly in their savings account for the first living child of the family, it fulfills the specific conditional care related to maternal and child health. The main objective of PMMVY is to provide compensation for wage loss so that the women can get sufficient rest during and after pregnancy. It is implemented using Anganwadi centers and ASHA/ANM workers. It is essential for the women to register her pregnancy at the government-approved health facility within 150 days from the date of the last menstrual period (LMP) for benefit. After registration, the women are provided mother and child protection (MCP) cards, which have their maternity-relevant details. Providing maternity benefits directly helps increase the proportion of the population receiving social protection and thereby achieves Target 4 of SDG 1.

2.10.7 Pradhan Mantri Awas Yojana—Gramin (PMAY-G) Rural Housing

Housing is universally and unanimously recognized as one of the basic human needs and is accepted as an important component of the government's poverty alleviation strategy. Initially, the Ministry of Rural Development launched Indira Awaas Yojana (IAY), which aimed to provide housing to families living below the poverty line (BPL) in rural areas. Since its inception, it has provided assistance for the construction of 360 lakh houses. "Housing for All" by 2022 became the priority of the government for financial year 2016–2017, and the rural housing scheme IAY has been restructured to Pradhan Mantri Awas Yojana—Gramin (PMAY-G). The main objectives of the scheme are to provide assistance for the construction of houses in rural areas along with unit assistance up to Rs. 1.20 lakh in plains and Rs. 1.30 lakh in hilly states and difficult areas.

The identification of beneficiaries is based on the Socio Economic and Caste Census (SECC) 2011 data that cover households for houseless or living in houses with kutcha walls and kutcha roof, with two rooms or less after excluding households falling under the automatic exclusion category and thereby directly focusing on achieving Target 5 of SDG 1.

2.10.8 Pradhan Mantri Gram Sadak Yojana

In 2000, Pradhan Mantri Gram Sadak Yojana (PMGSY) was initiated as part of a poverty reduction strategy to help states build an all-weather road with necessary culverts and cross-drainage structures, which is operable throughout the year. Its main objective is to provide connectivity, which in turn leads to the economic development of nearby people and areas. All-weather roads are built for the eligible unconnected habitations as per the core network with a population of 500 persons and above in plain areas as per the 2001 Census.

2.10.9 National Social Assistance Programme (NSAP)

The National Social Assistance Programme (NSAP), launched in 1995, is a social assistance program for poor households—for the aged (above 60 years), widows, and differently abled and in the case of death of the breadwinner—with the intention of ensuring that social protection to the beneficiaries is available throughout the country. It aims at ensuring minimum national standards in addition to the benefits that the states are providing or might provide in further. Under the NSAP, the following five schemes are operational in the country:

(i) Indira Gandhi National Old Age Pension Scheme (IGNOAPS): Under the scheme, assistance is provided to persons aged 60 years and above belonging to families living below the poverty line as per the criteria prescribed by the Government of India. Central assistance of Rs. 500 per month is provided to persons 80 years or older, and Rs. 200 per month is given to people who are in the age group of 60–79 years.

(ii) Indira Gandhi National Widow Pension Scheme (IGNWPS): Under this scheme, Rs. 300 per month is provided to widows belonging to families living below the poverty line and in the age group of 40–79 years, as per the criteria prescribed by the Government of India. After attaining the age of 80 years, the beneficiary is shifted to IGNOAPS for a pension of Rs. 500 per month.

(iii) Indira Gandhi National Disability Pension Scheme (IGNDPS): Under this scheme, financial assistance of Rs. 300 per month is provided to persons aged 18–79 years who have severe or multiple disabilities and belong to a below poverty line (BPL) family as per the criteria prescribed by the Government of India. After the beneficiary attains the age of 80 years, he would be shifted to IGNOAPS to obtain a pension of Rs. 500 per month.

(iv) National Family Benefit Scheme (NFBS): Under this scheme, a below poverty line (BPL) household is entitled to a lump sum amount of financial assistance on the death of primary bread earner aged between 18 and 59 years. The amount of financial assistance is Rs. 20,000.

(v) Annapurna: Under this scheme, 10 kg of food grains per month are provided free of cost to senior citizens who, although eligible under the Indira Gandhi National Old Age Pension Scheme (IGNOAPS), are not receiving the pension under IGNOAPS.

2.10.10 Social Security Schemes

The Government of India launched the following three ambitious social security schemes pertaining to the insurance and pension sectors for creating a universal social security system, targeted especially for the poor and the underprivileged:

2.10.10.1 Atal Pension Yojana (APY)

To address the longevity risks among the workers in the unorganized sector who are not covered under any statutory social security scheme, Atal Pension Yojana (APY) was launched in May 2015. The APY focuses on all Indian citizens in the unorganized sector who join the National Pension System (NPS). Any Indian citizen aged between 18 and 40 years can join through their savings bank account/post office savings account. Minimum pension of Rs. 1000 or Rs. 2000 or Rs. 3000 or Rs. 4000 or Rs. 5000 is guaranteed by the Indian government to the subscriber at the age of 60 years.

2.10.10.2 Pradhan Mantri Jeevan Jyoti Bima Yojana (PMJJBY)

The PMJJBY is a 1-year life insurance scheme which is renewable from year to year. It offers Rs. 2 lakhs coverage for death due to any reason and is available to people in the age group of 18–50 years (life cover up to 55 years of age) having a bank account who consented to join and enable autodebits.

2.10.10.3 Pradhan Mantri Suraksha Bima Yojana (PMSBY)

The PMSBY is a 1-year personal accident insurance scheme that is renewable each year, offering coverage for death/disability due to an accident. It is available to people in the age group of 18 to 70 years with a bank account who consent to join and enable autodebits. The risk coverage available will be Rs. 2 lakh for accidental death and permanent total disability and Rs. 1 lakh for permanent partial disability under this scheme.

Implementation of the above schemes enables affordability and targeting in favor of the poor and the underprivileged. Furthermore, these schemes would address the situation of low penetration of insurance in the country.

2.10.11 Schemes Launched to Provide Relief During the COVID-19 Pandemic

2.10.11.1 Pradhan Mantri Garib Kalyan Yojana

Under this scheme during the wake of the COVID-19 pandemic, 81 crore individuals are being provided with 5 kilos of rice or wheat and 1 kilo of chana every month. Cash transfers to 20 crore Jan Dhan accounts are a fresh stimulus. The beneficiaries would include 3 crore poor senior citizens, poor widows, and other groups. Approximately 16% of the EPFO subscribers are likely to be covered under this scheme. Under the Kisan Samman Nidhi program, an additional amount of Rs. 2000 was given as the front load to farmers. Mahatma Gandhi National Rural Employment Guarantee Act was also raised, and along with the daily wage, workers provide relief to the business. In February 2020, an additional working capital finance of 20% of the outstanding credit in the form of a term loan at a concessional rate of interest was announced. The government provided Rs. 20,000 crore subordinate debts for 2 lakh micro, small, and medium enterprises (MSMEs), which are non-performing assets (NPA) or are stressed. A fund with a corpus of Rs. 10,000 crore was set up by the government to provide equity funding support for MSMEs.

2.10.11.2 One Nation One Ration Card (ONORC)

The One Nation One Ration Card (ONORC) is an ambitious plan to ensure ceaseless delivery of subsidized food security to all people covered under the National Food Security Act 2013 (NFSA), irrespective of beneficiaries' physical location anywhere in the country. The program objective is to empower all NFSA beneficiaries to be self-reliant for their food security anywhere in the country through portability of the same existing ration cards.

The migrant beneficiary can easily locate shops in new areas using maps to reach the closest fair price shop and collect their subsidized food grains (in part or full) from any ePoS (electronic point-of-sale device)-enabled fair price shop in the country with Aadhaar/biometric authentication. Furthermore, their family members at home can also take balance/their requirement of food grains on the same ration card. This immensely helped migrant workers during COVID lockdowns.

Furthermore, ONORC has now also become a part of the "Prime Minister's Technology-Driven System Reforms under the Atma Nirbhar Bharat Abhiyan" due to its potential to empower migrants.

2.10.11.3 e-Shram Portal

The e-Shram portal is the first-ever National Database of Unorganized Workers (UWs), which was launched on August 26, 2021, and crossed the mark of 21.85 crores as on January 13, 2022 (https://register.eshram.gov.in). This database will have details of name, occupation, address, educational qualification, skill types, and family details for the optimum realization of their employability. The e-Shram portal seeded with Aadhaar will be used to deliver all the social security benefits of the central and state governments for the unorganized workers.

Under the PMSBY scheme, all eligible workers are registered on the e-Shram portal. They can avail an accidental insurance cover of 2 lakhs from the date of policy issuance. Various states have also started using the database to provide relief for the registered unorganized workers in lieu of the third COVID-19 wave in the state.

2.11 Way Forward

Globally as well, in India, it is well accepted that poverty is multidimensional and only through a concerted effort on its different dimensions can a real dent on poverty be made. Among many, there are two critical issues in the discourse on poverty in India. One relates to poverty measurement, and the second relates to effective poverty elimination. Poverty measures compare people in a society to assess the extent of unacceptable disadvantages that exist, yet any poverty measure is itself imperfect. The two most important factors of this imperfection are data limitations and the diversity of human lives being assessed more so in a vast country such as India.

Furthermore, perceptions of what defines basic human needs vary widely according to income, level of development, sociopolitical beliefs, and other factors. This is why views on how the poverty line should be defined vary widely. This makes the choice of a poverty line difficult.

Poverty line assessment if it were to be done in present circumstances cannot be based on minimum expenditure on subsistence basket as done in the past. The ongoing COVID-19 pandemic has put emphasis on the criticality and importance of certain "essentials"—access to quality healthcare, education and awareness, water and sanitation facilities, adequate nutrition, and the need for living spaces where social distancing can be practiced according to the guidelines laid out by the WHO. Recalibration of the poverty line should be done based on changes in income, consumption patterns, and prices. In India, poverty measurement of an individual has repeatedly led to contentious debates on the poverty line. Despite these shortcomings, conceptually having a poverty line and related poverty estimates helps to concentrate the public policy discourse around an agreed set of numbers as well as to track the progress in combating poverty.

Over time, priorities and needs have shifted with development in India. Today, the aspiring poor seek quality in education; betterment in health, housing, skills, and consumption; and not merely minimum food and shelter. Therefore, poverty is now not just about basic food to keep body and soul together but about living standards—sanitation, housing, piped water, electricity, education, health, financial inclusion, and jobs.

Current projects focusing on SDG and to develop the Multidimensional Poverty Index (MPI) spearheaded by NITI Aayog may be expected to provide poverty indices at national, state, and lower levels of granularity with a focus on multidimensionality to improve SDG 1, which in turn will pave the way for achieving Sustainable Development Goals.

Sustainable Development Goals are recognition of the fact that poverty cannot be addressed merely by increasing incomes. Development is a multidimensional construct similar to poverty. Thus, policy-makers in developing societies must address the intertwined and interrelated issues of development to address poverty. The twofold strategy of enabling the economy to grow rapidly (with high employment intensity) on a sustained basis and attacking poverty and addressing disparities through social welfare programs remains relevant along with bringing more efficiency in implementing the current schemes run by the government. Finally, SDGs is not just a program of the government; it is the need of the hour which can only be achieved if we, the people, join hands and come together as a nation so that even the future generations can reap the benefits of the sustainable developments.

Annexure

Case Study: Operation Chaturbhuj—A New Strategy Under MGNREGA Implemented in District Lakhimpur Kheri, Uttar Pradesh (http://planning.up.nic.in).

In the analysis of public complaints, it was found that in rural areas, cases of mutual rivalry and cross Police First information report (FIR) arise due to the encroachment of sector roads/earthen roads (Chak Roads), ponds, and public lands. In rural areas, even after the completion of "Chakbandi" or the measurement and demarcation of agricultural plots, roads are sometimes not built due to encroachment issues. Under Operation Chaturbhuj, the illegal encroachers of public roads and ponds were provided with a win-win option using Operation Chaturbhuj.

The aim of Operation Chaturbhuj is to provide livelihoods to thousands of job seekers, mitigate disputes after demarcations of roads, provide road connectivity after "Chakbandi," and boost the environment through plantations on either side of roads. This was possible due to the consolidated efforts of Rural Development, Revenue, Police, Agriculture, and other departments along with the passionate district administration.

Under this project, earthen road construction (chak roads) work was initiated in all villages where "Chakbandi" was completed within the last decade, but roads were not built. The villagers were sensitized about the socioeconomic benefits of having roads along with offering permanent solutions to the above-revenue/legal problems. Through this Operation, mostly, the small and marginal farmers received direct access to transportation rights from their farm to the market, who otherwise were very much dependent on others' land for passage. Agricultural equipment and machinery can now easily reach each field of villagers. This has reduced land disputes to a good extent. By constructing ponds and ensuring that it is being given on lease to the fishing community, water conservation efforts have received an impetus, along with providing employment to families. Plantation drives across the ponds also helped in environmental improvement. All this has created huge employment opportunities in the villages of Lakhimpur Kheri, that too during the COVID-19 pandemic when migrant laborers were coming in huge numbers. District Kheri has also been ranked number 1 in the state in providing employment during the COVID-19 pandemic under this campaign. After the success of the pilot campaign in the Gram Panchayat Parsa, Operation Chaturbhuj was expanded to other blocks as well. Since this campaign was developed under the built-in incentive structure and community participation, it was successful in providing a permanent solution.

References

Best Practices, Atma Nirbhar Uttar Pradesh Volume-1, Planning Department, Government of Uttar Pradesh from http://planning.up.nic.in/page/en/StateLevel.html

DBT Bharat Portal, DBT Mission Cabinet Secretariat, Government of India from https://dbtbharat.gov.in/state/state-ranking

e-SHRAM Portal, Ministry of Labor & Employment, Government of India from https://register.eshram.gov.in/#/web-dashboard

Explainer Note on National Multidimensional Poverty Index, NITI Aayog, Government of India, Press Information Bureau from https://pib.gov.in/PressReleasePage.aspx?PRID= 1775489

Gaur S, Rao NS (2020) Working paper on poverty measurement in India: a status update, Ministry of Rural Development, Government of India. https://rural.nic.in/sites/default/files/WorkingPaper_Poverty_DoRD_Sept_2020.pdf

Global Multidimensional Poverty Index Report (2018). On Oxford Poverty and Human Development Initiative from https://ophi.org.uk/gmpi-2018/

Global Multidimensional Poverty Index Report (2020) On Oxford Poverty and Human Development Initiative from https://ophi.org.uk/global-mpi-report-2020/

Global Multidimensional Poverty Index Report (2021) On Oxford Poverty and Human Development Initiative from https://ophi.org.uk/global-mpi-report-2021/

India Year Book (2021) Ministry of Information and Broadcasting, Government of India

Pradhan Mantri Jan Arogya Yojana from https://pmjay.gov.in/about/pmjay

Press Note on Poverty Estimates (2011–2012) Planning Commission, Government of India, July 2013

Ramachandran N (2019) India is making dramatic strides in addressing extreme poverty. Mint, November 11. https://www.livemint.com/opinion/columns/india-is-making-dramatic-strides-in-addressing-extreme-poverty-11573494674626.htm

SDG India Index Baseline report (2018) SDG India index and dashboard 2019–20 & 2020–21 on NITI Aayog, Government of India from https://www.niti.gov.in/reports-sdg

The World Bank (2000–2001) World development report: attacking poverty. The World Bank, Washington, DC. https://openknowledge.worldbank.org/handle/10986/11856

The World Bank (2022) Commission on Global Poverty. https://www.worldbank.org/en/programs/commission-on-global-poverty

Ravi Patni is an engineering graduate from VIT University, Vellore, and is currently working as research officer (Engineering) in Planning Department, Government of Uttar Pradesh. He is currently involved in socioeconomic data analysis and coordinating with central government to implement new initiatives and programs in the state.

3 Summary Measures of Population Health: Healthy Life Expectancy in India in the Context of Sustainable Development

Shewli Shabnam and Nandita Saikia

Abstract

Ensuring healthy lives and promoting well-being for all were recognized as an integral part of the United Nations 2030 Agenda for Sustainable Development. The present increase in life expectancy is primarily explained by the mortality reductions from chronic diseases at older ages. Therefore, debates arise on whether longer life means better health and if mortality indicators are enough to measure health. These considerations led to the development of Summary Measures of Population Health (SMPH) which combines both morbidity and mortality data to show the health status of a population as a single numerical index. In this context, we presented a brief history of the development of SMPH to understand how the nonfatal health outcomes have gained due recognition in measuring health. As healthy life expectancy (HLE) is one of the most popular and widely used SMPH, we estimated the healthy life expectancy of India using the Sullivan method based on the World Health Survey data (2003). We compared healthy life expectancies derived from self-rated general health, self-reported activity limitations, and self-reported functional limitations (involving social, psychological, and physical domains of health). In addition, we calculated the morbidity-free life expectancy (MFLE) of the males and females in India using the data of the National Sample Survey, round 75, conducted in 2017–2018. We found that healthy life expectancies, measured by all three aspects of health, were higher among males. In contrast, morbidity-free life expectancies based on the age-specific prevalence rate of ailing persons were higher among women. It reflects that MFLE differs from HLE as the latter includes much broader aspects of health. Therefore, researchers should carefully choose the appropriate health indicators for assessing health. As India is experiencing population ageing with substantial regional variation and the burden of noncommunicable diseases has already exceeded the burden of infectious diseases, we should use SMPH to assess population health in India. In our opinion, healthy life expectancy is a good indicator for evaluating population health in the context of SDG 3.

S. Shabnam (✉)
Department of Geography, Bidhannagar College, Government of West Bengal, Kolkata, India
e-mail: shewlijnu@gmail.com

N. Saikia
Department of Public Health and Mortality Studies, International Institute for Population Sciences, Mumbai, India

V. Dutta, P. Ghosh (eds.), *Sustainability: Science, Policy, and Practice in India*, Sustainable Development Goals Series, https://doi.org/10.1007/978-3-031-50132-6_3

Keywords

SDG 3 · Healthy life expectancy · Summary measures of population health · Morbidity-free life expectancy

3.1 Introduction

The aim of ensuring healthy lives and promoting well-being for all has been delineated by the United Nations as Sustainable Development Goal 3 (SDG-3), which must be achieved by 2030 (United Nations 2015). In this context, the vital question is how to measure a healthy life. Conventionally, data on mortality and morbidity has been used to understand the health status of a population (Murray 1994). Infant mortality rate, under-five mortality rate, life expectancy at birth, and incidence and prevalence rates of diseases are common indicators used to assess health status.

In the twentieth century, a continuous decline in death rates and a subsequent rise in life expectancies were observed in developed countries. The increase in the length of life was primarily associated with the decline of chronic ailments at older ages. It triggered an in-depth discussion on whether living longer signifies better health, and whether mortality indicators alone are sufficient to measure health. These deliberations fostered the development of Summary Measures of Population Health (SMPH) which combines both mortality and morbidity data to show the general health of a population as a single number (Field and Gold 1998). Some typical summary measures of population health include disability-free life expectancy (DFLE), active life expectancy (ALE), disability-adjusted life years (DALYs) and such others.

In this chapter, our objective is to present a brief history of the development and use of SMPH. It helps us understand how nonfatal health outcomes (due to disability and chronic diseases) have gained recognition in measuring health. In light of the increasing use of SMPH, one may be curious about the population health in India in terms of SMPH. As healthy life expectancy (HLE) and morbidity-free life expectancy (MFLE) are widely accepted SMPH, our second objective is to estimate the HLE and MFLE of India based on two nationally representative sample surveys. Our third objective is to explain the relevance of HLE in the context of Sustainable Development Goals.

3.2 Development of Summary Measures of Population Health

In 1964, Sanders first combined the data on mortality and morbidity to understand the standard of health of different communities (Sanders 1964). In the last 50 years, numerous researchers have contributed to the formulation, calculation, and use of SMPH. Broadly, SMPH can be grouped into two families based on a simple survivorship curve – (1) health expectancy and (2) health gap. Health expectancy is a generic term for those population health indicators "that estimate the average time (in years) that a person could expect to live in a defined health state" (Mathers et al. 2001b, p. 6). On the other hand, the health gap measures "the difference between the actual health of a population and some stated norm or goal for population health" (Murray et al. 2002, p. 17).

3.2.1 SMPH: Health Expectancies

The first use of health expectancies to measure population health dates back to the works of Sanders (Sanders 1964). Based on the Kit Carson County Morbidity Study conducted in America, Sanders concluded that the communities with adequate health care would register a greater prevalence of chronic diseases and lower death rates than those with inadequate health care. To evaluate the efficiency of health services in various communities, he developed a modified life table method based on (a) death rates and (b) functional effectiveness (as a substitute for morbidity). He then estimated productive man-years for each cohort of conceptions. A higher number of productive man-years per 100,000 conceptions indicated better health-care services.

The following year, Chiang developed mathematical models for describing the health state of a given population at a given time (Chiang 1965). Chiang tried to formulate general models to measure the frequency of illness, disease duration, and time lost due to death. Combining these three variables, he computed the average fraction of the year in which an individual remained healthy. He referred to that fraction as the mean duration of health, which was used as an index of the health of a population.

The first precise method of calculating health expectancy was developed by Sullivan in 1971. Based on the National Health Interview Survey data (1964) and the Vital Statistics of the United States (1965), he estimated the life expectancy free of disability and free of bed disability for White and all other persons by sex (Sullivan 1971). To calculate disability-free life expectancy (DFLE), he used period life tables where the variables were age-specific death and disability rates of a particular time. He found that in the mid-1960s in the USA, DFLE at birth was 61.6 years and 68.4 years for males and females, respectively. His analysis revealed that the DFLE at birth were 7.4 years higher for White males and 8.0 years higher for White females than their non-White counterparts.

In Japan, the social indicator movement in the 1960s led to the development of various useful socioeconomic indicators. The Council of National Living tried to integrate many of them and published trial results in 1974. The results included information on the changes in life expectancy (LE) and DFLE in Japan between 1966 and 1970. It was found that the rise in LE was slightly higher than the rise in DFLE during that period (The Council of National Living 1974).

Based on available data, researchers started calculating the health expectancies of Canada in the early 1980s. Using the information on (i) short-term disability and long-term activity restriction from the Canada Health Survey (1978–1979), (ii) long-term stay in hospitals from institutional records, and (iii) mortality data from vital statistics, Wilkins and Adams (1983) estimated life expectancies in different health states (e.g., life years lived institutionalized, life years under restricted activity but not involving institutionalization, and the like), and a sum of the indices was named quality-adjusted life expectancy (QALE). The estimated difference between LE and QALE was 5.0 years for males and 6.6 years for females.

The estimation of disability-free life expectancy for the United Kingdom was first published by Bebbington (1988). He calculated the expectation of life without disability (ELWD) for England and Wales for 1976, 1981, and 1985 based on the British General Household survey data. There were two questions in the survey schedule: "(i) Do you have any long-standing illness, disability or infirmity? (ii) Does this illness or disability limit your activities in any way?" (Bebbington 1988, p. 322). If the answer was "yes" to both questions, it confirmed activity limitation due to disability. He found that in 1985 in England and Wales, ELWD was 61.5 years for females and 58.7 years for males. Also, between 1976 and 1985, the rate of increase in life expectancy was higher than the rate of improvement in ELWD.

One major work on DFLE about the American population was published in 1989 by Crimmins et al. (1989). They tried to compare the DFLE of the United States between 1970 and 1980. For mortality data, they used the decennial life tables of the United States. To calculate the percentage of institutionalized persons, they used the census data. Information on disability for the noninstitutionalized population was collected from the National Health Interview Survey. Their study revealed that LE at birth rose about 3 years between 1970 and 1980 for males and females in the United States. However, in the case of DFLE, the increase was 0.7 years for males, and no change was observed for females. So, they concluded that improvement in medical sciences led to a rise in life expectancy, but people spent increasing proportions of their lives as bed-ridden dependents.

During the 1970s and 1980s, several studies on disability-free life expectancy pointed to the difficulties of getting comparable data over the long run. The International Network on Health Expectancy and the Disability Process (Réseau Espérance de Vie en Santé, or REVES) was established in 1989 to "conduct research and encourage the use of standardized methods for data collection and calculation of health expectancies" (Ined 2021). REVES started using DFLE for cross-national comparisons.

In 1993, information on DFLE was incorporated into the health database of the Organisation for Economic Co-operation and Development (OECD) (Mathers 2002). In 2011, a health-related report for OECD countries, "Health at a Glance," provided an estimation of DFLE (referred to as healthy life years in the report) at age 65 for European countries for the first time (OECD 2011). Besides, the Eurostat annually estimates DFLE/healthy life years (HLY) for European Union (EU) countries. Mortality data is obtained from Eurostat's demographic database for calculating healthy life years. Information on self-reported long-term activity limitations from the EU-SILC survey is used to estimate the disability. In 2016, the average HLY at birth was 64.2 years for females and 63.5 years for males in the EU (EUROSTAT 2019).

Health expectancy as a measure of population health has been used in many developing countries since the mid-1980s. Using the Sullivan method, Grab and colleagues in 1991, Wang in 1993, and Qiao in 1997 calculated the health expectancies of the people of China (Saito et al. 2003). Other research on health expectancies in China include the works of Gu et al. (2009), Zimmer et al. (2010), and Lu et al. (2018). Among other Asian countries, Tu and Chen (1994) calculated the DFLE and disease-free life expectancy of the adults in Taiwan for 1986 and 1991 using a double decrement life table model. LE and HLE of Thai elderly were calculated by Jitapunkul and Chayovan (2000) for 1986 and 1995.

In India, using the Sullivan method, a study on morbidity-free life expectancy was (MFLE) carried out by Thomas et al. (2014). They used the information on mortality rates from the Sample Registration System (SRS) and the data on morbidity from the 60th round of the National Sample Survey (NSS). They computed morbidity-free life expectancy, restricted activity-free life expectancy, and bed disability-free life expectancy for the elderly. In 2004, MFLE at age 60 was 11.2 years for males and 12.5 years for females in India. However, the percentages of MFLE to total LE were 67.3 for females and 68.1 for males. Bora and Saikia (2015) calculated the gender-specific DFLE in India based on self-reported health information. The data used for the study came from WHO Study on Global Ageing and Adult Health (SAGE) in India, 2007. They found a greater prevalence of severe and extreme disability among women in 14 out of 20 ADL measures. They observed that DFLE were higher among men irrespective of their ages. Sreerupa et al. (2019) compared the life expectancy and mobility-free life expectancy among the elderly in India between 1995–1996 and 2004 using the Sample Registration System and National Sample Survey data. Their study indicated the compression of morbidity in India. In another study, based on the data from the Census of India 2011 and the SRS abridged life tables of India 2011, Mishra et al. (2020) computed the DFLE in India. They found higher DFLE among females than males. The study revealed a significant regional disparity in DFLE in India.

Studies on health expectancies are also available from South America. Romero et al. (2005) estimated healthy life expectancies in Brazil using the World Health Survey (WHS) data conducted in 2003. They observed that the percentage of years lived in poor health was higher among females than males. Moreno et al. (2018) made a detailed study on DFLE among older adults in Santiago, Chile. The study also concluded that compared to the male population, females had a higher percentage of disabled years at older ages.

Studies on the comparison regarding DFLE among developing economies are limited. In one such study, Roberto Ham-Chande (2003) compared health expectancies among five Latin American and one Caribbean city (Sao Paulo, Montevideo, Santiago, Buenos Aires, Mexico City, and Bridgetown) in 2000. At age 65, the highest DFLE was estimated in Montevideo (Uruguay) and the lowest in Santiago (Chile). In another comparative study, Santosa et al. (2016) calculated the DFLE of six developing economies (China, Ghana, India, Mexico, the Russian Federation, and South Africa) based on the

WHO SAGE data (2007–2010). The lowest prevalence of disability was observed in China and the highest in India. In all six countries, women had higher life expectancies. Still, their percentage of DFLE at age 50 and over was lower than men.

3.2.2 SMPH: Health Gaps

The most popular measure of health gap, called disability-adjusted life years (DALY), was introduced to assess the Global Burden of Disease (GBD) in the "World Development Report 1993" (World Bank 1993). The rationale for developing DALY as an indicator of the burden of disease has been discussed in detail by Murray and Lopez in their monumental work the "Global Burden of Disease and Injury Series, Volume I" (Murray et al. 1996). Estimation of DALY combined "the years of life lost due to premature mortality (YLL) in the population and the equivalent 'healthy' years lost due to disability (YLD)" (Mathers et al. 2001b, p. 9).

The most important findings of the GBD study of the 1990s were published in four consecutive articles in the "Lancet" in 1997 (Murray and Lopez 1997a, b, c, d). The second article estimated disability-free life expectancy (DFLE) and disability-adjusted life expectancy (DALE) for eight world regions. DALE was defined as "the expectation of the equivalent number of healthy years of life at birth" (Murray and Lopez 1997b, p. 1349). It was found that DALE was the highest in established market economies and the lowest in sub-Saharan Africa.

The World Health Organization in the year 2000 published the ranking of its member states based on DALE. The next year, the WHO renamed DALE as health-adjusted life expectancy (HALE) following the feedback from the member states (Mathers et al. 2001a). The "World Health Report 2001" estimated HALE for 55 countries using an improved methodology, new epidemiological data for some diseases, and comparable data from 63 surveys across the world (World Health Organization 2001a).

Several studies presented estimations of DALY and DALE/HALE for different countries. One paramount national-level study on DALY was conducted in the Netherlands by the Dutch Burden of Disease Group in 1994. The three leading causes of YLL were found as ischemic heart disease, cerebrovascular disease, and cancers. On the other hand, mental disorders, alcohol dependence, and visual impairments were the leading causes of YLDs (Melse et al. 2000). Mathers et al. (2001c) calculated DALY and DALE for Australia in 1996. The research pointed to the high prevalence of mental disorders in Australian society (almost 30% of the morbidity burden in Australia). McKenna and his colleagues estimated DALY to identify the most important causes of disability and mortality in the United States. They concluded that in the mid-1990s, ischemic heart diseases accounted for the highest number of DALYs lost among males and females in the country (McKenna et al. 2005).

At the global level, the Global Burden of Disease Study (GBDS) has regularly produced estimates of DALYs. After the early 1990s, DALYs were estimated globally in 1999–2002 and 2004 (Das & Samarasekera 2012). GBD 2010 added a new crown to the efforts of quantifying the disease burden by analyzing 291 diseases and injuries in 21 world regions. It was found that noncommunicable diseases contributed to 78.6% YLDs in 2010 (Vos et al. 2012). High blood pressure was identified as the most critical risk factor contributing to the DALYs, followed by tobacco smoking and alcohol use (Lim et al. 2012). The information on the GBD is also available for 2016, where DALYs and HALE have been estimated for 195 countries and territories (GBD 2016 DALYs and HALE Collaborators 2017). According to this study, between 1990 and 2016, the contribution of NCDs to the total number of DALYs lost increased by 36.6%. However, the contribution of communicable, maternal, neonatal, and nutritional (CMNN) diseases decreased by 40.1%. At birth, the global HALE was 56.9 years in 1990 which increased to 63.1 years in 2016.

Literature on SMPH in terms of health gaps is very few in India. The India State-Level Disease Burden Collaborators recently made an enormous effort to estimate DALY for each state of India. They found that the age-standardized DALY rate dropped by 36.2% between 1990 and 2016 (India State-level Disease Burden Initiative Collaborators 2017). Menon et al. (2019) estimated the disease burden in India in 2017 using the data from 2017 UN death totals, Sample Registration System for 2010–2017, Million Death Study for 2010–2014, and YLD-YLL ratios from WHO Global Health Estimates. They found that for perinatal and nutritional conditions, chronic respiratory diseases and diarrhea, DALY rates were twice in rural areas than in urban areas. In contrast, DALY rates of ischemic heart disease were conspicuously higher in urban areas compared to its rural counterpart. A recent article assessed India's disease burden due to child and maternal malnutrition (India State-level Disease Burden Initiative Malnutrition Collaborators 2019). According to this study, malnutrition accounted for 68.2% of the total under-five mortality and 17.3% of all age DALYs in India in 2017.

From the literature review, it is evident that the use of various summary measures of population health to assess the health status of the population at the national and subnational levels has increased globally. The efforts to develop SMPH started in the 1960s, and different measures of health expectancies were derived in the next 30 years. DALY, an essential health gap measure, was formulated in the 1990s. Since then, the application of DALY has been increasing as it allows us to compare the magnitude of healthy life lost due to different diseases.

3.3 Healthy Life Expectancy (HLE) in India

Due to the effectiveness and popularity of SMPH, we are going to estimate the health status of the population of India using summary measures of population health. In this context, it is imperative to note that the procedure to calculate DALY followed by the Global Burden of Disease Study (GBDS) is not only complex; it needs an amalgamation of researchers and professionals and substantial financial assistance. In contrast, healthy life expectancy is a simple measure based on an abridged life table and the prevalence of disability by age groups (Sullivan 1971). Therefore, we have estimated the healthy life expectancy of India using the Sullivan method.

According to the World Health Organization (WHO), "health is a state of complete physical, mental and social well-being and not merely the absence of disease or infirmity" (WHO 1946). The World Health Survey (WHS), 2002–2004, conducted by the World Health Organization in collaboration with 70 countries, provides information on various dimensions of health, such as the respondent's perception of overall health status, activity limitation, and functional health (which involves social, psychological, and physical domains of health). Therefore, WHS data on self-rated health (SRH) and self-reported disabilities (SRD) were widely used to depict the population's health status in various countries. For our analysis, we also used the data of WHS because it captured the information on all aspects of health defined by the WHO.

Healthy life expectancy is the average number of years expected to be spent in good health at a particular age, assuming fixed age-specific mortality and disability rates (Stiefel, Perla and Zell 2010). Based on an extensive literature review, it is found that the measure of self-rated health based on a five-point scale "has demonstrated stability, consistency and good test-retest reliability and is strongly related to a wide set of health outcomes, including general morbidity, reported symptoms, health care utilization and mortality" (Hardy et al. 2014, p. 320). However, studies on HLE using self-rated health and self-reported disabilities have recently gained momentum in India. Singh et al. (2013) studied the male-female differences in SRH/SRD among the older population in India using the data of NSSO Round 60. Their study revealed that after covariate adjustment, a higher percentage of women reported poor SRH than men. Similar findings were reported by several other studies (Roy and

Chaudhuri 2008; Dhak 2009; Bora and Saikia 2015; Panday and Ladusingh 2013; Oksuzyan et al. 2018). In continuation with the ongoing research, in the present chapter, we have computed HLE in India based on the data on self-rated health (SRH) and self-reported disabilities (SRD) obtained from the WHS India 2003. In addition, we have also analyzed the morbidity-free life expectancy (MFLE) of the males and females in India using the 75th round of the National Sample Survey conducted in 2017–2018.

3.4 Methodology

The World Health Survey conducted in India in 2003 covered six major states: West Bengal, Maharashtra, Rajasthan, Uttar Pradesh, Karnataka, and Assam. The selection criteria of the states and the details of the sample design are available in the report "Health System Performance Report: World Health Survey 2003, India" (Arokiasamy et al. 2006). Following the report, we have pooled the data of these states for all India-level analyses, as the above six states covered almost half of the population in India. The survey covered 10,279 households. The questionnaire on health information covered a total sample of 9994 individuals aged 15 years and above.

Mathematically, healthy life expectancy (HLE) is expressed as.

$$e'_x = 1/l_x \sum_{x}^{w} \left(1 -_n \pi_x\right) nLx$$

where e'_x = HLE at age x; l_x = number of survivors at age x; w = oldest age category; nLx = the total number of person-years lived between exact ages x and x + n; ${}_n\pi_x$ = prevalence of poor health between the ages x and x + n; and $(1-{}_n\pi_x)$ = age-specific rate of being healthy.

We have estimated healthy life expectancies considering different aspects of health. In the first case (Case 1), we have estimated HLE based on self-rated general health. Here, the respondents were asked to rate their overall health status on the survey day. The answers were available on a five-point Likert scale: "very good," "good," "moderate," "bad," and "very bad." Then, we dichotomized the responses as "poor self-rated health," combining "bad" and "very bad" and "good self-rated health" combining other answers. The age-specific prevalence rates (ASPR) of poor health were subtracted from 1, to get the age-specific prevalence rate of being healthy.

In the second case (Case 2), the estimation of HLE was based on activity limitation. The respondents were asked: "Overall, in the last 30 days, how much difficulty did you have with work or household activities?" The responses were provided on a five-point Likert scale: "none," "mild," "moderate," "severe," and "extreme/cannot do." The reactions were pooled to obtain two categories: "activity with difficulties," which combined the responses "severe" and "extreme/cannot do," and "activity without difficulties," which combined the other responses. In this case, we subtracted the proportion of individuals in each age group who performed the activities with difficulty from 1 to obtain the age-specific rate of being healthy.

In the third case (Case 3), the estimation of HLE was based on the severity of functional limitations involving the social, psychological, and physical domains of health. "The International Classification of Functioning, Disability and Health (ICF)" has classified functioning as an essential component of health and defined functioning as an umbrella term for body functions, body structures, activities, and participation (WHO 2001b). Considering the multidimensional aspect of functioning, the World Health Survey in India included 16 questions to capture an individual's health. The questions were grouped under eight domains (mobility, self-care, pain and discomfort, cognition, interpersonal activities, vision, sleep and energy, and affect). Appendix Table 3.2 presents an overview of the health domains and questions regarding functional difficulties. These domains are incorporated in several popular health status-related surveys, such as the Short Form 36 (SF 36), the EuroQol 5D, and the Health Utilities Index Mark 3 (HUI 3) (Moussavi et al. 2007).

A composite score was computed based on the answers to the 16 questions on functional health. Since the item responses were based on a five-point ordered categorical scale ("none" = 1, "mild" = 2, "moderate" = 3, "severe" = 4, and "extreme/cannot do" = 5), factor analysis was done using the principal component method to obtain factor score. The analysis was performed with the help of SPSS version 22. The score in the first principal component was transformed to a (0, 1) scale, where 0 corresponds to the best health (without any functional limitation) and 1 indicates the worst health. The proportion of disability by age group was calculated as the arithmetic means of the scale's value in that age group. The rate of being healthy is derived by subtracting that value from 1.

Finally, we applied age-specific prevalence rates of being healthy (derived from Cases 1, 2, and 3) on SRS life tables of India for 2001–2005 to estimate healthy life expectancies.

3.4.1 Results

Figure 3.1 shows HLE in India based on self-rated general health, self-reported activity limitations, and self-reported functional health. It is found from Fig. 3.1 that healthy life expectancies derived from the information on general health and activity limitations produced almost similar results (Case 1 and Case 2). Healthy life expectancies in Case 3 based on multidimensional aspects of health were lower than in other cases. However, after age 60, the differences in healthy life expectancies derived from different health dimensions were minimized (Appendix Table 3.3).

Note: Case 1, self-rated general health; Case 2, self-reported activity limitation; and Case 3, self-reported functional limitations involving social, psychological, and physical domains of health.

Now, let us look at the gender differentials in HLE based on different aspects of health. Although the differences were minimal, healthy life expectancies were lower among the females in all three cases (Case 1, general health; Case 2, activity limitation; and Case 3, functional health) than males (Table 3.1).

From our analysis, we may say that general health rating is highly influenced by a person's activity limitation, social interaction capacity, and mental condition. The results are so similar for Cases 1, 2, and 3 that healthy life expectancies based on self-rated general health can be used as a proxy to HLE estimated involving functional health or activity limitations.

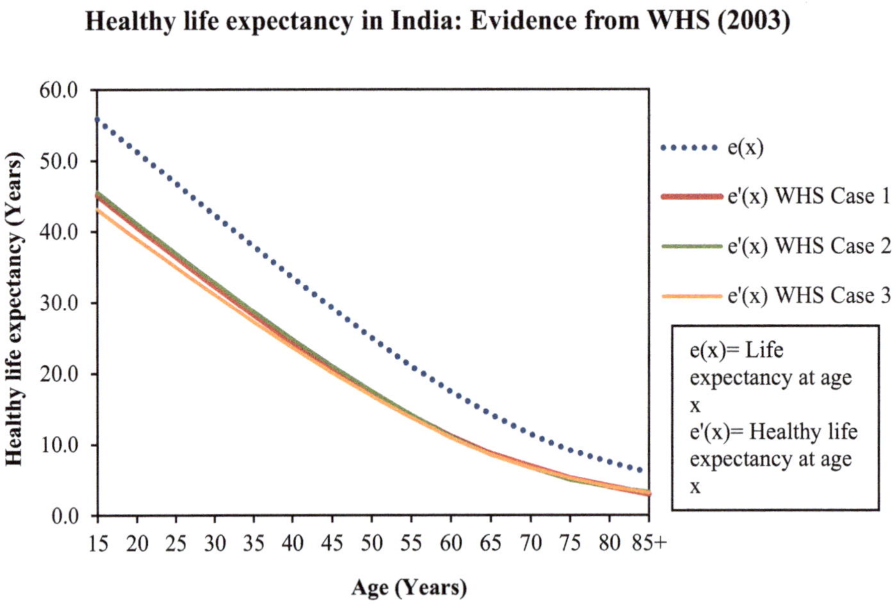

Fig. 3.1 Healthy life expectancies by different aspects of health in India

Table 3.1 Life expectancies and healthy life expectancies among males and females in India 2003

Age	Male e(x)	Female e(x)	e′(x) Case 1		e′(x) Case 2		e′(x) Case 3	
			M	F	M	F	M	F
15	54.3	57.7	46.1	44.2	46.4	44.6	44.1	42.1
20	49.7	53.2	41.6	39.9	42.0	40.3	39.9	38.1
25	45.2	48.8	37.4	35.8	37.8	36.2	35.8	34.2
30	40.7	44.4	33.1	31.7	33.5	32.1	31.8	30.5
35	36.3	39.8	28.9	27.8	29.4	28.1	27.9	26.7
40	32.0	35.4	25.0	23.9	25.5	24.2	24.2	23.2
45	27.8	30.9	21.1	20.2	21.6	20.4	20.6	19.6
50	23.8	26.6	17.6	16.8	18.0	16.9	17.2	16.4
55	20.0	22.4	14.3	13.7	14.8	13.6	14.2	13.4
60	16.5	18.6	11.4	10.9	11.8	10.4	11.4	10.5
65	13.4	15.1	9.0	8.4	9.1	7.8	8.9	8.1
70	10.9	12.2	7.2	6.5	7.3	5.9	7.2	6.1
75	8.7	9.8	5.4	5.1	5.6	4.4	5.7	4.7
80	7.1	7.8	4.7	3.5	4.8	3.4	4.8	3.4
85+	5.7	6.2	3.7	2.6	4.5	2.6	4.1	2.7

Source: Computed by the author based on the SRS life table, male, and females, India (2001–2005) and the WHS India 2003

e(x) = life expectancy at age x, e′(x) = healthy life expectancy at age x

M male, *F* female

Case 1: based on self-rating of general health

Case 2: based on self-reporting on activity limitation

Case 3: based on functional health involving social, psychological, and physical domains of health

3.5 Morbidity-Free Expectancy (MFLE) in India

India is experiencing a rapid health transition with a significant and rising burden of chronic diseases (Reddy et al. 2005). In this context, information on morbidity-free life expectancy (MFLE) is important. The literature review section shows that health outcomes vary with gender. Therefore, we have estimated MFLE for the males and females in India.

3.5.1 Methodology

We collected the data on death rates from the Sample Registration System (SRS) life tables of India for 2014–2018 (Office of the Registrar General & Census Commissioner, India 2020). Information on morbidity has been computed from the unit-level data of the 75th round (July 2017–June 2018) of the National Sample Survey (NSS). The household social consumption on health schedule (25.0) of the 75th round, collected data on the nature, duration, status, and spell of the ailment of household members during 15 days before the survey (National Statistical Office 2019). We found that 39,902 persons reported at least one spell of ailment, and they constitute our samples.

Status of ailment in terms of the time of initiation and time of end/continuation till the survey date has been grouped into four categories. These are:

Status 1: Started more than 15 days ago and is continuing (on survey date).

Status 2: Started more than 15 days ago and has ended (before survey date).

Status 3: Started within 15 days and is continuing (on survey date).

Status 4: Started within 15 days and has ended (before survey date).

Status 1 and 3 of any ailments are used to estimate point prevalence, and Status 3 and 4 have been used to calculate incidence. In our analysis, the prevalence (point prevalence at the time of the survey) of ailment has been calculated as:

$$\text{Prevalence of ailment} = (\text{Number of persons reporting any ailment at the time of survey / Total population})^{*}\,100{,}000$$

We have applied the Sullivan method to estimate MFLE. The mathematical expression of MFLE is similar to HLE. To obtain MFLE for 2017–2018, we applied statistics of proportion ailing (i.e., the prevalence of ailment) by age and sex on SRS life tables of India for 2014–2018.

3.6 Results

Life expectancies and morbidity-free life expectancies of males and females in India in 2004 are shown in Fig. 3.2.

It is observed from Fig. 3.2 that both life expectancies and morbidity-free life expectancies remained higher for women than men at any stage of life. MFLE at birth was 63.1 years for males and 64.4 years for females, and at the age of 60, the MFLE were 13 years and 14 years for males and females, respectively (Appendix Table 3.4). However, the life expectancy curves show a wider gender difference than those showing MFLE, indicating that the morbidity prevalence was higher among women than men. Based on the data of NSS round 60 conducted in 2004, Thomas et al. (2014) also found higher MFLE among women. Various biological, socioeconomic, and psychological factors are responsible for the higher morbidity rates among women (Gove and Hughes 1979; Emsile, Hunt and Mackintyre 1999; Bambra et al. 2021). In the previous section, we observed that healthy life expectancies (based on self-reported general health, activity limitation, and functional health) were lower among females than males. Therefore, our findings suggest that MFLE differs from HLE as the latter includes much broader aspects of health.

3.7 Limitations of the Study

A few limitations of the present study can be pointed out. We have used the information on self-rated health (SRH) from the World Health Survey (WHS) India 2003 to calculate healthy life expectancy (HLE). Several studies pointed to the lacunae in using SRH in determining the socioeconomic inequalities in health (Sauerborn et al. 1996; Subramanian et al. 2009; Jain et al. 2012). For example, there is a tendency to underreporting ill-health among the poor. Many poor

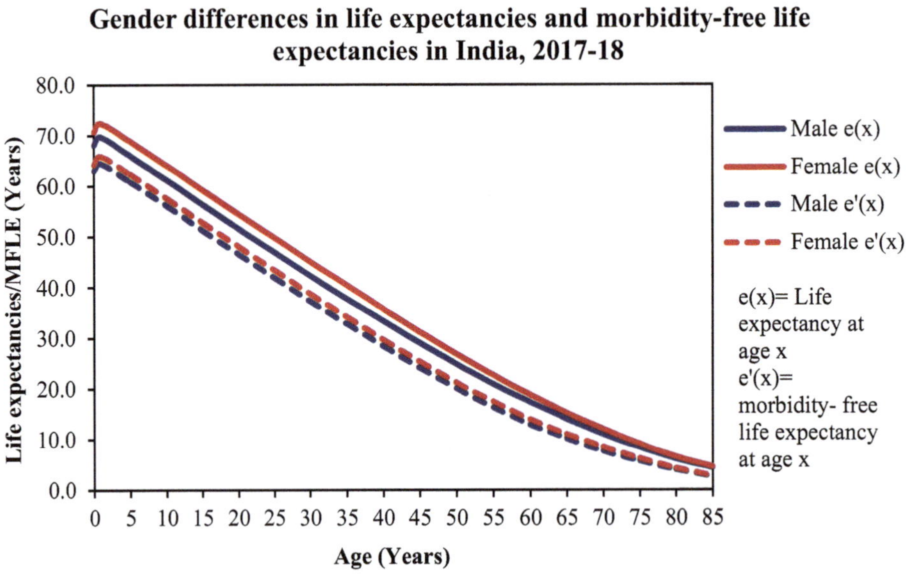

Fig. 3.2 Life expectancies and morbidity-free life expectancies by sex: India, 2017–2018

people cannot afford health services; consequently, they fail to recognize their ailment. Sen (2006) observed that in India, Kerala had a much higher reported morbidity rate than Bihar. According to him, Kerala had a higher older population, and the reporting of morbidity in Kerala was more due to its high literacy rate and extensive public health facilities. In the NSS data, the prevalence of any disease/condition is also based on self-reporting. The self-reported information may vary from the clinically tested and laboratory-confirmed results of any disease.

3.8 Conclusion

The Agenda 2030 has delineated several targets to achieve the goal of SDG 3 (United Nations 2015). The targets are related to reproductive, maternal, newborn, and child health, communicable and noncommunicable diseases, injuries and violence, and health system. Except for the health system, the indicators of other targets are mostly related to reducing mortality and morbidity rates. SDG 3 indicators help to analyze the performance of a country in each category. To ensure healthy lives for all by 2030, the SDG 3 targets must be fulfilled. However, for cross-national and interstate comparisons of healthy lives, it is difficult to get the data of all the indicators, and comparing them is a cumbersome job. Furthermore, one country/region may have a higher infant mortality rate but a lower prevalence of noncommunicable diseases. So the ranking of a country changes depending on the choice of the indicator. To overcome these constraints, we can use the summary measures of population health which "combines both mortality and morbidity data to represent overall population health as a single number" (Field and Gold 1998, p. 4). With the increasing burden of noncommunicable diseases at older ages, the use of SMPH has substantially increased as it incorporates the impact of nonfatal health outcomes in health planning and priority setting. SMPH is also crucial in assessing socioeconomic inequality in health "because people in lower socio-economic groups do not only live shorter lives but also spend a larger proportion of their life in ill-health" (Mackenbach 2002; p. 1777).

In India, declining fertility rates, reduction in mortality, and increasing survival at older ages are causing population ageing. The share of the elderly is projected to reach 20% in 2050 (United Nations Population Fund 2012). Therefore, the prevalence of morbidity at older ages will increase in society. The burden of noncommunicable diseases has already exceeded the burden of infectious diseases in India (India State-level Disease Burden Initiative Collaborators 2017). Also, different regions of India are at various stages of demographic transition. Population ageing is occurring faster in the southern states of India compared to the north-central states. In this situation, more research is needed to assess people's health status in India using SMPH. Therefore, we estimated healthy life expectancies in India. As the developed countries witnessed population ageing and the rise in the burden of chronic diseases long before the developing countries, the use of SMPH has become widespread in developed countries. The European Commission and the OECD regularly publish data on the healthy life years of the countries of the European Union. The Government of India should also take the initiatives to publish India's health status using healthy life expectancy or other SMPH.

In 2003 in India, healthy life expectancy at the age of 20 was 40.8 years, 41.2 years, and 39.1 years based on self-rated general health, self-reported activity limitation, and self-reported functional health, respectively. At age 60, the corresponding values were 11.2 years, 11.2 years, and 11.0 years. Such close values indicate that when people assess their health, they perceive health in totality. Therefore, HLE derived from self-rated general health could be a substitute for HLE based on activity limitations or functional health, particularly among the elderly. In this context, we strongly argue to include the question of self-rated general health in large-scale health surveys, like NFHS.

Self-rated health is subjective and contextual depending on socioeconomic status, health conditions of peers of similar age, and culture, among other factors (Tissue 1972; Baron-Epel et al.

2005). However, it is irrefutable that the basis of rating lies in the biological and psychological condition of the individuals (Jylha 2009). Further, it is often not feasible for a researcher to get information on medically diagnosed illness or disability, particularly in developing countries. In such cases, they must rely on the respondents' reporting about a disease or general health (Suchman et al. 1958). Cullati et al. (2018) assessed the construct validity of the SRH item of WHS India 2003 and concluded that the self-rated general health was a reliable indicator for determining the health status of the population of India.

We found that when health was measured in terms of the prevalence of ailment, women had higher MFLE. However, when health was expressed in terms of physical, mental, and social well-being, women experienced lower healthy life expectancy. Therefore, researchers should carefully choose the appropriate SMPH for their study. Particularly, those working on women's health, must explore the physical, social, and psychological aspects of health. Our findings on gender differentials in HLE are important because it reveals that an increase in the length of life should not be the primary objective of public health policymakers. It is essential to emphasize the quality of life.

Appendix

Table 3.2 Health domains and questions regarding functional difficulties

World Health Survey India 2003	
Health domains	Questions regarding functional difficulties
Mobility	Overall, in the last 30 days, how much difficulty did you have with moving around?
	In the last 30 days, how much difficulty did you have in vigorous activities, such as running 3 km (or equivalent) or cycling?
Self-care	Overall, in the last 30 days, how much difficulty did you have with self-care, such as washing or dressing yourself?
	In the last 30 days, how much difficulty did you have in taking care of and maintaining your general appearance (e.g., grooming, looking neat and tidy, etc.)?
Pain and discomfort	Overall, in the last 30 days, how much of bodily aches or pains did you have?
	In the last 30 days, how much bodily discomfort did you have?
Cognition	Overall, in the last 30 days, how much difficulty did you have with concentrating or remembering things?
	In the last 30 days, how much difficulty did you have in learning a new task (e.g., learning how to get to a new place, learning a new game, learning a new recipe, etc.)?
Interpersonal activities	Overall, in the last 30 days, how much difficulty did you have with personal relationships or participation in the community?
	In the last 30 days, how much difficulty did you have in dealing with conflicts and tensions with others?
Vision	In the last 30 days, how much difficulty did you have in seeing and recognizing a person you know across the road (i.e. from a distance of about 20 meters)?

Source: (WHO, 2002) Individual Questionnaire of World Health Survey, 2002)

Table 3.3 Life expectancies and healthy life expectancies in India, 2003

Age	e(x)	e′(x) Case 1	e′(x) Case 2	e′(x) Case 3
15	55.9	45.2	45.6	43.2
20	51.4	40.8	41.2	39.1
25	46.9	36.6	37.1	35.1
30	42.4	32.4	32.9	31.2
35	38.0	28.4	28.8	27.4
40	33.6	24.5	24.9	23.7
45	29.3	20.7	21.1	20.2
50	25.1	17.2	17.6	16.9
55	21.1	14.1	14.3	13.9
60	17.5	11.2	11.2	11.0
65	14.3	8.8	8.6	8.6
70	11.5	6.9	6.7	6.8
75	9.2	5.3	5.0	5.2
80	7.5	4.1	4.0	4.1
85+	6.0	3.0	3.4	3.2

Source: Computed by the author
e(x) values are taken from SRS life table, India, 2001–2005
e′(x) is computed from the unit-level data of the WHS India 2003
e(x) = life expectancy at age x, e′(x) = healthy life expectancy at age x
Case 1: Based on self-rating of general health
Case 2: Based on self-reporting on activity limitation
Case 3: Based on functional health involving social, psychological, and physical domains of health

Table 3.4 Life expectancies and morbidity-free life expectancies among males and females in India, 2017–2018

	e(x)		e′(x)	
Age	M	F	M	F
0	68.2	70.7	63.1	64.4
1	69.8	72.5	64.5	66.0
5	66.1	69.0	60.9	62.5
10	61.3	64.2	56.2	57.7
15	56.5	59.4	51.4	52.9
20	51.7	54.6	46.7	48.2
25	47.0	49.9	42.0	43.5
30	42.4	45.2	37.4	38.8
35	37.8	40.5	32.8	34.3
40	33.4	35.8	28.4	29.7
45	29.1	31.3	24.2	25.4
50	24.9	26.9	20.1	21.3
55	21.0	22.8	16.4	17.5
60	17.4	18.9	13.0	14.0
65	14.1	15.3	10.2	11.1
70	11.1	12.1	7.7	8.6
75	8.5	9.2	5.8	6.3
80	6.2	6.6	4.1	4.4
85+	4.5	4.7	2.8	2.9

Source: Computed by the author
e(x) values are taken from SRS life table, male and female, India, 2014–2018
e′(x) is computed from the unit-level data of NSS, round 75, 2017–2018
e(x) = life expectancy at age x, e′(x) = morbidity-free life expectancy at age x
M male, *F* female

References

Arokiasamy P, Guruswamy M, Roy TK, Lhungdim H, Chatterji S, Nandraj S (2006) Health system performance assessment, world health survey, 2003, India. International Institute for Population Sciences (IIPS), World Health Organization (WHO), and World Health Organization (WHO)-India-WR office

Bambra C, Albani V, Franklin P (2021) Covid-19 and gender health paradox. Scand J Public Health 49:17–26. Retrieved from https://journals.sagepub.com/doi/pdf/10.1177/1403494820975604 on 11.07.2022

Baron-Epel O, Kaplan G, Haviv-Messika A, Tarabeia J, Green MS, Kaluski DN (2005) Self-reported health as a cultural health determinant in Arab and Jewish Israelis, MABAT – national health and nutrition survey 1999-2001. Soc Sci Med 61(6):1256–1266. https://doi.org/10.1016/j.socscimed.2005.01.02

Bebbington AC (1988) The expectation of life without disability in England and Wales. Soc Sci Med 27(4):321–326. https://doi.org/10.1016/0277-9536(88)90265-1

Bora JK, Saikia N (2015) Gender differentials in self-rated health and self-reported disability among adults in India. PLoS one 10(11):e0141953. Retrieved from https://journals.plos.org/plosone/article/file?id=10.1371/journal.pone.0141953&type=printable on 23.06.2017

Chiang C L (1965) An index of health: mathematical models. Public health service publication no. 1000 – series 2, no 5. National Centre for health statistics. Retrieved from https://stacks.cdc.gov/view/cdc/12902 on 01.01.2022

Crimmins EM, Saito Y, Ingegneri D (1989) Changes in life expectancy and disability-free life expectancy in the United States. Popul Dev Rev, 15 (2):235–267. Retrieved from https://doi.org/10.2307/1973704 on 10.06.2017

Cullati S, Mukhopadhyay S, Sieber S, Chakraborty A, Burton-Jeangros C (2018) Is the single self-rated health item reliable in India? A construct validity study. BMJ Glob Health 3(6):e000856. Retrieved from https://gh.bmj.com/content/bmjgh/3/6/e000856.full.pdf on 30.06.2020

Das P, Samarasekera U (2012) The story of GBD 2010: a super-human effort. Lancet, 380(9859):2067–2070. Retrieved from https://www.thelancet.com/action/showPdf?pii=S0140-6736%2812%2962174-6 on 10.10.2016

Dhak B (2009) Gender differences in health and its determinants in the old-aged population in India. J Biosoc Sci 41(5):625–643. https://doi.org/10.1017/S002193200999006X

Emsile C, Hunt K, Mackintyre S (1999) 'Gender' or 'job' differences? Working conditions amongst men and women in white-collar occupations. Work Employ Soc 13(4):711–729. Retrieved from https://doi.org/10.1177/09500179922118196 on 07.06.2015

EUROSTAT (2019) Healthy life years statistics. Retrieved from https://ec.europa.eu/eurostat/statisticsexplained/index.php/Healthy_life_years_statistics on 10.9.2019

Field MJ, Gold MR (eds) (1998) Summarizing population health: directions for the development and application of population metrics. National Academies Press (US), Washington DC. https://doi.org/10.17226/6124

GBD 2016 DALYs and HALE Collaborators (2017) Global, regional, and national disability-adjusted life-years (DALYs) for 333 diseases and injuries and healthy life expectancy (HALE) for 195 countries and territories, 1990–2016: A systematic analysis for the Global Burden of Disease Study 2016. Lancet 390(10100):1260–1344. Retrieved from https://www.thelancet.com/action/showPdf?pii=S0140-6736%2817%2932130-X on 23.02.2018

Gove WR, Hughes M (1979) Possible causes of the apparent sex differences in physical health: an empirical investigation. Am Sociol Rev 44:126–146. Retrieved from https://doi.org/10.2307/2094822 on 12.12.2017

Gu D, Dupre ME, Warner DF, Zeng Y (2009) Changing health status and health expectancies among older adults in China: gender differences from 1992 and 2002. Soc Sci Med 68(12): 2170–2179. Retrieved from https://www.ncbi.nlm.nih.gov/pmc/articles/PMC2727476/pdf/nihms106966.pdf on 21.09. 2021

Hardy MA, Acciai F, Reyes AM (2014) How health conditions translate into self-ratings: a comparative study of older adults across Europe. J Health Soc Behav 55(3):320–341. Retrieved from https://www.ncbi.nlm.nih.gov/pmc/articles/PMC4669051/pdf/nihms-730117.pdf on 01.01.2021

India State-level Disease Burden Initiative Collaborators (2017) Nation within a nation: variations in epidemiological transition across the states of India, 1990-2016 in the global burden of disease study. Lancet 390(10111):2437–2460. Retrieved from https://www.thelancet.com/action/showPdf?pii=S0140-6736%2817%2932804-0 on 20.12.2017

India State-level Disease Burden Initiative Malnutrition Collaborators (2019) The burden of child and maternal malnutrition and trends in its indicators in the states of India: the Global Burden of Disease Study 1990–2017. Lancet Child Adolesc Health 3(12):855–870. Retrieved from https://www.thelancet.com/action/showPdf?pii=S2352-4642%2819%2930273-1 on 14.04.2020

Ined (INSTITUT NATIONAL D'ÉTUDES DÉMOGRAPHIQUES) (2021) About REVES. Retrieved from https://reves.site.ined.fr/en/home/about_reves/ on 12.10.2021

Jain K, Goli S, Arokiasamy P (2012) Are self-reported morbidities deceptive in measuring socio-economic inequalities? Indian J Med Res 136(5):750–757. Retrieved from www.ncbi.nlm.nih.gov/pmc/articles/PMC3573595/pdf/IJMR-136-750.pdf on 23.06.2016

Jitapunkul S, Chayovan N (2000) Healthy life expectancy of Thai elderly: did it improve during the soap-bubble economic period? J Med Assoc Thailand 83(8):861–864. PMID: 10998838

Jylha M (2009) What is self-rated health and why does it predict mortality? Towards a unified conceptual model. Soc Sci Med 69:307–316. https://doi.org/10.1016/j.socscimed.2009.05.013

Lim SS, Vos T, Flaxman AD, Danaei G, Shibuya K, Adair-Rohani H, et al (2012) A comparative risk assessment of burden of disease and injury attributable to 67 risk factors and risk factor clusters in 21 regions, 1990–2010: a systematic analysis for the Global Burden of Disease Study 2010. Lancet, 380(9859): 2224–2260. Retsrieved from https://www.ncbi.nlm.nih.gov/pmc/articles/PMC4156511/pdf/nihms589916.pdf on 13.12.2021

Lu B, Liu X, Lim J, Yang M (2018) Changes in the morbidity prevalence and morbidity-free life expectancy of the elderly population in China from 2000 to 2010. J Econ Ageing 13(C):113–121. Retrieved from https://doi.org/10.1016/j.jeoa.2018.05.002 on 20.09.2021

Mackenbach JP (2002) Socio-economic inequalities in health in developed countries: the facts and the options. In: Detels R, McEwen J, Beaglehole R, Tanaka H (eds) Oxford textbook of public health, 4th edn. Oxford University Press, Oxford, pp 1773–1790

Mathers CD (2002) Health expectancies: an overview and critical appraisal. In: Murray CJL, Salomon JA, Mathers CD, Lopez AD (eds) Summary measures of population health. World Health Organization, Geneva, pp 177–204

Mathers CD, Murray CJL, Lopez AD, Salomon JA, Sadana R, Tandon A, et al (2001a) Estimates of healthy life expectancy for 191 countries in the year 2000: methods and results. World Health Organization. Global Programme on Evidence for Health Policy Discussion Paper No. 38

Mathers CD, Vos T, Lopez AD, Salomon J, Ezzati M (eds) (2001b) National burden of disease studies: a practical guide, Edition 2.0. Global program on evidence for health policy. World Health Organization, Geneva

Mathers C, Vos ET, Stevenson CE, Begg ST (2001c) The burden of disease and injury in Australia. Bull World Health Organ 79(11):1076–1084. Retrieved from https://www.ncbi.nlm.nih.gov/pmc/articles/PMC2566696/pdf/11731817.pdf on 09.09.2021

McKenna MT, Michaud CM, Murray CJL, Marks JS (2005) Assessing the burden of disease in the United States using disability-adjusted life years. Am J Prev Med 28(5): 415–423. Retrieved from https://www.ajpmonline.org/action/showPdf?pii=S0749-3797%2805%2900067-X on 11.09.2015

Melse JM, Essink-Bot ML, Kramers PG, Honeymans N (2000) A national burden of disease calculation. Dutch disability-adjusted life-years. Dutch Burden of Disease Group. Am J Public Health 90(8):1241–1247. Retrieved from https://www.ncbi.nlm.nih.gov/pmc/articles/PMC1446331/pdf/10937004.pdf on 04.06.2021

Menon GR, Singh L, Sharma P, Yadav P, Sharma S, Kalaskar S, et al (2019) National burden estimates of healthy life lost in India, 2017: an analysis using direct mortality data and indirect disability data. Lancet Global Health 7(12):E1675–E1684. Retrieved from https://www.thelancet.com/action/showPdf?pii=S2214-109X%2819%2930451-6 on 03.01.2022

Mishra RS, Mishra R, Mohanty SK (2020) Gender differential and regional disparity of disability-free life-expectancy among disable in India. Clin epidemiol Global Health 8(3):818–827. Retrieved from https://doi.org/10.1016/j.cegh.2020.02.007 on 31.04.2021

Moreno X, Albala C, Lera L, Leyton B, Angel B, Sanchez H (2018) Gender, nutritional status and disability-free life expectancy among older people in Santiago, Chile. PLoS One, 28, 13(3):e0194074. Retrieved from https://www.ncbi.nlm.nih.gov/pmc/articles/PMC5874002/pdf/pone.0194074.pdf on 04.04.2019

Moussavi S, Chatterji S, Verdes E, Tandon A, Patel V, Ustun B (2007) Depression, chronic diseases, and decrements in health: results from the world health surveys. Lancet 370:851–858. Retrieved from https://www.thelancet.com/action/showPdf?pii=S0140-6736%2807%2961415-9 on 03.03.2014

Murray CJL (1994) Quantifying the burden of disease: the technical basis for disability-adjusted life years. Bull World Health Organ 72(3):429–445. Retrieved from https://www.ncbi.nlm.nih.gov/pmc/articles/PMC2486718/pdf/bullwho00414-0105.pdf on 23.05.2014

Murray CJL, Lopez AD (1997a) Mortality by cause for eight regions of the world: global burden of disease study. Lancet 349(9061):1269–1276. Retrieved from https://www.thelancet.com/action/showPdf?pii=S0140-6736%2896%2907493-4 on 02.02.2013

Murray CJL, Lopez AD (1997b) Regional patterns of disability-free life expectancy and disability-adjusted life expectancy: global burden of disease study. Lancet 349(9062):1347–1352. Retrieved from https://www.thelancet.com/action/showPdf?pii=S0140-6736%2896%2907494-6 on 02.02.2013

Murray CJL, Lopez AD (1997c) Global mortality, disability and contribution of risk factors: global burden of disease study. Lancet 349(9063):1436–1442. Retrieved from https://www.thelancet.com/action/showPdf?pii=S0140-6736%2896%2907495-8 on 02.02.2013

Murray CJL, Lopez AD (1997d) Alternative projections of mortality and disability by cause 1990–2020: global burden of disease study. Lancet 349(9064):1498–1504. Retrieved from https://www.thelancet.com/action/showPdf?pii=S0140-6736%2896%2907492-2 on 02.02.2013

Murray CJL, Lopez AD, World Health Organization, World Bank, Harvard School of Public Health (1996) The global burden of disease: a comprehensive assessment of mortality and disability from diseases, injuries, and risk factors in 1990 and projected to 2020, global burden of diseases and injury series Volume I. Harvard University Press, Cambridge, MA

Murray CJL, Salomon JA, Mathers CD (2002) A critical examination of summary measures of population health. In: Murray CJL, Salomon JA, Mathers CD, Lopez AD (eds) Summary measures of population health. World Health Organization, Geneva, pp 13–40

National Statistical Office (NSO) (2019) Key indicators of social consumption in India: Health, NSS 75th Round, (July 2017–June 2018). Ministry of Statistics and Programme Implementation, Government of India. Retrieved from http://164.100.161.63/sites/default/files/publication_reports/KI_Health_75th_Final.pdf on 13.09.2020

OECD (2011) Health at a glance 2011: OECD indicators. Retrieved from https://www.oecd.org/els/health-systems/49105858.pdf on 12.12.2018

Office of the Registrar General & Census Commissioner, India (2020) SRS based abridged life tables 2014–18. Ministry of Home Affairs, Government of India. Retrieved from https://censusindia.gov.in/nada/index.php/catalog/42517/download/46143/SRS-Abridged_Life_Tables_2014-2018.pdf on 13.09.2020

Oksuzyan A, Singh PK, Christensen K, Jasilionis D (2018) A cross-national study of the gender gap in health among older adults in India and China: similarities and disparities. Gerontologist 58(6):1156–1165. Retrieved from https://www.ncbi.nlm.nih.gov/pmc/articles/PMC6455934/pdf/gnx111.pdf on 23.10.2020

Pandey A, Ladusingh L (2013) Socio-economic correlates of gender differential in poor health status among older adults in India. J Appl Gerontol 34(7):879–905. https://doi.org/10.1177/0733464813481850

Reddy SK, Shah B, Varghese C, Ramadoss A (2005) Responding to the threat of chronic diseases in India. Lancet 366(9498):1744–1749. Retrieved from https://www.thelancet.com/journals/lancet/article/PIIS0140-6736(05)67343-6/fulltext on 05.06.2013

Roberto H-C (2003) Ageing and health expectancies in urban Latin America. In: Robine J-M, Jagger C, Mathers CD, Crimmins EM, Suzman RM (eds) Determining health expectancy. Wiley, Chichester, pp 319–333

Romero DE, Leite IC, Szwarcwald CL (2005) Healthy life expectancy in Brazil: applying the Sullivan method. Cad Saὑde Pὑblica 21:S7–S18. Retrieved from https://www.scielo.br/j/csp/a/B8kSFGTfGzFnW3W85x8tP3B/?format=pdf&lang=en on 23.07.2021

Roy K, Chaudhuri A (2008) Influence of socio-economic status, wealth and financial empowerment on gender differences in health and healthcare utilization in later life: evidence from India. Soc Sci Med 66(9):1951–1962. https://doi.org/10.1016/j.socscimed.2008.01.015

Saito Y, Qiao X, Jitapunkul S (2003) Health expectancies in Asian countries. In: Robine J-M, Jagger C, Mathers CD, Crimmins EM, Suzman RM (eds) Determining health expectancy. Wiley, Chichester, pp 289–317

Sanders BS (1964) Measuring community health levels. Am J Public Health 54(7):1063–1070. Retrieved from https://www.ncbi.nlm.nih.gov/pmc/articles/PMC1254928/pdf/amjphnation00174-0025.pdf on 01.01.2022

Santosa A, Schroders J, Vaezghasemi M, Ng N (2016) Inequality in disability-free life expectancies among older men and women in six countries with developing economies. J Epidemiol Community Health, 70(9):1–7. Retrieved from https://www.ncbi.nlm.nih.gov/pmc/articles/PMC5013163/pdf/jech-2015-206640.pdf on 13.07. 2020

Sauerborn R, Adams A, Hien M (1996) Household strategies to cope with the economic costs of illness. Soc Sci Med 43:291–301. Retrieved from https://doi.org/10.1016/0277-9536(95)00375-4 on 17.09.2020

Sen A (2006) Health achievement and equity: external and internal perspectives. In: Anand S, Peter F, Sen A (eds) Public health, ethics, and equity. Oxford University Press, New Delhi, pp 263–268

Singh L, Arokiasamy P, Singh PK, Rai RK (2013) Determinants of gender differences in self-rated health among older population: evidence from India. SAGE open, April–June:1–12. Retrieved from https://journals.sagepub.com/doi/pdf/10.1177/2158244013487914 on 04.01.2021

Sreerupa RS, Shweta I, Ajay S, Saito Y, Malhotra R (2019). Living longer: for better or worse? Changes in life expectancy with and without mobility limitation among older persons in India between 1995–96 and 2004. Int J Popul Stud 4(2): 23–34. Retrieved from https://article.accscience.com/journal/IJPS/4/2/10.18063/ijps.v4i2.761/761-3066-5-PB.pdf on 02.06.2021

Stiefel MC, Perla RJ, Zell BL (2010) A healthy bottom line: healthy life expectancy as an outcome measure for health improvement efforts. Milbank Q 88(1):30–53. Retrieved from https://www.ncbi.nlm.nih.gov/pmc/articles/PMC2888015/pdf/milq0088-0030.pdf on 02.02.2020

Subramanian SV, Subramanyam MA, Selvaraj S, Kawachi I (2009) Are self-reports of health and morbidities in developing countries misleading? Evidence from India. Soc Sci Med 68:260–265. Retrieved from https://www.ncbi.nlm.nih.gov/pmc/articles/PMC2652643/pdf/nihms92773.pdf on 09.05.2016

Suchman EA, Phillips BS, Streib GF (1958) An analysis of the validity of health questionnaires. Soc Forces 36(3):223–232. Retrieved from https://doi.org/10.2307/2573809 on 17.06 2021

Sullivan DF (1971) A single index of mortality and morbidity. HSMHA Health Rep 86:347–354. Retrieved from https://www.ncbi.nlm.nih.gov/pmc/articles/PMC1937122/pdf/hsmhahr00004-0061.pdf on 01.01.2022

The Council of National Living (1974) Social indicators of Japan. Cited in Saito, Y., Qiao, X., & Jitapunkul, S. (2003). Health expectancies in Asian countries. In: Robine J-M, Jagger C, Mathers CD, Crimmins EM,

Suzman RM (eds) Determining health expectancy. Wiley, Chichester, pp 289–317
Thomas MB, James KS, Sulaja S (2014) Does living longer mean living healthier? Exploring disability-free life expectancy in India. Working paper 322. The Institute for Social and Economic Change, Bangalore. Retrieved from http://www.isec.ac.in/WP%20322%20-%20Benson%20Thomas%20et%20al.pdf on 11.09.2016
Tissue T (1972) Another look at self-rated health among the elderly. J Gerontol 27(1):91–94. Retrieved from https://doi.org/10.1093/geronj/27.1.91 on 13.12.2021
Tu EJ, Chen KJ (1994) Recent changes in healthy life expectancy and their implications for medical costs in Taiwan. In: Mathers C, McCallum J, Robine J-M (eds) Advances in health expectancies. Australian Institute of Health and Welfare (AGPS), Canberra, pp 367–382
United Nations (2015) Transforming our world: the 2030 agenda for sustainable development, A/RES/70/1. Retrieved from https://sustainabledevelopment.un.org/content/documents/21252030%20Agenda%20for%20Sustainable%20Development%20web.pdf on 30.08.2020
United Nations Population Fund (2012) Report on the status of elderly in selected states of India, 2011. Retrieved from https://india.unfpa.org/sites/default/files/pub-pdf/AgeingReport_2012_F.pdf on 07.08.2021
Vos T, Flaxman AD, Naghavi M, Lozano R, Michaud C, Ezzati M et al (2012) Years Lived with Disability (YLDs) for 1160 sequelae of 289 diseases and injuries 1990–2010: a systematic analysis for the Global Burden of Disease Study 2010. Lancet, 380(9859):2163–2296. Retrieved from https://www.ncbi.nlm.nih.gov/pmc/articles/PMC6350784/pdf/nihms-1005038.pdf on 13.12.2021
Wilkins R, Adams OB (1983) Health expectancy in Canada, late 1970s: demographic, regional, and social dimensions. Am J Public Health 73(9):1073–1080. Retrieved from https://www.ncbi.nlm.nih.gov/pmc/articles/PMC1651059/pdf/amjph00644-0063.pdf on 18.10.2021
World Bank (1993) World development report 1993: investing in health. Oxford University Press, New York
World Health Organization (2001a) Mental health: new understanding new hope. World Health Organization. Retrieved from https://apps.who.int/iris/bitstream/handle/10665/42390/WHR_2001.pdf?sequence=1&isAllowed=y on 13.04.2017
World Health Organization (2001b) International classification of functioning, disability and health (ICF). World Health Organization, Geneva
World Health Organization (WHO) (1946) Constitution of the World Health Organization. Retrieved from https://apps.who.int/gb/bd/PDF/bd47/EN/constitution-en.pdf?ua=1 on 23.04.2020
Zimmer Z, Keneda T, Tang Z, Fang X (2010) Explaining late life urban vs rural health discrepancies in Beijing. Soc Forces 88(4):1885–1908. Retrieved from https://www.ncbi.nlm.nih.gov/pmc/articles/PMC2998765/pdf/nihms225586.pdf on 20.09.2021

Shewli Shabnam is currently an Assistant Professor in Geography at Bidhannagar College, Government of West Bengal, Kolkata. She has keen interests in gender issues, morbidity and healthcare system, cultural geography, and social demography. She has published several research papers and book chapters.

Nandita Saikia is presently the Professor and Head of Department of Public Health and Mortality Studies of the International Institute for Population Sciences, Mumbai. She is one of the prominent demographers in India, and her research interest lies in formal demography, mortality and causes of death, inequalities in health, and undocumented immigration. She has published several articles in reputed international journals. She is an associate editor of several international journals.

Quality Education: Foundation for 16 SDGs

4

Apoorva Bhatnagar and Triguna Singh

Abstract

Development has always been an essential agenda globally. However, is development enough? Adding sustainability to the idea of development connects the present with the future, aiming at being responsible today for a better tomorrow. The United Nations Brundtland Commission (1987) defined sustainability as "meeting the needs of the present without compromising the ability of future generations to meet their own needs." Quality education is the fourth and fundamental goal out of the 17 Sustainable Development Goals given by the United Nations, which mark inclusivity, equity, and quality education to all by 2030. Nobel Laureate Amartya Sen rightly says "If we continue to leave vast sections of the people of the world outside the orbit of education, we make not only the world less just but also less secure." Education is the foundation stone for accomplishing almost every goal in the sustainable development list. The idea is to provide effective primary education and increase skills both technical and vocational for better employment as well as entrepreneurial opportunities for men as well as women. Gender disparity is an avid concept, especially in developing countries, leading to unequal access to quality education. Eliminating this flaw would garner equal opportunities for both men and women.

Quality education is not just about literacy and numeracy but also about inculcating sustainable practices in the lifestyle itself, building awareness about diverse cultures, and promoting peace, nonviolence, and equality. Moreover, it shifts the focus to global citizenship to a better and bigger learning environment with an increase in the number of scholarships to developing and least developed nations. On the supply side of the education industry, it highlights the need for well-qualified teachers to join hands and ensure quality education for all. Considering the impact of the COVID-19 pandemic, progress in education has slowed down. The first wave led to the complete lockdown in India, and schools and colleges had to shift to the online mode of learning. This sudden transition insisted on the need to develop infrastructure and training methods for online learning modules. There is the need for better infrastructure, such as laptops, Internet facilities, quality study material, and access to online libraries, especially in developing and least developed nations, to drive past the impact

A. Bhatnagar (✉)
Balaji Institute of Modern Management,
Sri Balaji University, Pune, India
e-mail: apoorva10.bhatnagar@gmail.com

T. Singh
City Montessori School, Lucknow, India
e-mail: triguna.singh19@gmail.com

V. Dutta, P. Ghosh (eds.), *Sustainability: Science, Policy, and Practice in India*,
Sustainable Development Goals Series, https://doi.org/10.1007/978-3-031-50132-6_4

caused by this pandemic. The goal of this chapter is to get the reader acquainted with the importance of SDG 4, i.e., quality education as the basis for achieving the remaining 16 SDGs.

Keywords

Quality education · SDGs · SDG 4 · Sustainable development · United Nations · SDG interactions · India

4.1 Introduction

Sustainability is all about empowering today without forgetting the future. Quality education is concerned with inclusive growth along all sectors and the provision of quality education. SDG 4 aims to nullify the gap between formal and informal learning: public and private education (The Energy Progress Report 2019). It focuses on integrating global and local education and encourages sustainable learning to meet the demands of the competitive world. The inclusivity and equity indicators of SGD 4 have directed attention to bringing equal right to education to both genders. It tends to increase youth as well as adult literacy. Bringing inclusive learning to fore would enable growth in the remaining 16 SDGs (The Energy Progress Report 2019). This integration would entail the strength of quality education in capturing the essence of all other sustainable goals and meeting the agenda in 2030 (Lawrence 2018). Quality education is an imperative factor that leads to greater well-being and health, as the World Happiness Report 2020 by the UN clearly indicates that citizens of Finland have the highest level of contentment and satisfaction, which could be because Finland radically overhauled its education system years ago (Helliwell et al. 2020). This paper explores how the development of Sustainable Development Goal (SDG) 4, i.e., quality education, can lead to the growth of 16 other SDGs.

Education is an indispensable instrument in the development of society. Sustainable Development Goals were built upon Millennium Development Goals that aimed to reduce poverty. The introduction of SDGs widened the base for multilateral policies globally. India has been a developing country, and investing in education would lead to development in all aspects, including poverty upliftment, hunger reduction, improvement in health and well-being, gender equality, climate change and mitigation, industry, innovation, and infrastructure. Mahatma Gandhi had visualized education as a means of awakening the nation's conscience to injustice, violence, and inequality entrenched in the social order. Quality education has always been a challenge for the nation. Focusing on the development of SDG 4 in India would empower citizens, promote equality, bring prosperity, and help to make sustainable choices. SDG 4 holds a lifelong perspective on learning and will help India come to the global stage in the education panorama.

4.2 Quality Education (SDG 4) and 16 SDGs

Quality Education and End Poverty in All Its Forms Everywhere (SDG 1)

The multidimensional poverty index, which is an international poverty line, defines the poor as a person who has no access to basic amenities such as education and basic health facilities and leads to a poor standard of living (Choudhury and Chauhan 2013; Bhukta 2020). SDG 1, which is no poverty, aims to eradicate poverty in all forms according to the national dimensions of each country. It is concerned with the provision of resources equally to all sections of society, an earning means and strategies that focus on gender-sensitive development and pro-poor growth. Years of research have shown the correlation between education and income-earning capacity (Chae 2022). The provision of quality education would lead to better productivity and promising career prospects (Lawrence et al. 2020). Quality education tends to be advantageous for eradicating poverty in terms of proliferating skills and competencies, which would lead to employment capabilities, thus blurring the economic gap between rich and poor. According

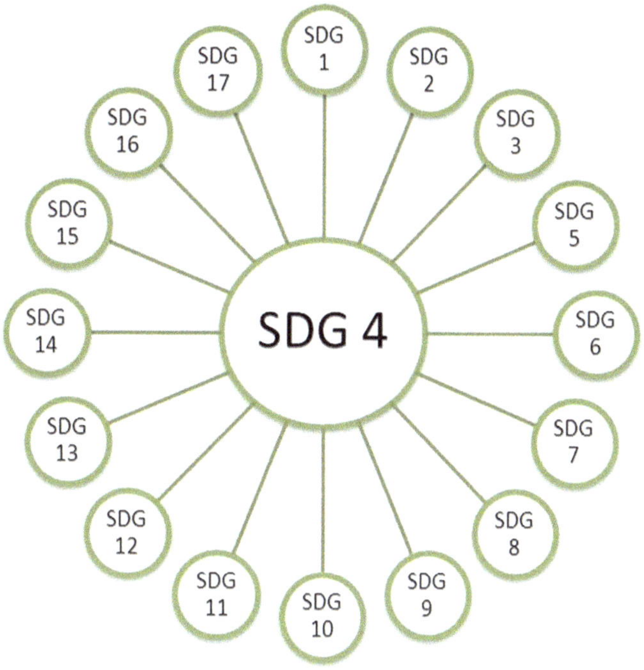

Fig. 4.1 SDG 4 and its interconnection with other SDGs

to an estimate given by the UNESCO, if students of less developed and developing countries acquire basic reading skills, then approximately 171 million people could move to the above poverty line segment. Moreover, it is further suggested that encouraging the adult population to complete their secondary education would result in reducing the global poverty rate to less than 50%. The National Education Policy can help fight poverty, as it focuses on primary and secondary education and emphasizes the need for vocational and technical training to enhance job prospects (Kumar et al. 2021). Figure 4.1 shows that SDG 4 is the foundation of all other SDGs.

4.3 Quality Education and Food Security, Improved Nutrition, and Sustainable Agriculture (SDG 2)

SDG 2 aims to make nutritious food accessible to all sections of society, especially vulnerable sections, to tackle the widespread problem of malnutrition among infants, lactating women, and older people. It even focuses on research and development and adoption of sustainable agricultural practices to safeguard the environment and enhance agricultural infrastructure. Several studies have indicated the interrelation among all the SDGs, which reflects the interdependence of quality education (SDG 4) and zero hunger (SDG 2) (Griggs et al. 2014; Lawrence et al. 2020). Advancing the education system to the less earning section of society and encouraging people to obtain at least basic education and providing them with meal facilities in return would reduce hunger issues while simultaneously increasing literacy rates. It has been stressed to make provisions to retain students once they are enrolled, especially those who are coming from lower economic backgrounds. Providing quality education along with meal programs at school and colleges would help in handling the dual challenges of providing quality and reducing hunger in implementing Sustainable Development Goals (Anyolo et al. 2018). Lawrence et al. (2020) show an indirect relation to target 2.5 of SDG 2, which states "by 2020, maintain the genetic diversity of seeds, cultivated plants and farmed and domesticated animals and their related wild species, including through soundly managed and diversified seed and plant banks at the national, regional and international levels, and promote access to and fair and equitable sharing of benefits arising from the utilization of genetic resources and associated traditional knowledge, as internationally agreed." Furthermore, Saravanakumar et al. (2020) suggest that problems with nutritional security exist due to the existence of social evils such as gender disparity, illiteracy, and caste discrimination.

4.4 Quality Education and Health and Well-Being for All at All Ages (SDG 3)

Health and well-being, which is SDG 3, was made to ensure universal healthcare by reducing the maternal mortality ratio and neonatal mortality rate. Furthermore, it aims to combat diseases such as AIDS and malaria and strengthen the system for drug abuse treatment, tobacco control, and early monitoring and warning systems for global health issues. Awareness is a major part of quality education, and major health risks could be mitigated by educating people from a very young age on sexual and reproductive healthcare, family planning, and the harms of drugs (Lawrence et al. 2020). Education could help alleviate the level of information as well as the standard of living, which would stimulate health consciousness in society. The level and quality of education of parents have an impact on the decisions made for the health of children (Nilsson et al. 2018), especially in developing countries such as India. Moreover, better education leads to more awareness about sanitation, nutrition, and disease prevention (Haldar and Hembram 2020).

4.5 Quality Education and Gender Equality (SDG 5)

SDG 5 is gender equality, which deals with the culmination of practices that encourage gender discrimination in any form. It ensures an end to violence against women, any type of exploitation, forced marriage, genital mutilation, and even child marriage. Quality education would ensure the participation of women in decision-making, politics, and policy-making. Furthermore, education would bring recognition to the unpaid services of women and provide them with equal opportunities to grow and become empowered (Fonseca et al. 2020). The UNESCO has been continuously working to address the gender gap through education within the education industry as well as other fields beginning from early childhood to higher education. It aims to create empowering opportunities for the development of women and girls by 2030. Lawrence et al. (2020) suggest an indirect relationship between quality education and gender equality. There is an urgent need to include issues related to gender, implement gender-sensitive education in syllabi, and train educators (Patel et al. 2020). Ending gender-based violence and addressing the issue of gender equality at an early stage require the development of role models and immense training for both faculty and staff (ibid).

4.6 Quality Education and Clean Water and Sanitation (SDG 6)

Clean water and sanitation (SDG 6) deals with equal access to drinking water with efficient utilization of water, as the planet has been facing freshwater scarcity for quite a long time. In addition, these goals aim to culminate activities that cause pollution due to hazardous activities by industries, untreated waste disposal, and the problem of open defecation (United Nations 2015). Empowering people with quality education would essentially make them aware of water and sanitation management (Lawrence et al. 2020). This would reduce pollution and encourage the sustainable usage of water as a resource (Londoño et al. 2020), especially in developing nations such as India, which face water issues although it is surrounded by water on the three sides and is the land of two of the major rivers.

4.7 Quality Education and Sustainable Energy (SDG 7)

SDG 7 ensures the availability of sustainable energy at affordable prices amidst the currently ongoing global energy crisis. The goal is international cooperation to develop accessible renewable and clean energy to reduce damage to the environment and develop the infrastructure needed to supply clean energy. Education would entail the development of such technology at affordable rates (Lawrence et al. 2020) in developing nations such as India. Moreover, people in developing nations need to change their behavior and awareness about the usage and benefits of renewable and sustainable energy at the basic level. Quality education could bring about a change in awareness level as well as acceptance of clean energy for usage. Furthermore, education would teach skills, impart knowledge, and impact the technological and financial decisions made in the field of clean energy (Nilsson et al. 2018).

4.8 Quality Education and Decent Work and Economic Growth (SDG 8)

SDG 8 is concerned with decent work and economic growth. This goal ensures productive employment for everyone with promising economic growth. It encourages development-oriented policies by increasing consumption and production, keeping in mind the damage to the environment (United Nations 2015). SDG 8 stimulates youth to acquire better education and training to obtain better employment opportunities (United Nations 2021). Quality education would uplift the creative side of people, which affects revenue-earning capacity by creating diverse avenues of employment and economic growth (UN ESCAP 2019; Lawrence et al. 2020). Education would acquaint people with international labor standards, impacting the reduction in the exploitation rate (Fonseca et al. 2020).

4.9 Quality Education and Resilient Infrastructure, Inclusive and Sustainable Industrialization, and Innovation (SDG 9)

SDG 9 aims to build infrastructure that passes the test of time and sustainable industrial growth to stimulate economic development. It encourages innovation by stressing the need for research and development and promoting small-scale industries in developing nations such as India through the accessibility of affordable credit facilities and suitable infrastructure (United Nations 2021). Education impacts knowledge that seeds the plant of curiosity and gives birth to innovation. Thus, education has a huge impact on infrastructure development (Pradhan et al. 2017; Lawrence et al. 2020). SDG interactions add a holistic view to the sustainability dimension (Nilsson et al. 2018).

4.10 Quality Education and Reducing Inequalities Within and Among Countries (SDG 10)

Reducing inequalities is the goal of SDG 10. It ensures the elimination of discriminatory practices that promote inequalities among gender, age, race ethnicity, origin, and religion or in any other form. It stimulates social and political inclusion of all citizens irrespective of their demographics (United Nations 2021). Education is a tool that brings awareness and culminates discriminatory practices based on unscientific and orthodox beliefs (Smith 2019). SDG 4 would help society transform and uplift from age-old traditions of discriminatory practices by making victimized people aware of their rights and responsibilities (Lawrence et al. 2020). Furthermore, the targets of SDG 4 and SDG 10 are found to be correlated, and closer analysis reveals a trade-off and cobenefit between them (Pham-Truffert et al. 2020).

4.11 Quality Education and Sustainable Cities and Communities (SDG 11)

SDG 11 is a sustainable city and community concerned with the availability of basic amenities such as housing, public transportation facilities, and better road safety. Moreover, the goal is to work in the field of sustainable urbanization and improvised settlement planning due to the ever-increasing migrating population (United Nations 2015). It also addresses disaster management, waste management, and the inclusion of green spaces to make communities sustainable for living (ibid). Previous studies suggest that education brings civic sense to the public, which means the development of skills needed for understanding the environmental impact of disasters and waste and following rules for road safety to avoid deaths caused by them (Nakicenovic 2019; Lawrence et al. 2020). Education is a powerful tool for mitigating challenges pertaining to the environment, society, and economy.

4.12 Quality Education and Sustainable Consumption and Production Patterns (SDG 12)

SDG 12 ensures sustainable consumption and production by the efficient utilization of natural resources, keeping in mind the needs of the present as well as the future. This goal encourages reduction, recycling, and reuse to encourage sustainability in lifestyle (United Nations 2015). Moreover, for sustainable consumption and production, it is essential to develop the country's technological front to implement strategies that reduce wasteful consumption and eliminate the usage of fossil fuels (United Nations 2021). All this is possible with the provision of quality education, as it rationalizes the needs and consumption activity of the consumer, which indirectly impacts production (Oluwadamilola et al. 2018; Lawrence et al. 2020). Target 8 of SGD 12 directly correlates with SDG 4, as it states that "By 2030, ensure that people everywhere have the relevant information and awareness for sustainable development and lifestyles in harmony with nature" (United Nations 2015).

4.13 Quality Education and Combat Climate Change and Its Impacts (SDG 13)

Global warming has been in its alarming stage, which makes SDG 13, i.e., combat climate change and its impacts to be an essential goal. This goal integrates measures to adapt to the changing climate due to global warming and measures to mitigate disasters and make people aware of early warning systems (United Nations 2015). It also deals with mechanisms to develop strategies to reduce the impact of climate change, strengthen the green climate fund, and capitalize on the fund to spread awareness through learning courses (United Nations 2021). Target 3 of SDG 13 which states that "improve education, awareness-raising and human and institutional capacity on climate change mitigation, adaptation, impact reduction and early warning" directly correlates with SDG 4 (Lawrence et al. 2020; Pham-Truffert et al. 2020). Climate education should be integrated with innovative technologies to address sustainability-related climate change issues, and youth must be empowered with this knowledge, as they have the potential to bring change (Franco et al. 2020). There is the need to make people aware of the rampant usage of fossil fuels, especially by poor, unaware, and uneducated people. Unrestricted use of fossil fuels and deforestation are not only the cause of global warming but are also causing shortages of resources needed for the existence of life on earth.

4.14 Quality Education and Sustainable Use of Oceans, Seas, and Marine Resources (SDG 14)

Oceans form the major part of Earth. Therefore, SDG 14 addresses the conservation of underwater life-forms and the sustainable usage of resources extracted from oceans. According to the United Nations (2021), there is an increase in dead zones, which are areas where there is insufficient oxygen to support marine life. Dead zones emerge due to pollution, which stimulates the growth of algae. These algae are responsible for sucking up oxygen, and the underwater area becomes a dead zone where very few organisms can survive. The largest dead zone is found in the Gulf of Oman connecting the Arabian Sea and Persian Gulf. The Bay of Bengal has also developed a dead zone half the size of Bangladesh (Bristow et al. 2016). The goal is to minimize marine pollution and manage marine resources sustainably to reduce adverse impacts on underwater life. It also focuses on reducing illegal practices such as overfishing or unregulated fishing to safeguard marine resources. Target 14a of SDG 14, which states that "Increase scientific knowledge, develop research capacity and transfer marine technology, taking into account the Intergovernmental Oceanographic Commission Criteria and Guidelines on the Transfer of Marine Technology, in order to improve ocean health and to enhance the contribution of marine biodiversity to the development of developing countries, in particular small island developing States and least developed countries," directly correlates with SDG 4 (Lawrence et al. 2020). Imparting knowledge about underwater resources and their safety is an important issue, as this knowledge would help in building alternative strategies to conserve water resources and life below water. A study suggested using disposed-off fish skins to produce fish that would ensure sustainable usage of underwater resources (Palomino 2020). Disposed-off fish skins are a by-product of the sea food industry (ibid).

4.15 Quality Education and Sustainable Use of Terrestrial Ecosystems, Manage Forests, Combat Desertification, and Halt and Reverse Land Degradation and Biodiversity Loss (SDG 15)

SDG 15 works to safeguard life on land by ensuring sustainable use and conservation of forests and terrestrial life-forms, wetlands, and freshwater ecosystems. It ensures the safety of biodiversity and encourages sustainable development measures to prevent the extinction of threatened species and take serious actions against illegal practices of poaching and trafficking of wildlife and flora (United Nations 2015). Conservation of biodiversity is only possible when local communities are taught to lead sustainable lifestyles with minimum intervention to ecosystems (Lawrence et al. 2020). Arana et al. (2020) stressed on making people aware of the usage of organic material such as biodegradable fibers and crop waste fibers, using discarded fishing nets for nylon fibers, sustainable production of yarns, and reusing and recycling the waste material to reduce the pollution caused by them. In the Indian context, schools and colleges have mandatory environmental science courses to help students understand the importance of saving the environment and leading sustainable lives by making responsible choices. Quality education would help bring sustainable solutions to conserve terrestrial ecosystems, thus making SDG 4 an essential requirement for SDG 15.

4.16 Quality Education and Peaceful and Inclusive Societies for Sustainable Development, Justice for All, and Inclusive Institutions (SDG 16)

SDG 16 has a wide scope, as it endorses the elimination of illegal practices that pose danger to the peace of nations, such as violence, abuse, trafficking, bribery, corruption, and exploitation. It ensures that the rule of law prevails and that every citizen gets justice and is allowed participation in decision-making along with protection of fundamental rights (United Nations 2015). Target 16.10, which states "ensure public access to information and protect fundamental freedoms, in accordance with national legislation and international agreements," and Target 16.a, which states "strengthen relevant national institutions, including through international cooperation, for building capacity at all levels, in particular in developing countries, to prevent violence and combat terrorism and crime," correlate directly with SDG 4 (Lawrence et al. 2020). Education would make people aware about their fundamental rights, which would eventually reduce the exploitation and corruption that take place due to the lack of awareness. Furthermore, education brings an understanding of coexistence that reduces conflict situations, leading to a peaceful environment. Good governance brings about the transformation in society and motivates them to lead sustainable livelihoods, ensuring justice to all (Franco and Derbyshire 2020), especially in the case of developing nations such as India, and education would prove to be a huge contributor for the same.

4.17 Quality Education and Global Partnership for Sustainable Development (SDG 17)

SDG 17 stresses the integrated efforts of all nations in implementing Sustainable Development Goals and initiates coordination among different United Nations units to enhance international cooperation. Furthermore, this goal aims to develop and distribute environmentally friendly technologies to build capacity and bring macroeconomic stability to developing nations such as India (United Nations 2015). SDG 4 is important in implementing SDG 17 because quality education would make communities understand the gravity of cooperating with each other to implement these goals and benefit globally. Target 17.6 of SDG 17 is even associated with SDG 4, as it states "Enhance North–South, South–South and triangular regional and international cooperation on and access to science, technology and innovation and enhance knowledge sharing on mutually agreed terms, including through improved coordination among existing mechanisms, in particular at the United Nations level, and through a global technology facilitation mechanism" (Lawrence et al. 2020). This target stresses the need for knowledge sharing and accessible scientific and technological information to reduce the knowledge gap among nations (ibid). The importance of education is clear by the fact that sustainability depends upon accountability and even companies are moving toward a sustainable performance approach to map multi-stakeholder partnerships (Franco and Abe 2020).

4.18 Conclusion

Quality education as SDG 4 came out to be the backbone for fulfilling the objectives of the remaining 16 SDGs. A holistic approach is needed to accelerate sustainable development, and quality education has been an indispensable tool in integrating the essence of all Sustainable Development Goals. Making education a priority in policy-making would shift society from making irresponsible choices to making more conscious choices, which would help the ecosystem to flourish in sustainable ways. Sustainability is no longer a choice; it has become a need of the hour. Education is imperative as a direct or indirect target of all Sustainable Development Goals associated with the SDG 4 indicators. Nutritional requirements (SDG 2) must be fulfilled to end the problem of hunger. Providing basic meal facilities at the school level would encourage students to learn and prepare themselves for the future and at the same time obtain the provision of meals. Furthermore, sustainability can be galvanized by reducing the challenge of poverty (SDG 1), which has an integrated relation with employment (SDG 8), and employability could be increased through quality education.

In addition, the surge in global warming has been a matter of concern, and making people aware of the practices that lead to global warming (SDG 13) and developing infrastructure to reduce the same is possible with the accessibility of quality education. Both lives on land (SDG 15) and under water (SDG 14) have been facing existential threats due to pollution and irresponsible waste disposal and deforestation. Education not only brings awareness about wrong practices but also helps to develop innovative solutions to such challenges. Sustainability is about reminding the present generation about the needs of future generations, which demands educating the general public about sustainable consumption and sustainable production (SDG 12). In addition, sustainable consumption and sustainable production, peace and justice (SDG 16), and civic sense (SDG 11) are also very important components of democratic society, and education plays a huge role in helping people understand their rights and duties and remedies in case of exploitation.

Gender equality (SDG 5) would empower equal participation in decision-making, policy-making, and even harnessing opportunities that were previously understood to be male-dominated areas only. With education comes understanding of the concept of gender neutrality and equal distribution of income and wealth (SDG 10) and that development takes place when society uplifts all the sections irrespective of their caste, creed, sex, or economic status. Furthermore, education enhances the possibility of research and development to boost innovation (SDG 9) and encourage the transition to better technology and resources such as clean energy (SDG 7) and clean drinking water (SDG 6). In addition, better health facilities and awareness about reproductive, maternal, and child health (SDG 3) would increase the life expectancy and productivity of people as a whole. For development to take place sustainably, it becomes essential for all stakeholders to voluntarily join hands (SDG 17) and work on sustainable development integration and capacity building. Sustainability should not be a goal; it should be a way of life, inculcated in such a manner that people automatically make responsible choices.

References

Anyolo EO, Kärkkäinen S, Keinonen T (2018) Implementing education for sustainable development in Namibia: school teachers perceptions and teaching practices. J Teach Educ Sustain 20(1):64–81

Arana C, Franco IB, Joshi A, Sedhai J (2020) SDG 15 life on land. In: Actioning the global goals for local impact. Springer, Singapore, pp 247–264

Bhukta A (2020) No poverty: how much, how far. In: Sustainable development goals. Springer, Cham, pp 1–11

Bristow LA, Dalsgaard T, Tiano L, Mills DB, Bertagnolli AD, Wright JJ, Hallam SJ, Ulloa O, Canfield DE, Revsbech NP, Thamdrup B (2016) Ammonium and nitrite oxidation at nanomolar oxygen concentrations in oxygen minimum zone waters. Proc Natl Acad Sci 113(38):10601–10606

Brundtland, G. H. (1987). Our Common Future World Commission On Environment And Developement

Chae H (2022) Income or education? Community-level antecedents of firms' category-spanning activities. Strateg Manag J 43(1):93–129

Choudhury G, Chauhan C (2013) Measuring poverty: how the world counts its poor. Hindustan Times. Accessed 4 Dec 2021

Fonseca LM, Domingues JP, Dima AM (2020) Mapping the sustainable development goals relationships. Sustainability 12(8):33–59

Franco IB, Abe M (2020) SDG 17 partnerships for the goals. In: Actioning the global goals for local impact. Springer, Singapore, pp 275–293

Franco IB, Derbyshire E (2020) SDG 16 peace, justice and strong institutions. In: Actioning the global goals for local impact. Springer, Singapore, pp 265–274

Franco IB, Tapia R, Tracey J (2020) SDG 13 climate action. In: Actioning the global goals for local impact. Springer, Singapore, pp 219–228

Griggs D, Smith MS, Rockström J, Öhman MC, Gaffney O, Glaser G, Shyamsundar P (2014) An integrated framework for sustainable development goals. Ecol Soc 19(4):49

Haldar SK, Hembram S (2020) Health progress in India with respect to millennium development goals: are health targets of SDGs achievable? An empirical study at subnational level. In: Sustainable development goals. Springer, Cham, pp 41–55

Helliwell JF, Layard R, Sachs J, De Neve JE (2020) World happiness report 2020. Sustainable Development Solutions Network, New York

Kumar K, Prakash A, Singh K (2021) How National Education Policy 2020 can be a lodestar to transform future generation in India. J Public Aff 21(3). https://doi.org/10.1002/pa.2500

Lawrence AW (2018) Some governance and peaceful coexistence issues for sustainable advancement: a Nigerian perspective. Author House, Bloomington

Lawrence AW, Ihebuzor N, Lawrence DO (2020) Macrolevel studies of direct and indirect relationships between SDG 4 and the 16 SDGS. Mod Econ 11(06):11–76

Londoño NAC, Velasco JO, García FC, Franco IB (2020) SDG 6 clean water and sanitation. In: Actioning the global goals for local impact. Springer, Singapore, pp 85–104

Nakicenovic N (2019) The world in 2050: transformative pathways toward sustainability. In: 2019 Annual meeting, February. AAAS, Washington, DC

Nilsson M, Chisholm E, Griggs D, Howden-Chapman P, McCollum D, Messerli P, Stafford-Smith M (2018) Mapping interactions between the sustainable development goals: lessons learned and ways forward. Sustain Sci 13(6):1489–1503

Oluwadamilola O, Akinyemi O, Adediran O (2018) Human capital development and inclusive growth: implications for achieving SDG-4 in Nigeria. Afr Popul Stud 32(1):4088–4096

Palomino E (2020) SDG 14 life below water. In: Actioning the global goals for local impact. Springer, Singapore, pp 229–246

Patel A, Banerjee S, Hans A (2020) Gender equality and the sustainable development goals: focusing on the adolescent girls in India. In: Sustainable development goals. Springer, Cham, pp 79–93

Pham-Truffert M, Metz F, Fischer M, Rueff H, Messerli P (2020) Interactions among sustainable development goals: knowledge for identifying multipliers and virtuous cycles. Sustain Dev 28(5):1236–1250

Pradhan P, Costa L, Rybski D, Lucht W, Kropp JP (2017) A systematic study of sustainable development goal (SDG) interactions. Earth's Future 5(11):1169–1179

Saravanakumar V, Malaiarasan U, Balasubramanian R (2020) Sustainable agriculture, poverty, food security and improved nutrition. In: Sustainable development goals. Springer, Cham, pp 13–39

SDG, U (2019) Sustainable development goals. The energy progress report. Tracking SDG, 7

Smith WC (2019) One indicator to rule them all: how SDG 4.1. 1 dominates the conversation and what it means for the most marginalized. In: Annual review of comparative and international education 2018. Emerald, Bingley

UN ESCAP (2019) Investing in sustainable infrastructure for all. UN ESCAP, Bangkok

United Nations (2015) Transforming our world: the 2030 agenda for sustainable development. https://sustainabledevelopment.un.org/post2015/transformingourworld

United Nations (2021) The sustainable development goals report 2021. United Nations, New York

Apoorva Bhatnagar is working as a Senior Assistant Professor of Marketing at Balaji Institute of Modern Management, Sri Balaji University, Pune. She has her PhD in Marketing from Gurukul Kangri University, Haridwar, Uttarakhand. Her area of expertise includes marketing, celebrity endorsements, advertising, consumer behavior, personality psychology, sustainable marketing, and entrepreneurship. She has expertise in data analytics softwares like SPSS and AMOS. She has several research papers and has also coauthored a book More Than Marketing Sustainable Marketing.

Triguna Singh is a psychologist attached with City Montessori School, Lucknow, Uttar Pradesh. Her research interests include child psychology and education. She has conducted various seminars on child psychology and education. She has gained considerable experience in the field of counseling.

5 Mapping the Provisions for Inclusive and Equitable Quality Education in the Indian Education Policy Documents: NEP 1968 to NEP 2020

Chasul Phogat

Abstract

Among 17 Sustainable Development Goals of the United Nations, the fourth goal focuses on quality and inclusive education for all. The focus of the current study is on the keyword "all." According to research and studies, the meaning of inclusive education is not limited to including learners with special needs into the mainstream, but it has a wider scope of making education and the environment inside a classroom more culturally appropriate so that "all" learners can find themselves included. Inside a classroom, learners and instructors do not interact in a vacuum; they interact via their demographic characteristics such as culture, language, religion, caste, language, and so on. Education would be considered as inclusive when learners despite of their caste, status, religion, and gender would feel included. Previous research indicates that caste plays an important role in the existing economic divide and maintaining the status quo. The focus of the study is to investigate how many initiatives are taken by the government to provide quality education to the learners from the scheduled caste and scheduled tribes. According to the Indian Constitution, Article 45, every child aged up to 14 years has the right to get free and compulsory education. Similarly, the national education policies prepare a roadmap for what these children would receive under "the free and compulsory education." Research suggests that policies are always context-specific; hence, it must be prepared in accordance with the needs of the citizens with a maximum coverage. The current study has analyzed the national education policies, National Curriculum Framework 2005, and national-level campaigns to locate the initiatives taken to include all learners, making education accessible to everyone. Since NEP 2020 has been drafted after the SDGs were announced by the UN, the focus is more upon locating the changes made in NEP 2020 by keeping the fourth goal in focus, i.e., equitable and inclusive education for all.

Keywords

Inclusive education · Culture · Caste · Economic divide · Indian Constitution · National education policies · National Curriculum Framework 2005

5.1 Introduction

Sustainable Development Goals (SDGs) were formulated as a continuation of the Millennium Development Goals (MDGs) and are being

C. Phogat (✉)
Bennett University, Greater Noida, India
e-mail: Chasul.phogat@bennett.edu.in

V. Dutta, P. Ghosh (eds.), *Sustainability: Science, Policy, and Practice in India*, Sustainable Development Goals Series, https://doi.org/10.1007/978-3-031-50132-6_5

implemented globally. There are 17 Sustainable Development Goals with 169 targets set by the United Nations in 2015 for the year 2030. The fourth goal, i.e., "ensuring equitable quality education and promoting lifelong learning opportunities for all" (UN 2015), is the focus of discussion of the current study. Education is always considered as a stepping stone for attaining the goal of growth and development in any society apart from creating aware and responsible citizens. Both physical and mental well-being are equally important for an individual to grow, and education is immensely crucial for the mental growth of individuals. Without being an educated person, it becomes difficult for a person to become aware about his/her rights. The Constitution of India provides the basic right of equality to all its citizens, and under Article 21-A, the right to education is a fundamental right of all children between the age group of 6 and 14. Similarly, many steps have been taken by the state to ensure that every citizen gets equal opportunities for education. Despite the various programs such as *Sarva Shiksha Abhiyan* (education for all), *Beti Bachao, Beti Padhao* (save the girl child, educate the girl child), *Samagra Shiksha Abhiyan* (integrated scheme for school education), and *Padhna Likhna Abhiyan* (scheme of adult education) initiated by the government, the fourth goal of education for all is still not attained! Education policies have been made since 1968, but despite the countless efforts, India is on rank 121 among 193 nations. The results are an indicator of poor performance when it comes to quality education, and data indicates that the ranking is deteriorating since past years (United Nations 2022). The argument proposed by the study is the education system is not inclusive or favorable toward the marginalized people in the country. The term marginalized has been used to refer to the scheduled caste and scheduled tribes in the present study. "Many great policies were made on education in the country since the time India got freedom, but they remained good only on paper because no strategies were ever designed for their implementation" said M. Sisodia (The Hindu, Sept. 11 2022a). The duality of policy-making takes its roots in the way a policy is made. In contrast, it is highly contextualized, and policy implementation is context-dependent (Yang 2007). India, being a multicultural, multilingual country, becomes crucial to understand the different subcultures exist here. To make the implementation process smoother, context-specific studies are required. In the present study, one aspect of the culture has been analyzed. Since caste plays a dominant role in Indian culture, the study has tried to investigate how much attention has been given to the educational upliftment of the scheduled caste and scheduled tribes in the education policies, the curriculum framework, and the national-level campaigns for education and literacy.

For the growth of any society, it becomes crucial that all its citizens get equal opportunities. "Ending poverty and other deprivations must go hand-in-hand with strategies that improve health and education, reduce inequality, and spur economic growth" (United Nations 2015). The achievement objectives are of improving health and education for everybody by reducing inequality, and when there is equal access to opportunities for all, economic growth would be a result. The study attempts to answer the research question of whether the national education policies of India are promising and striving to provide equal opportunities to "all" or not. Inclusive education does not limit its meaning to including children with special needs, but the scope is much wider. The children from remote areas, rural areas, slums, minorities, and tribal communities should get equal opportunity to access the educational facilities which any person living in a developed area is getting. "Inclusion is not just an educational philosophy; it is more importantly a process towards the practical changes that must be brought about to help children learn to their full potential while recognizing that all children are different and have different individual learning needs and speeds rather than 'special' needs" (Saxena 2012, p. 148). There must be no discrimination based on caste, creed, religion, sex, and ethnicity inside a classroom. India being a multicultural and multilingual country accommodates people from various cultures and subcultures and religions. As per Pew Research Data (2021),

Hindus take up 79.8% of India's population. Hinduism follows a rigid system of caste, and it is among the world's oldest form of surviving social stratification (BBC 2019). The structure has been synthesized in a way that no individual can follow Hinduism without associating themselves with a caste. As B.R. Ambedkar said, if someone is a Hindu, automatically the person falls under a specific category of caste from the day she/he is born, from a mother's lap till her/his death bed. Discrimination exists even if a whole society is facing a natural calamity such as an earthquake United Nations World Conference 2001). A society progresses when each one of its members get an equal opportunity to grow and excel. India, being a diverse country, consists of various cultures, languages, religions, and castes. Defined by Manusmriti, the caste hierarchy includes four major categories: the first three categories are considered as upper castes, whereas the fourth category is considered as low or backward caste. There is a fifth category as well, which are outcastes; they are not even counted as humans. People falling under this fourth caste are addressed as the scheduled castes, scheduled tribes, and other backward classes. Mahatma Gandhi gave them the name of *Harijans*, but the terminology evokes sympathy, and with progression of time, they are addressed as Dalits. In the hierarchy, the most upper caste is of *Brahamanas* who are considered as the purest of all; they are the only people allowed to complete the *puja* ceremonies inside a temple or inside people's houses. An optimal learning environment for every student inside a classroom makes education inclusive in real terms. In 1950, with the strenuous efforts of Dr. Ambedkar, a provision of quota was announced for Dalits so that they can have access to education and employment. Although there is a provision of reservation for minorities, scheduled castes (SC), scheduled tribes (ST), and other backward castes (OBC) in the Indian Constitution, there are any special efforts made to fulfill their needs keeping their historical and cultural background in mind! This study investigates the space provided in the educational policies, the national-level education campaigns, and the National Curriculum Framework 2005 to fulfill the special needs of children from these marginalized communities, i.e., SC and ST communities. Social justice and social inclusion become the first step in the direction of achieving the goal of equality for all.

5.2 The Trajectory of Caste-ism in Educational Institutions

Restriction from getting education is an age-old, forced tradition on Dalits by Manusmriti. The text prohibits education for Dalits. There were severe punishments if Dalits wanted to study Sanskrit. They were not supposed to even listen to Sanskrit which was the elite language spoken and understood by the upper caste only. Gradually, people from Dalit communities started to go to school. The environment inside the schools was completely hostile toward these Dalit students. Covertly, the teachers used to abuse them and beat them up for no reasons. All children used to sit on mats, where these few children from the Dalit community used to sit on the floor in the corners of the classroom. In his autobiography, *Joothan*, Om Prakash Valmiki writes how he was always abused by his teachers. For 3 days, constantly, he was sweeping the school premises when all his classmates were sitting inside the classroom and was studying (Valmiki 2007). Students from these communities were harassed physically and mentally by both the teachers and students equally. "The Hindus of Kavitha ordered the untouchables not to insist upon sending their children to the common village school maintained by Government. What sufferings the untouchables of Kavitha had to undergo, for daring to exercise a civic right against the wishes of the Hindus" (Ambedkar 2021, p. 12). Such was the plight of these people who had no access to necessities of life. They were denied of all their rights and were forced to live a life of forced and unpaid labor. It resulted in very low self-esteem and confidence among them. Even today, there are numerous cases of atrocities against Dalit community. Dalits are created by the god to serve the upper castes. Even in today's modern world, people are treated respectfully only based on their caste. A person from a Dalit community no matter whatever post she/he has acquired would be mocked if belongs to a Dalit community. In his report, Mahatma Jyotirao Phule wrote that there should be separate schools for these children so that they can study. The problems of these Dalit children can only be understood by the Dalit teachers. In *Apne-Apne Pinjre*, Mohandas Naimishray writes that there was a school in their locality and all the children from their community used to go to study there. No upper caste teacher wanted to teach them since it became a Dalit school (all students who were enrolled in the school belong to *Chamar* community which is considered as low caste according to Hinduism). "Even if the teachers came to teach us, they would come with a negative body language such as shrinking their noses, putting a handkerchief on their mouths, standing on a certain distance" (Namishrai 2018).

In the eighteenth century, Mahatma Jyotirao Phule and his wife Savitribai Phule worked on a massive scale for the rights and education of Dalit people. The couple was the first one in India to open schools and organizations for the upliftment of Dalits, women, and Dalit women. Savitribai Phule was the first woman teacher of India. She along with her friend Fatima Sheikh used to teach women and made them aware about their rights. The couple belonged to the *Mali* community. Jyotirao Phule being an experienced victim of the Brahmanical ideology worked for the upliftment of Dalit people, women of all castes, and widows. He along with his wife dedicated his entire life in educating women and Dalit. He was the first person who used to demand separate schools for Dalits. He recommended that Brahmins should not teach these children because the former cannot understand the problems of the latter and the objective of social equality would not be fulfilled. When the first Education Committee was drafting their rules, suggestions were taken from Jyotirao Phule, and he recommended that there should be special schools for the education of Dalit community people and their teachers must be Dalit (Arya 2021). In the next century, Dr. Ambedkar carried forward Phule's work to a next level. He was the one who fought for the rights of the depressed classes. Dr. Ambedkar followed the ideology of Jyotirao Phule. Dr. Ambedkar also insisted upon the significance of education for the Dalits. Being a person from the Mahar community, he also experienced various incidents of untouchability and embarrassment. The life of his mission was to demolish the caste system. In his book *Annihilation of Caste*, he explains how the economic divide is the result of the caste system and until the caste system is demolished there is no scope for social justice and social equality (Ambedkar 2021). Education is a weapon which would help individuals to fight against inequality and discrimination. In independent India, the Dalit students were facing discrimination. "They were forbidden to keep certain domestic animals, to use certain metals for ornaments; were obliged to wear a particular type of dress, to eat a particular type of food, to use a particular type of footwear and were forced to occupy the dirty, dingy and unhygienic outskirts of villages and towns for habitation where they lived in dark, insanitary and miserable smoky shanties or cottages" (Keer 2019). Dalit women used to give tax if they wanted to cover their breasts in the Kerala State of southern India. This tax was known as breast tax (BBC 2016).

The incidents of Dalit atrocities are a common phenomenon in the Indian society. A grave incident took place in August 2022 in Rajasthan. A 9-year-old Dalit boy was beaten up by his upper caste teacher mercilessly, and after 24 days, the boy died (The Wire 2022b). The boy mistakenly drank water from an earthen pot which was kept for the upper castes. In October 2021, Dalit students were forced to queue up and eat separately during the midday meal in Uttar Pradesh. There are clear instructions for them to keep their utensils separate. They are beaten on minor mistakes (The Indian Express 2021). Such incidents clearly showcase that caste discrimination exists in the educational institutions even today. Under such circumstances, how can a person coming from a "different" background can cope up with the pressures she/he going to face (Rege 1998)? In higher education institutions where students like Payal Tadvi and Rohith Vemula commit suicides, there is a dire need to make changes in the educational policies to create a safe, secure, and healthy environment inside the educational institutions. "Privatisation of higher education in India at the expense of public education which has adversely affected the overall prospects for oppressed social groups to break their vicious cycle of exploitation. The lack of the policy of reservation to ensure social justice and representation of the oppressed combined with exorbitant fees makes such private education institutions inaccessible and exclusive" (The Wire 2021). On the other hand, government is initiating schemes for paying the victims of atrocities from SC and ST communities but not able to curb such incidents. A culture needs to be developed where everyone is treated with respect and dignity. Such culture cannot be established without including the principles of social justice and equality in the educational institutions. Educational organizations are the place where great think tanks are produced. Since the schools and colleges do not exist in vacuum and the learners and teachers there are also a by-product of the culture itself, the policies, curriculum, and everything must focus upon the underrepresented people.

5.3 Special Provisions for Quality Education: Recommendations to William Hunter in Pre-independent India

In October 1882, a commission named Hunter Commission was appointed to investigate the progress in education after the Woods Dispatch Commission 1854. Hunter Commission, with an aim to improve the quality of education, called for various suggestions from people who were working on grassroots level in the education field. Jyotirao Phule, working actively for the education of the underprivileged children and women along with his wife Savitribai Phule, submitted his recommendations to the Commission. This was the first attempt made by any Indian citizen to not only speak and write but also highlight the special needs of Dalit children in the education reports. These reports were the basis on which the first draft of the National Education Policy 1968 was synthesized. In pre-independent India, education was primarily given by Christian Missionaries. Jyotirao Phule along with his wife Savitribai Phule was the first person in India to open schools for Dalits and Dalit women. Savitribai Phule, Jyotirao Phule's wife, was the first woman teacher who taught women even after receiving life-threatening backlash from upper caste men. Jyotirao Phule's recommendations included his views not only on quality education but also on quality education for Dalit children and women. Being a victim of caste atrocities, Jyotirao Phule knew that the only way for Dalits to progress and live a life of dignity is education. His idea of quality education was the inclusion of scholarships and special aids for brilliant students. He recommended that technical education and moral education must be the focus of the education system. He clearly mentioned in his recommendations that education is completely under the control of upper caste people; it is by the upper castes and for the upper castes. One of his major suggestions was to appoint Dalit teachers in the government schools, especially in rural India so that the discrimination faced by the Dalit students in schools can be reduced. He also proposed that a Dalit teacher would be able to understand and empathize more with the problems faced by these Dalit students. But after reading the literature available, the researcher proposes that it was possible to appoint Dalit teachers since the availability of Dalit teachers is a big question. The Dalit community was suppressed by the upper castes in such a way that they did not have access to necessities such as water and food. He emphasized that education must be mandatory for children till 12 years of age. In each village, there should be a separate school for Dalits if the population is high. His ideas for quality education focus a lot upon quality teachers. He emphasized upon the appointment of qualified teachers in the indigenous schools in the villages (Arya 2021).

5.4 Provisions for Quality Education for "All" in National Educational Policies and Governmental Campaigns: Does Congruity Exist Between the Plans and Implementation at the Ground Level?

The first National Education Policy 1968 was synthesized taking the previous reports as the foundation. "Suitable programmes should be developed to reduce the prevailing wastage and stagnation in schools and to ensure that every child who is enrolled in schools successfully completes the prescribed course" (Ministry of Human Resource and Development, NEP 1968, pp. 39–40). The heading "Free and Compulsory Education" (Ministry of Human Resource and Development, NEP 1968, p. 40) says "Strenuous efforts should be made to equalise educational opportunity." Considering it an ambiguous statement, the researcher proposes that an education policy must guide and instruct about what efforts should be taken. Going further under Sections b and d, the policy document states "Intensive efforts are needed to develop education among the backward classes and especially the tribal people." Page number 44, Section 16, repeats the same pattern: "The educational interests of the minorities should be promoted." There is no clear methodology proposed by the policy-makers to address the needs of "all." On the other hand, the NEP 1986 has devoted an entire section (Section 4) for the education for children who belong to the scheduled tribes and scheduled castes. Part III, 1986 Education Policy, states that suitable steps would be taken by the government to ensure that all the learners despite of their gender and caste would receive the same education. Part III focuses upon the education of women as well as education for the children from scheduled caste and scheduled tribes. The policy document mentions that various scholarships and schemes should be run by the government to uplift the education level of women (Department of Higher Education, Ministry of Education 1986). There are such schemes as *Beti Bachao, Beti Padhao* (save the girl child, educate the girl child); *Shiksha Karmi Yojna* for people living in remote areas, especially girls living in remote areas; *Sarva Shiksha Abhiyan* (education for all); *Kasturba Gandhi Balika Vidyalayas*; and *Samagra Shiksha Abhiyan*, a recent campaign by the government that subsumes the three earlier schemes into it (Ministry of Education 2021); however, *Samagra Shiksha Abhiyan* focuses upon including the SC and ST children into the mainstream. Although Section IV in NEP 1986 discusses about providing scholarships to students who belong to SC and ST communities, the discrimination faced by them inside the educational institutions is an indicator of a lack of social justice (Thorat 2021). NEP 1986 even pays attention toward recruiting people from the marginalized community (as suggested by Jyotirao Phule in his recommendations to the Hunter Commission). NEP 1986, Section 4, discusses in detail about all the steps which should be taken in the direction of equality in education. But the question remains the same even after 35 years of NEP 1986: Why are we still lagging in achieving equality in education? Section 4 even emphasized upon the utilization of schemes such as NREP and RLEGP to ensure the employment status of SC and ST people so that the children in these families can secure their future. The section focuses upon the use of tribal languages in the initial stage of learning for tribal children. The policy also mentions about opening of "Ashram schools" for the tribal students. But according to PRS Legislative Research, the data reveals that only 616 Ashram schools were functional in 2014 (PRS Legislative Research 2014). As per 2019 data, there are total 1018 functional Ashram schools in India out of the 1205 sanctioned (Renuka Singh, Union Minister, Lok Sabha). In 1997–1998, Eklavya Model Residential Schools under Article 275(1) were opened to provide quality education to tribal children. "By 2022, 462 Eklavya Model schools would be opened" (Ministry of Tribal Affairs 2020). The center allocates 30 Lakh INR for constructing these schools, and the state spends 30 Lakh annually for the maintenance. "Only 244 schools could be made functional" (Ministry of Tribal Affairs 2022).

"The global education development agenda reflected in the Goal 4 (SDG4) of the 2030 Agenda for Sustainable Development, adopted by India in

2015, seeks to 'ensure inclusive and equitable quality education and promote lifelong learning opportunities for all' by 2030. Such a lofty goal will require the entire education system to be reconfigured to support and foster learning, so that all of the critical targets and goals (SDGs) of the 2030 Agenda for Sustainable Development can be achieved" (Introduction Section, Ministry of Human Resource and Development, NEP 2020). The key phrase in the above-quoted lines from the Introduction Section of NEP 2020 is "the entire education system to be reconfigured to support and foster learning." While the fourth SDG is emphasizing upon "equity and inclusiveness" of education, NEP 2020 is not able to penetrate the keyword itself. Fostering learning is an essential element but the keyword "all" is missing. A similar trend can be observed in the reforms made in contemporary times by the state governments also. A report published on September 21, 2022, in *The Hindu* mentions that the government is trying to reform the education system which is in alignment with NEP 2020. To make education more inclusive, the Andhra Government is trying to develop programs which focus more upon building critical thinking and problem-solving skills. The objective of the plan is not aligning with the methodology the government is applying in this scenario. "The aim is to ensure that all children have access to quality education within an equitable and inclusive classroom environment, taking into account their diverse backgrounds and different academic abilities, thereby making them active participants in the learning process" (Varma 2022, The Hindu).

Four pages are devoted to "inclusive education" in NEP 2020 in Chapter 6. In Part 6.2, the policy discusses in detail that attention should be given to the underrepresented people such as students from the scheduled caste and tribal caste, those who are living in remote rural areas, transgender communities, and so on. Section 6.2.2 mentions that steps would be taken to bridge the gap in terms of access, participation, and learning outcomes among students from underrepresented sections. The post-matric and pre-matric schemes by the government are all based on an old model. Section 6.4 mentions about the initiatives taken by the government such as distributing bicycles for transportation to women, but no such initiative is there for people from SC and ST communities. Section 6 discusses a lot about women education, about Kasturba Gandhi Vidyalaya; it pays attention to the gender part and to the people with special needs, but the focus on SC and ST people is lacking.

5.5 Conclusion

The national education policies should provide an exclusive space to explain the inclusion element. Not only there is a need to write about inclusive education into the national education policy but also there is a need to include the concept of inclusive education in the teacher education section. It is the need of the hour to lace student teachers with the key concepts of social justice. The training which student teachers acquire before their actual years of teaching acts as a foundation for their future years of teaching. The teachers are required to be sensitized about various ways to include the marginalized in the mainstream. There should be precocious guidelines from curriculum design to teachers' training not only about inclusive education but also about how to include all learners in the teaching-learning process. The pedagogy and the content must be both designed and created in a way that no learner feels excluded. The curriculum of school students, higher education students, and teacher students must be designed in a way that nobody feels excluded inside a classroom. Teachers should be sensitized toward the caste issue. They should be trained in a way that they can empathize to their problems. There should be a provision of continuous professional development programs which exclusively train the pre-service teachers about the caste problems.

References

Ambedkar BR (2021) Annihilation of caste, 6th edn. R Sons Books, Delhi, pp 10–37

Arya JL (2021) Kranti Joti Savitri Phule, 2nd edn. Vaani Prakashan, Delhi, pp 116–121

BBC (2019) "What is India's Caste System?" BBC News. June 19, 2019. Retrieved July 20, 2022, from https://www.bbc.com/news/world-asia-india-35650616

Department of Higher Education, Ministry of Education (1986) National Policy of Education 1986. Government of India. Retrieved April 22, 2022, from https://www.education.gov.in/en/higher_education

Department of Higher Education Ministry of Education (1992) National Policy of Education 1968. Government of India. Retrieved April 22, 2022, from https://www.education.gov.in/en/higher_education

Department of Higher Education Ministry of Education (2020) National Policy of Education 1968. Government of India. Retrieved April 22, 2022, from https://www.education.gov.in/en/higher_education

Hari A (2021) Casteism is rampant in higher education institutions, but is 'Wilfully neglected' study. The Wire, October 8. https://thewire.in/education/casteism-rampant-higher-education-institutions-wilfully-neglected-study

Keer D (2019) Dr. Ambedkar: life and mission, 5th edn. Popular Prakashan, Mumbai, pp 47–87

Kramer S (2021) Key findings about the religious compositions of India. Pew Research Center, September 21. Retrieved March 16, 2022, from https://www.pewresearch.org/fact-tank/2021/09/21/key-findings-about-the-religious-composition-of-india/

Ministry of Education (2021) Samagra Sikhsha Scheme. Department of School Education and Literacy Ministry of Education, Government of India, August 16. Retrieved April 22, 2022, from https://samagra.education.gov.in/

Ministry of Human Resource and Development (1968) National Policy of Education 1968. Retrieved June 20, 2022, from https://www.education.gov.in/en

Ministry of Human Resource and Development (2020) National Education Policy 2020. Retrieved June 18, 2022, from https://www.education.gov.in/en

Ministry of Tribal Affairs (2020) EMRS. Ministry of Tribal Affairs, Government of India, November 13. Retrieved April 22, 2022, from https://tribal.nic.in/EMRS.aspx

Ministry of Tribal Affairs (2022) 41st Report, Standing Committee on Social Justice and Empowerment, 2022–2023. Ministry of Tribal Affairs, Government of India. Retrieved December 25, 2023, from https://loksabhadocs.nic.in/lsscommittee/Social%20Justice%20&%20Empowerment/17_Social_Justice_And_Empowerment_41

Mishra A (2021) We were told to sit apart, beaten up: Dalit children of Amethi School. The Indian Express, October 1. https://indianexpress.com/article/india/we-were-told-to-sit-apart-beaten-up-dalit-children-of-amethi-school-7544896/

Namishrai M D (2018) 4th Edition, Vani Prakashan, Delhi, pp 15–31

PRS Legislative Research (2014) Working of Ashram Schools in Tribal Areas. Retrieved July 25, 2022, from https://prsindia.org/files/policy/policy_committee_reports/1393414764_SCR%20Summary-Working%20of%20Ashram%20Schools%20in%20Tribal%20Areas.pdf

Rege S (1998) Dalit women talk differently: a critique of 'difference' and towards a Dalit feminist standpoint position. Econ Polit Wkly 33:WS39–WS46

Saxena V (2012) Contemporary trends in education, 1st edn. Pearson, London, pp 145–178

Staff Reporter (2016) The woman who cuts off her breasts to protest a tax. BBC News, July 28. Retrieved March 11, 2022, from https://www.bbc.com/news/world-asia-india-36891356

Staff Reporter (2022a) Several state laws need to be tweaked for NEP 2020: Sisodia. The Hindu, September 11, New Delhi

Staff Reporter (2022b) Rajasthan: nine-yr-old-Dalit boy passes away after alleged assault by school teacher. The Wire, August 14. https://thewire.in/caste/rajasthan-nine-yr-old-dalit-boy-passes-away-after-alleged-assault-by-schoolteacher

Thorat S (2021) Discrimination on the campus. The Hindu, January 20. https://www.thehindu.com/opinion/lead/Rohith-Vemula-suicide-Dalit-and-higher-education-Discrimination-on-the-campus/article62116029.ece

United Nations (2015) Transforming Our World: The 2030 Agenda for Sustainable Development. Retrieved December 27, 2023, from https://sdgs.un.org/sites/default/files/publications/21252030%20Agenda%20for%20Sustainable%20Development%20web.pdf

United Nations (2022) *The sustainable development goals report 2022*. United Nations Department of Economic and Social Affairs. Retrieved July 30, 2022, from https://sdgs.un.org/goals

United Nations World Conference (2001) A Report by Human Rights Watch for the United Nations World Conference Against Racism, Racial Discrimination, Xenophobia and Related Intolerance. Durban, South Africa, September 2001. Retrieved July 25, 2022, from https://www.hrw.org/reports/2001/globalcaste/caste0801.pdf

Valmiki O P (2007) Joothan, 1st Edition, Bhatkal & Sen, Delhi, pp 09–20

Varma S P (2022) Reforms and the task of getting teachers on board. *The Hindu*. September 21, 2022. Retrieved December 30, 2022, from https://www.thehindu.com/opinion/op-ed/reforms-and-the-task-of-getting-teachers-on-board/article65915190.ece

Yang R (2007) Comparing policies. In: Bray M, Adamson B, Mason M (eds) Comparative education research. Approaches and methods. Springer, Cham, pp 241–262

Chasul Phogat is currently an Assistant Professor of English at Bennett University, Greater Noida. She has completed her PhD from BITS Pilani, India. Her research interests include women studies, Dalit studies, and intercultural communication.

Impact of the Language of Instruction for STEM Subjects on Student Engagement, Motivation, and Further Learning: A Case Study in Maharashtra, India

6

Darshan Gaikwad, Jyotsna Akurathi, and Mukund Nagarajan Rao

Abstract

Quality education under the paradigm of sustainable development is a challenging policy domain, especially in a diverse country like India. With multitudes of cultures, languages, and dialects, the medium of instruction in the field of education poses a challenge, especially for subjects like science, technology, engineering, and mathematics. The New Education Policy (NEP) 2020 also argues for imparting technical education in regional languages. Earlier research also suggests that concepts are better communicated and understood in native languages. Yet, a rudimentary analysis of the number of student attendees in Marathi medium municipal schools in Mumbai, for example, shows a decrease of about 66% between the years 2010–2011 and 2019–2020. Language of instruction is deeply connected to quality and inclusivity in education. It also determines the future motivation and opportunities that students receive. Hence, this challenge warrants a deeper probe and analysis of the motivations and aspirations of stakeholders in the education ecology.

This study takes the State of Maharashtra as a case to examine the key factors that determine preferences for language of instruction within different stakeholder groups. It aims to give an overview of the development of language-of-instruction-centered policies in the state through interactions with stakeholders to gather their inputs and insights. It further attempts to consolidate them into actionable recommendations.

Keywords

Language of instruction · English · Marathi · Regional · STEM · Maharashtra · India · SDG 4

6.1 Introduction

Sustainable Development Goal (SDG) 4 seeks to ensure inclusive and equitable quality education and promote lifelong learning opportunities for all under the Agenda 2030 of the United Nations. Education is a stimulant for socioeconomic progress and holds immense potential in the quest for sustainable development. The UNESCO has declared the years 2022–2032 as the decade for indigenous languages (UNESCO 2019). The

D. Gaikwad (✉) · J. Akurathi · M. N. Rao
Indian Institute of Management, Rohtak, India
e-mail: ipm01darshang@iimrohtak.ac.in; ipm01jyotsnaa@iimrohtak.ac.in; ipm01mukundr@iimrohtak.ac.in

V. Dutta, P. Ghosh (eds.), *Sustainability: Science, Policy, and Practice in India*, Sustainable Development Goals Series, https://doi.org/10.1007/978-3-031-50132-6_6

international drift toward indigenous roots is in stark contrast with the shift of the local populace toward acquiring global languages.

While the world has made rapid strides over the years in ensuring access to education, a lot more is left to be achieved in light of the fact that nearly a fifth (260 million) of the world's children were out of school in 2018. Further, more than half of all children and adolescents fail to meet the minimum proficiency standards in mathematics and reading (Education – United Nations Sustainable Development n.d.). The pandemic has only worsened the situation with increased disruptions and dropout rates across geographies. Thus, there is an urgent need for a review of the current situation to ensure that education as a tool can be used to empower people and help us on the path toward sustainable development.

6.2 Education in India: A Linguistic Overview

Due to the colonial rule in India, the introduction of English led to the consideration of the language as a major foreign language in the country. Eventually, English has become the medium of instruction in schools and other educational institutions. With these functional changes and the benefits offered through the language, its status has turned from a foreign to a global language in India (Vijayalakshmi and Babu 2014).

6.3 Structure

At the school level, India has a mix of government-run, government-aided, and privately run not-for-profit schools that fulfill the education needs of the population. Higher education is classified under a complex web of various heads in India but can broadly be classified under central university, state university, private university, Institute of National Importance, and deemed-to-be-university based on their management (Sharma and Sharma 2015; Shah 2015).

India has a total of 1.5 million schools of which two-thirds are government-run, 39,391 colleges, and 981 universities with 37.4 million people enrolled in higher education with a gross enrolment ratio of 24.5%. Moreover, due to the ongoing pandemic, only a third of India's school children have enrolled in some form of online education. Of these students, only 32.5% are pursuing their online education through live online classes (Annual Status of Education Report 2020) (Fig. 6.1).

6.4 STEM

Science, technology, engineering, and mathematics (STEM) is a sought-after field in India. Nearly one-third of the eight million odd people graduating in India each year pursue a degree in STEM. With higher penetration of the STEM courses in India, there are significant need and importance of the development of English language skills in engineering students as means to excel in their professional career as engineers of the country (Clement and Murugavel 2015; Ghanta 2020; Gupta 2013) (Fig. 6.2).

While a majority of the STEM courses are being delivered in the English language, there is a significant role that the language plays in India. English is used as a mode of communication in professional settings like offices and other workplaces. While Indian employers from the job market demand engineers with proficiency in speaking English, even the global workplaces are posing a demand for people with effective communication skills in English (Bairagi Patrai and Ashok Kumar Mohanty 2016). The dual nature of English acting as a means of comprehending their studies and a means of communication in their professional careers has been underlined in this paper.

6.5 Laws and Policies

The Right of Children to Free and Compulsory Education Act or Right to Education Act (RTE) recognizes education as a fundamental right and mandates free and compulsory education for students between the ages 6 and 14 years. In India,

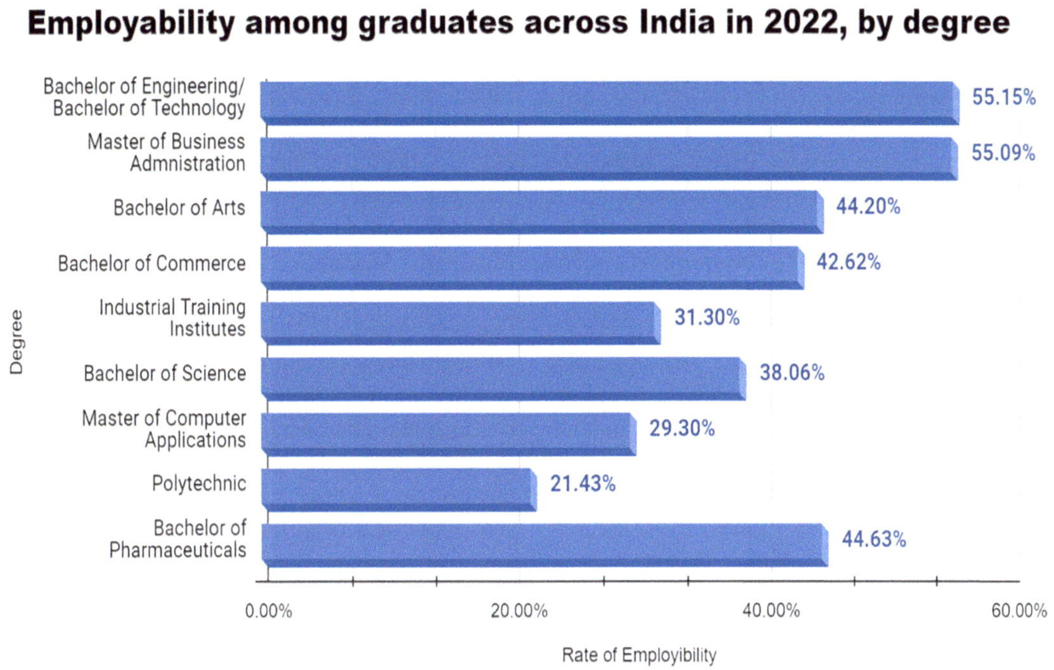

Fig. 6.1 Employability of Indian graduates in 2022, by degree. (*Note*: From Statista. (January 21, 2022). Employability among graduates in India 2022, by degree. https://www.statista.com/statistics/738255/employability-among-graduates-by-degree-india/)

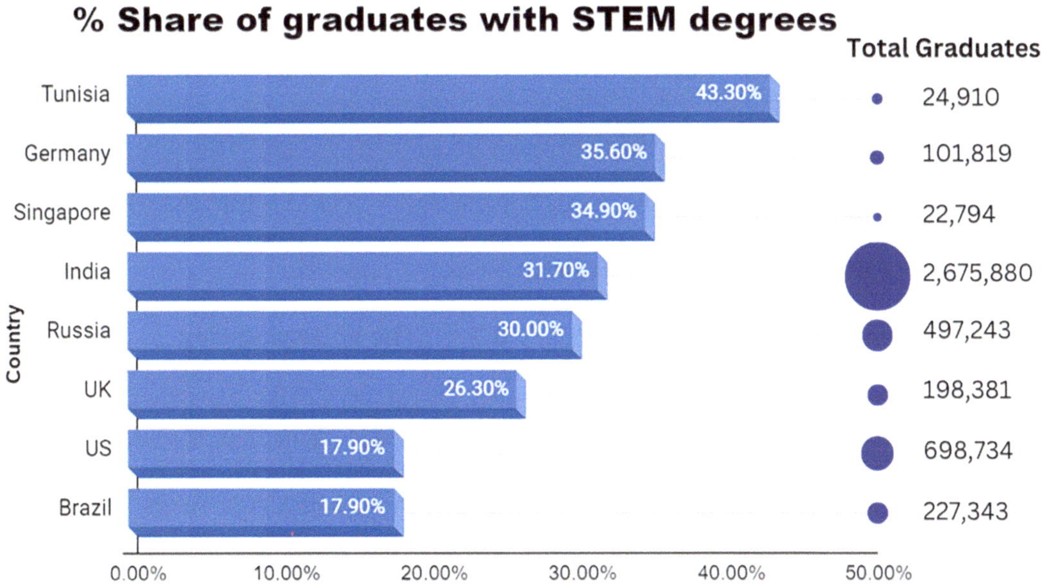

Fig. 6.2 Share of STEM graduates in 2018, by country. (*Note*: From Statista Infographics, by Buchholz, K. (2020, September 16). Where Most Students Choose STEM Degrees. https://www.statista.com/chart/22927/share-and-total-number-of-stem-graduates-by-country/)

education is included in the concurrent list, meaning both state and central governments can legislate on the same.

On the policy front, India has had three national education policies over the years. The first national education policy was introduced in 1968, followed by one in 1986 which was further revised in the year 1992. The most recent and comprehensive education policy was introduced in the year 2020 which provides a framework for elementary education to higher education and vocational training with an aim to transform India's education system (Nanda 2020). In addition, there are several government-sponsored schemes aimed at the education sector such as the Midday Meal Scheme, Rashtriya Madhyamik Shiksha Abhiyan, and Girls' Hostel Scheme among many others.

6.6 Student Enrolment and Job Opportunities

The importance of quality education has been sufficiently established. However, the means and methods of imparting quality education are highly debated topics, especially the one on the medium of instruction in a complex and diverse country like India. English merely as a "library language" (Meganathan 2019) does not translate to profitable job opportunities (Vijayalakshmi and Babu 2014). Several researchers have emphasized on the need for adopting the mother tongue of the child as the preferred mode of instruction (Khan 2016). Researchers have highlighted the problems and challenges that come naturally to these students, who are not very proficient in English (Pinke 2018). It places a high cognitive load on the students, to match their peers, and learning a technical and STEM courses might become a painful experience (Soosai Raj et al. 2018).

According to the 2011 census, linguistically more than 19,500 languages or dialects are spoken as mother tongues in India (Indian Express 2018). Out of these, there are a total of 121 languages and 270 mother tongues that are spoken by more than 10,000 people at a pan India level (Census of India 2011). While English and Hindi are recognized as the official languages of the Indian union, there are 22 languages under the eighth schedule of the Indian Constitution. The eighth schedule contains the official languages of India, namely, Assamese, Bengali, Gujarati, Hindi, Kannada, Kashmiri, Konkani, Malayalam, Manipuri, Marathi, Nepali, Oriya, Punjabi, Sanskrit, Sindhi, Tamil, Telugu, Urdu, Bodo, Santhali, Maithili, and Dogri (Department of Official Languages n.d.) (Fig. 6.3).

Millions of graduates pass out of the universities in India. However, only about 20% of them are absorbed into the various industries across the country. Such a huge gap among the unemployable workforce is at the heart of the country's employment crisis (Bhattacharyya 2019). Adding to this is the language barrier. Today, English as a language is an essential life skill – an ability that enables a person to be successful and lead a productive and satisfying life. As a result, there is an urgent need, for students particularly pursuing STEM courses, to stay prepared to enter into the job market, thereby enabling them with the job-oriented English skills by the appropriate institutions (Maninder and Mahesh 2014; Singh 2018). The linguistic diversity of India brings to the forefront several questions regarding the medium of instruction that are adopted by various educational institutions. Currently, the medium of instruction for courses at various institutions in India varies from English and Hindi to the regional language of the particular state.

6.7 The Widening Gap

The golden rule of language teaching has been critically stated: "thou shall not teach language without also teaching the culture of the language." Through this rule, several researchers have attempted to find a correlation between the English language competency of students and the cultural background and emotional intellect of the student (Kadavakollu and Jayakar 2020). Such studies have concluded that aspects like the social, cultural, and economic background of the students pursuing these courses are to be evalu-

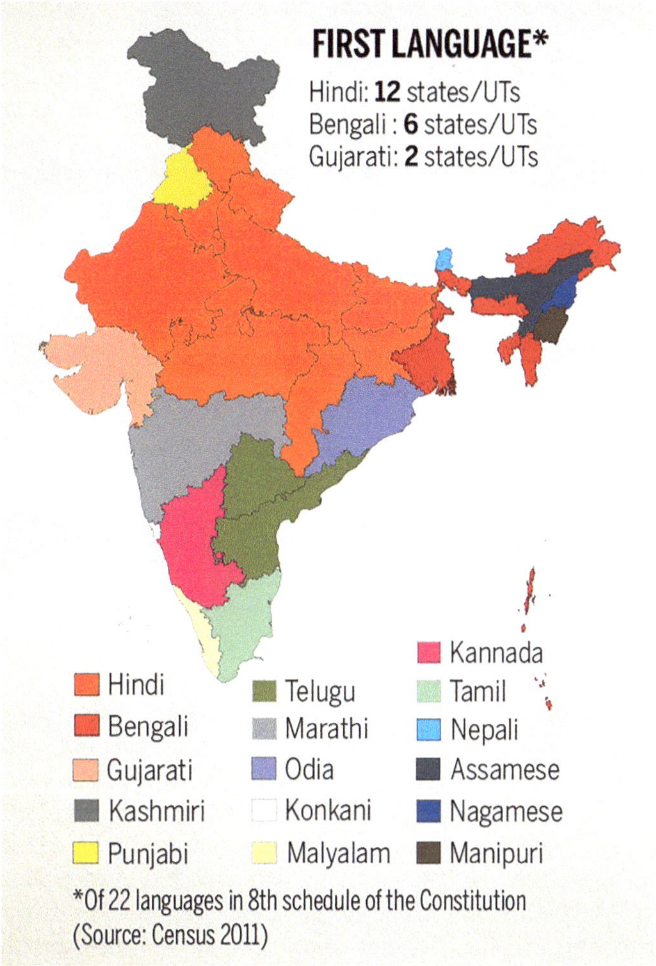

Fig. 6.3 First language across India by state. (*Note*: From The Times of India, February 21, 2020). Will your grandchildren speak your language? https://timesofindia.indiatimes.com/india/will-your-grandchildren-speak-your-language/articleshow/71167365.cms)

ated and the curriculum is to be designed accordingly.

However, in a diverse country such as India, despite the fact that states are linguistically divided, there exist several languages and dialects spoken within a particular state. In fact, many a times, there is a considerable variation even within a state (Shashidhar 2018). For example, in the State of Nagaland, there are 14 languages and 17 dialects in use with Konyak, being the most widely spoken language in the state. However, data suggests that Konyak is spoken among only 46% of the population. Similarly, several states show tremendous regional variation in terms of dialects spoken which is often conveniently subsumed into the parental language. The report says, "Himachal Pradesh, for example, has only 1.3 effective languages with 86% of the population identifying Hindi as their mother tongue. Breaking it up into dialects, though, it turns out the state has nearly 6 effective languages, with the largest being Pahari (a dialect of Hindi), which is spoken by 32% of the state's population. Similarly, 78% of Bihar's population speak Hindi (when measured at the language level), giving it 1.6 effective languages, but broken down to the dialect level, Hindi is only spoken by 26% of the population (the other prominent dialect in the state being Bhojpuri)."

According to Chaise LaDousa, a linguist anthropologist, "the mother tongue" of a child emerges as something that has to be upheld and praised, in contrast to the mode of delivery at school which ends up being something that has to be dismissed (LaDousa 2010). Moreover, the concept of teaching in the child's mother tongue is based on the fact that the child can better grasp the content taught in a language familiar to them. In such a situation with extreme linguistic diversity, the basis for deciding a language as the mode of instruction is a difficult choice to make. In the case of multiple modes of instruction, the logistics and feasibility of the same pose a problem in a developing country like India. At times, Germany, Japan, Korea, etc. are cited as successful examples of countries who have facilitated education in the regional language. However, the homogenous linguistic nature of these countries is often ignored which is a big consideration in a country like India.

Secondly, several technical terms in the English language do not have any equivalents in the regional language. Studies have indicated that activity-based teaching like presentations and group discussions have been reported to be effective methods of teaching the course to the students (Subha et al. 2019; Shaby 2019; Shah 2010). However, English language learning competency and communicative skills of the students were not up to the mark. One of the major needs of the students is to translate the concepts of STEM into their own regional language, mentally. This becomes difficult when listening to lectures in live at their respective classrooms; thereby increasing the cognitive load on the students' minds. Eventually, the major problem arising from this is the reduced comprehension and difficulty in conceptualizing verbal questions due to their poor fluency in English that prevents students from asking clarifying questions (Guo 2018). Oftentimes, the context, scientific explanation and the essence of the point may be lost in translation especially for technical subjects. Hence, ensuring quality translation with locally recognizable, understandable, and usable equivalents is of paramount importance.

The third issue that arises is that of transitioning to other languages/English at a later stage. This especially gains significance given the fact that higher education in India is tilted toward English as the medium of instruction. Thus, those transitioning from a vernacular language to English in such cases face issues in staying abreast with the demands of higher education (Anuradha and Rengaraj 2017). The issue of interstate migration for both education and jobs poses one more serious challenge in terms of adaptability. Language may pose as a barrier, thereby undoing the advantages that social mobility brings along with it. Several researchers have identified that students have reported having displayed a lack of confidence, with increased cognitive load and the inability to secure a good opportunity in the fast-paced English-speaking job market (Livingston 2012). A student completing their schooling in their respective home state in the regional language may be at a competitive disadvantage when moving to another state for education or in search of a job.

6.8 Research Methodology

This study aims to analyze the key factors that language of instruction has an effect on, focusing on the case of the Indian State of Maharashtra. Maharashtra has seen various policy changes in terms of language of instruction that needs to be used especially for primary and secondary education in government-run or aided schools. In contrast, there have been very little to no shifts in the languages of instruction for education in technical fields from the 11th standard and above, post the formation of the state in 1960. The language of instruction for these subjects has been limited to English.

Under the New Education Policy 2020, the All India Council for Technical Education announced offering engineering courses to students in eight regional languages. These include Hindi, Marathi, Tamil, Telugu, Gujarati, Kannada, Bangla, and Malayalam. Nineteen universities in the country were in preparation to offer technical education in regional languages. The council had opened 1230 seats for this initiative, out of which 255 seats have been occupied till January 2022.

This step is likely to have a lot of ramifications for many individuals across the country and requires a deeper analysis of the impact that language of instruction in technical fields has for students. Hence, a semi-structured interview was considered the best method of data collection for this study. The collected data was then thematically analyzed to identify vital strains and factors that language of instruction in STEM fields employ.

6.9 Sample Selection

Technical education is a large domain with an equally large stakeholder base. It was crucial to select a sample that represents all major stakeholder groups and can give an overall well-rounded insight into the topic. Hence, a stratified convenience sampling technique was employed to sample *25 individuals*. There were primarily four bases of sampling:

1. *Technical faculty*: Native Marathi-speaking individuals who currently teach STEM subjects in various places in the state can give a first-hand account of the challenges and features of using a particular language of instruction.
2. *Current STEM students*: Students currently learning engineering or related fields give significant details about the impact the language of instruction has on their learning and future prospects.
3. *Working professionals*: Individuals who natively speak Marathi, have completed their education in STEM subjects, and are currently working in different spaces in and out of the state are most appropriate to provide a retrospective account of their experience of language use in the job market and availability of opportunity.
4. *Linguistic experts*: Experienced people in the field of languages are also a vital source of understanding the challenges, prerequisites, and benefits of various languages of instructions.

Individuals from all of the abovementioned subgroups were interviewed for their insights and experience. Interviews were conducted in English and Marathi, depending on respondent preference. Twenty stakeholders were virtually or physically interviewed from the above backgrounds. All respondents spoke Marathi as their first language.

Respondents were encouraged to elaborate on the reasons and motivations behind their responses to each question to better understand their position. Preferences by interviewees were a binary; that is, they predominantly preferred either English or the state regional language as the medium of instruction.

6.10 Results

A detailed thematic coding of the interviewee responses reveals four key elements within the quality education paradigm on which a particular language of instruction needs to be analyzed. These are:

Comprehension abilities: The ability to gain accurate and deep conceptual understanding of the subject matter was of great concern to all the respondents. Hence, many cited comprehension abilities as one of their primary reasons of supporting a particular preference.

Career opportunities: Another crucial factor respondents explicitly identified was that of the impact language of instruction has on career opportunities and utility. Respondents had varying outlook toward this factor, but they deemed it crucial nonetheless.

Inclusion and reach: Language of instruction has a significant impact on inclusion and reach according to many respondents. Respondents also mentioned that this has been detrimental to their career decisions as well.

Institutional setup: The quality of educational material and instructors available is detrimental in selecting a particular language to many respondents.

6.10.1 Comprehension

Respondents from varying stakeholder groups had different outlooks on this factor. Individuals with a background in teaching and linguistics emphasized thoroughly on the importance of a native language in learning new concepts. To note:

> Learning anything new in a language one is more familiar with is always preferable than the alternative. This is because the learner already possesses an inherited corpus of words and concepts that can be utilised. This may not be the case with second languages because the conceptual grounding of very basic terms is relatively weaker. Hence, regional languages can be used to teach and learn any field of knowledge. (Rangnath Pathare, Faculty of Physics and acclaimed Marathi novelist)

Other academic respondents also gave similar thoughts on the topic. They unanimously referred to examples of countries where the national language was used for all kinds of communication including STEM education. (We can cite relevant research to support contrast the point made here.)

On the other hand, working professionals and current students had largely different opinions to the ones mentioned above. Most of these stakeholder groups were indifferent or did not favor a regional language of instruction even from a comprehension perspective:

> Language overall did not play that important a role in my higher education in science and engineering. I believe I would not be any better or worse learning the same subject in any other language. But because all of my science education till date has been in English, I would prefer to continue that. (Respondent, third year student of metallurgy and material sciences at IIT Bombay)

> I was enrolled in a Marathi medium government school up to 10th standard. Then onward, I chose the science stream for my 11th and 12th, JEE preparation, and am currently learning everything in English during Engineering. It was slightly difficult for me to cope in the beginning, but I managed. Although I could see some of my friends did struggle with the English curriculum. (Respondent, fourth year student, IIT Bombay)

Working professionals and students recognized the difficulties faced by students who don't have enough exposure to English to be able to cope up with a higher education in STEM fields and even cited the challenges experienced by themselves and people they know. But as a solution, they preferred remedial lessons in English or lowering the age at which students formally start learning English in vernacular medium schools over a higher STEM education in regional languages.

6.10.2 Career Opportunities

This factor has great influence across stakeholder groups but elicits different solutions from each of them. Working professionals highlighted the utility of a technical education in the English language most noticeably in the context of the career opportunity it affords them, compared to a regional language education. This is because of the prevalence of English as a language of communication internationally as well as within different language speakers within India. Getting a formal education in English is therefore more convenient for students as it enables a smoother transition for them during interstate work or international education. A higher education instructed in English is also perceived to increase the radii of their employability by many. The following two interview excerpts reflect the utility of studying in English:

> Given that a large chunk of the world speaks English including many people in India, it offers me a much larger field to operate in than learning those subjects in Marathi does. And like it or not, speaking fluent English significantly improve my chances at an interview.

> After completing my master's in chemical engineering, I secured employment in a Gujarat-based firm. I used to speak in broken Hindi with my colleagues and superiors, but all the official channels of communication were English. Three years later, I was working for a German firm where even casual communication was English.

Respondents who were more inclined toward the promotion of STEM education in regional languages had a distinct perception of the challenges they might face in career opportunities. But most of them opined that conversational English enough to effectively communicate with non-native language speakers can be acquired easily enough, without necessitating a formal education in the English language.

A crucial aspect highlighted by respondents that were cautious of STEM education in regional languages was that of national-level education institutes like IITs, NITs, IIITs, IIMs, NLUs, and other such institutions where students from across the length and breadth of the country, speaking different native languages, come to learn technically sophisticated subjects. Because of the multilinguistic nature of the student cohort studying here, English is the primary language of instruction. These institutes are some of the most sought-after places of higher education in the country. Respondents felt that if a student wants to learn STEM subjects in their regional languages, they will have to forgo their chances of entering some of the institutions mentioned above. They believe this to be too large of an opportunity cost.

6.10.3 Inclusion and Reach

Languages of instruction play a pivotal role in determining societal sections for whom higher STEM education is accessible. All respondents showed concern regarding maximizing inclusion and reach of STEM education. This reason was echoed by many respondents who supported regional language education. They believe that students from rural and tribal areas who attend district council-run schools, which use Marathi as the medium of instruction up to standard tenth, will be the primary beneficiaries of further STEM education in regional languages. Vinay Sahasrabuddhe, Rajya Sabha MP, highlighted this:

> It is not primarily a question about which language is more effective or useful, because this is not a zero sum game. Getting an opportunity to pursue any branch of knowledge in one's native language is a right that needs to be upheld. Whether to exercise the right or not remains entirely up to any individual. Expanding the language base for engineering subjects affords a more diverse student base to pursue the field.

In contrast, some respondents felt that individuals pursuing STEM higher education in regional languages may experience disenfranchisement in their professional lives. Moreover, these respondents specifically pointed to national-level systems of education which use English as the medium of instruction. Students and professionals who migrate to states need a common language to integrate in new environments. Respondents believe that for a country as linguistically diverse as India, it is better to learn technical subjects in a more widely known language:

> I come from a part of Maharashtra where we speak a very local dialect of Marathi with major differences between standard Marathi and what we speak. So for me, there wasn't a major difference in the difficulties of learning science as a subject in Marathi till standard tenth and then learning it further in English was very little. (Respondent, M Tech Civil engineering, BITS Pilani)

6.10.4 Institutional Setup

Offering alternatives to various languages in higher education poses a unique set of challenges. Respondents had a wide array of perspectives on each of these.

One of the more highlighted aspects of the institutional setup was that of resource planning. Respondents from all stakeholders' groups focused on this aspect. Establishing a vast field of education into a new language is a resource-intensive task. Employing and paying qualified educators and instructors, formulating educational material in regional languages and making it accessible, establishing qualification standards, ensuring overall quality control, and the resource diversion this move would require were major concerns from stakeholders. Creating a lucid vocabulary for many scientific and technical terminologies in STEM disciplines is an important factor brought forth by stakeholders:

> It is indeed very important to make higher STEM education in regional languages available. But even more important is to ensure that it is updated and meets quality standards. (Dr. Anagha Mandavkar, faculty of Marathi, DG Ruparel College, Mumbai)

Respondents working in the field of regional language development mentioned previous efforts done regarding the institutionalization of regional languages in higher education:

> It's not like this is a totally unprecedented step or initiative. Ironically, it was in colonial India that attempts to translate mother-level mathematics and medicine-related texts were made by one captain, Jarvis. He translated several mathematics-related books in Marathi, wrote original texts in Marathi, and got them published as well. Similarly, a medicine school was opened in Maharashtra that taught and issued degrees in Marathi as a primary language. (Respondent, who works with the State Marathi Language Department, Maharashtra)

(The claims made in the above quote were corroborated with another journalist source: Potnis 2018.)

Some respondents did raise concern regarding the widespread institutional capacity building that needs to be undertaken on a national level. Doing so would require much of already scarce resources which they feel can be put to better use. The transition phases from systems that support one language of instruction to those that support multiple ones are likely time-consuming. Interviewees believed that this can be damaging to the academic development and the journey of students in the short term.

These findings indicate a complex policy framework that can take care of inclusion, efficiency, and quality by way of the language of instruction in STEM fields.

6.11 Policy Recommendations

The use of regional languages as the sole medium of instruction brings along with it a host of issues. Finding equivalents in regional languages for technical terms, ensuring quality translation of content, the transition from school education where learning happens in a regional language to higher education which is almost completely conducted in English, opportunities on the professional front in a highly globalized world, and interstate migration for educational and job purposes are some of the issues. This, however, cannot discount the findings of various studies and the UNESCO guidelines that stress on the importance of the use of regional language to help deliver the content of the subject better.

In such a scenario, it would be useful to adopt a bilingual approach as also touched upon in the NEP 2020. Approaches such as dual delivery of content in both English and in a regional language by means of using English words for technical terms and contexts while delivering the content in the regional language can be further explored. In view of this, the following policy actions will help in better outcomes:

1. Establishing a national-level quality control board including translation board for STEM subjects that can ensure quality translation of text into regional languages. Further, the option of providing combined multilingual textbooks for students needs to be developed.
2. Training and capacity building in terms of resources, both human and material, to ensure that an adequate number of qualified regional language teachers and quality materials are available.
3. Providing the option of learning in regional languages for higher education would go a long way in reducing the inequalities between those accustomed to learning in an English medium setup and those learning in regional languages. This, however, needs to be accompanied by the simultaneous teaching of English, especially in cases where it is a necessity for professional advancement and to prevent disenfranchisement.

6.12 Conclusion

With about 37 million students enrolled in higher education in India, the language of delivery plays a significant role when these students enter the job market. This problem deepens when it comes to the graduates of STEM education. The competency of the graduates in the job market poses a challenge not just to the employees themselves but also to the employers and the country as language incompatibility leads to loss of efficiency in the job market. This chapter aims to analyze the key elements that impact the preferences toward a particular language: comprehension abilities, career opportunities, inclusion and reach, and institutional setup. These aspects have been studied from the perspective of different stakeholders involved in the education, linguistics, and STEM job sectors. While the results indicated a complex policy framework, it is crucial for all the stakeholders to act upon the solutions to reduce the gap between the language competency of graduates and job opportunities. It is recommended to make bilingual education with L1 as the primary mode of instruction and L2 (English) as a supplementary medium. This approach is in alignment with the National Education Policy 2020 and can prove to be beneficial for the education sector and employability for STEM students in India.

While the researchers have attempted to fill the gap in the existing literature, there still exists significant scope for further research in the present research. The following limitations have arisen due to the restrictions with regard to the ongoing pandemic and potential of the researchers:

1. Research gap in other regional languages of India.
2. The review of the related literature indicates the gap in the research of regional languages and education in languages other than Tamil. The existing literature focuses on the English competency of students who have been imparted with technical education in the Tamil language. Hence, further studies can be performed in other regional languages, using the framework used in this paper.
3. The qualitative nature of the present research lacks concrete evidence to show a significant relationship between the language of instruction and professional competency.

The present study focused on a qualitative methodology, structured in the form of interviews with the stakeholders. However, quantitative research that adopts a survey questionnaire methodology could be the scope for further research studies on this idea. This will help identify patterns and correlations between the language of instruction and the job competency of students with education imparted in regional languages.

Annexure: Questionnaire

Section A: Personal Information

Name: ______________________________

Age: ________________

Level of education: __________________

Medium of education up to std. 10th: ___________

Medium of education during 11th and 12th: _____________

Medium of education in undergraduate studies: ____________

Medium of education in postgraduate studies: ______________

Profession: ___________________

Position at workplace: __________________

Years of experience: ______________

Section B: About Language and STEM Education

(For STEM instructors)

1. What has been your experience with teaching (relevant STEM subject) in an institution situated in Maharashtra?
2. What to you is the role of language in teaching STEM subjects?
3. Do you see students from different linguistic educational backgrounds deal with the course material differently?
4. Is there a need for institutional change in STEM education pertaining to language?

For STEM learners and subsequent professionals:

1. Did you find a change in your learning experience in your primary education, and when you started to pursue higher education in STEM? If yes, could you say more about it?

2. Do you feel that the medium of instruction has some influence on your technical education?
3. Was there a particular language requirement in the job description at the spaces you were looking to work at? If yes, was it a challenge for you?
4. For you, what impact has language of instruction had on your career opportunities and prospects?

For language instructors and specialists:

1. How significant the role languages play in learning through various stages of formal education?
2. Are their benefits to getting educated in languages spoken more widely?
3. What according to you the relationship between language, STEM education, and inclusivity?

References

Anuradha V, Rengaraj M (2017) Students' attitude towards English language learning and their academic achievement among first year engineering graduates: a case study. Indian J Sci Technol 10(29):1–6. https://doi.org/10.17485/ijst/2017/v10i29/116503

ASER, P (2020) Annual Status of Education Report 2020 wave 1. Annual Status on Education Report

Bairagi P, Mohanty AK (2016) Importance of english for engineering students: an evaluation of the prevalent teaching-learning system in the indian context.Int. J. English Lit. (IJEL) 6(4):21–31

Bhattacharyya A (2019) Bridging the gap between education and employment. The Sunday Guardian Live, June 10. https://www.sundayguardianlive.com/culture/bridging-gap-education-employment

Census of India (2011) Language – Census of India Website. Office of the Registrar General & Census Commissioner, India. India. https://www.censusindia.gov.in/2011Census/C-16_25062018_NEW.pdf. Accessed 31 Jan 2022

Clement A, Murugavel T (2015) English for employability: a case study of the English language training need analysis for engineering students in India. English. Lang Teach 8(2). https://doi.org/10.5539/elt.v8n2p116

Department of Official Languages (n.d.) Languages Included in the Eighth Schedule of the Indian Constitution | Department of Official Language | Ministry of Home Affairs | GoI https://rajbhasha.gov.in/en/languages-included-eighth-schedule-indian-constitution

Education – United Nations Sustainable Development (n.d.) United Nations. https://www.un.org/sustainabledevelopment/education/. Accessed 31 Jan 2022

Ghanta S (2020) Enhancing English teaching methodologies for speaking skills of engineering students. Int J Psychosoc Rehabil 24(04):2647–2652. https://doi.org/10.37200/ijpr/v24i4/pr201372

Guo PJ (2018) Non-native English speakers learning computer programming. Proceedings of the 2018 CHI conference on human factors in computing systems. https://doi.org/10.1145/3173574.3173970

Gupta D (2013) Teaching English to engineering students in India. J Educ Pract 4(11):131–135

Indian Express (2018) July 1. https://indianexpress.com/article/india/more-than-19500-mother-tongues-spoken-in-india-census-5241056/. Accessed 31 Jan 2022

Kadavakollu T, Jayakar J (2020) Challenges to vernacular learners (L1) in acquisition of English language (L2) at U.G. level in Hyderabad region. J Educ Pract 4(14):14–19

Khan MS (2016) The impact of Native Language use on Second Language vocabulary learning by Saudi EFL students. Engl Lang Teach 9(5):134. https://doi.org/10.5539/elt.v9n5p134

LaDousa C (2010) On mother and other tongues: sociolinguistics, schools, and language ideology in northern India. Lang Sci 32(6):602–614. https://doi.org/10.1016/j.langsci.2010.08.001

Livingston G (2012) Using communication skills laboratory in enhancing speaking skills of engineering students. Language in India strength for today and bright hope for tomorrow 12:781–788

Maninder KK, Mahesh K (2014) Teaching english language at engineering and technology institutes in india: Problems and remedies. Int j Engl: lit lang. skills 4(2):15–22

Meganathan R (2019) Research in English language education in India. *National Council of Educational Research and Training.* https://www.academia.edu/44728758/Research_in_English_Language_Education_in_India. Accessed 4 Feb 2022

Nanda PK (2020) Systemic reform and outcome based education, hallmarks of new education policy, July 29. Retrieved May 3, 2022, from https://www.livemint.com/education/news/systemic-reform-and-outcome-based-education-two-hallmarks-of-new-education-policy-11596032795159.html

Pinke (2018) A study of English language teaching in India. J Adv Sch Res Allied Educ 15(6):75–79. https://doi.org/10.29070/15/57720

Potnis S (2018) Captains jarvis contribution for marathi. Loksatta, July 23. https://www.loksatta.com/navneet/captains-jarvis-contribution-for-marathi-1718602/. Accessed 1 Feb 2022

Shaby D (2019) Enhancing oral communication through task based language teaching: an experimental study

on under graduate students. Think India 22(3):277–282. https://doi.org/10.26643/think-india.v22i3.8201
Shah S (2010) Best practices in teaching and learning of technical institutes in India. Int J Eng Technol 2(4):374–377. https://doi.org/10.7763/ijet.2010.v2.150
Shah PJ (2015) Regulatory Structure of Higher Education in India (Ref: F-35110-INC-1). https://www.theigc.org/sites/default/files/2015/11/Shah-2015-Working-paper.pdf. Accessed 31 Jan 2022
Sharma S, Sharma P (2015) Indian higher education system: challenges and suggestions. CORE Scholar. https://corescholar.libraries.wright.edu/cgi/viewcontent.cgi?article=1179&context=ejie. Accessed 30 Jan 2022
Shashidhar K (2018) Nagaland is the most diverse state in India, language-wise. Mint, July 11. https://www.livemint.com/Politics/N5QQIaSOB5GVUMP0IM3f6K/Nagaland-is-the-most-diverse-state-in-India-language-wise.html. Accessed 31 Jan 2022
Singh E (2018) Developing speaking skills in the English language courses using ICT for the engineering students: an Indian context. Int J Engl Learn Teach Skills 1(2):121–125. https://doi.org/10.15864/ijelts.1204
Soosai Raj AG, Ketsuriyonk K, Patel JM, Halverson R (2018) Does native language play a role in learning a programming language? Proceedings of the 49th ACM technical symposium on computer science education. https://doi.org/10.1145/3159450.3159531
Subha S, Abirami K, Ambikavathy M, Eswaran D, Manikandan K (2019) A study on the English language competency of the first year Tamil medium engineering students. Indian J Sci Technol 12(32):1–3. https://doi.org/10.17485/ijst/2019/v12i32/142336
UNESCO (2019) Rationale of the IDIL2022-2032. https://www.unesco.org/en/decades/indigenous-languages/about/rationale#:~:text=Following%20the%202016%20Resolution%2071,Permanent%20Forum%20on%20Indigenous%20Issues
Vijayalakshmi M, Babu M (2014) A brief history of English language teaching in India. Int J Sci Res Publ 4(5):1–15

Darshan Gaikwad is a student of Public Policy at the University of Edinburgh. He has completed his bachelor's degree in Business Adminstration from the Indian Institute of Management, Rohtak. His research interests include education policy, water governance, and climate cooperation.

Jyotsna Akurathi is an MBA student at Indian Institute of Management, Rohtak. She has completed her bachelor's degree in Business Adminstration from the Indian Institute of Management, Rohtak as a part of the Integrated Program in Management (IPM). Her research interests include learning and education, public policy and women and gender studies.

Mukund Nagarajan Rao is an MBA student at Indian Institute of Management, Rohtak. He has completed his bachelor's degree in Business Adminstration from the Indian Institute of Management, Rohtak as a part of the Integrated Program in Management (IPM). His research interests include education policy, water governance, sports administration and policy and urban development.

Clean Water and Universal Sanitation in an Era of Sustainable Development: Understanding the Challenges and Prospects for SDG 6 in the Ganga River Basin

7

Sya Buryn Kedzior

Abstract

In a recent progress report by the WHO and UNICEF (Progress on household drinking water, sanitation and hygiene, 2000–2020. https://www.who.int/publications/i/item/9789240030848, 2021), India was lauded for advancement toward SDG 6 and improvements in its provision of basic water and sanitation services for a burgeoning population. Indeed, the report cites improvement in rates of basic water services by 5 percentage points and in basic sanitation services by an impressive 15 percentage points over only the previous 5 years. Improved provision of basic sanitation, combined with aggressive social programs (Jain et al., Int J Environ Res Public Health 17(4):1384, 2020), has contributed to another sharp decline in open defecation rates from 29% to 15% over the same period. But these improvements begin to pale when examining far lower rates of access to *safe and clean* water supplies, piped water services, safe sanitation, and wastewater treatment across the country, along with often sizable gaps in access between social groups and subnational regions. What emerges from this report is an incredibly complex picture of significant, though hampered, progress and many questions about India's prospects for meaningful and substantiative progress toward achieving SDG 6.

Drawing on more than a decade of research on water access and water quality in the Ganges River Basin, I explore these questions through an in-depth examination of changing conditions in India's largest watershed. Home to a population of more than 400 million people, along with significant centers of agricultural and industrial production, India's Ganges River Basin has long been the focus of government efforts to improve water and sanitation access while addressing related goals of improving water quality, sanitation, and ecosystem health. Most recent of these is the Namami Gange Programme, a conservation and pollution abatement effort that serves as one of the three flagship programs of the Union Government for achieving SDG 6, along with the national Swachh Bharat Abhiyan (Clean India Campaign) and the National Rural Drinking Water Programme. Together, these initiatives seek to develop the Ganga Basin as a model for water and sanitation management throughout the nation. By exploring the dynamics surrounding clean water and sanitation in this subregion, I highlight the challenges associated with providing

S. B. Kedzior (✉)
Department of Geography and Environmental Planning, Towson University, Towson, MD, USA

Environmental Science and Studies Program, Towson University, Towson, MD, USA
e-mail: skedzior@towson.edu

V. Dutta, P. Ghosh (eds.), *Sustainability: Science, Policy, and Practice in India*, Sustainable Development Goals Series, https://doi.org/10.1007/978-3-031-50132-6_7

sustainable access to water and sanitation across India and identify some of the potential pathways and prospects for achieving the aims of SDG 6.

Keywords

Clean water · Ganga River Basin · Sanitation · SDG 6 · India

7.1 Introduction

The cost of poor access to clean water and adequate sanitation is felt acutely across India. It has the highest number of premature pollution-related deaths in the world with nearly 700,000 annual deaths attributed to water pollution alone (GAHP 2019). Even when adjusted for population, India is second only to North Korea in per capita pollution-related deaths (ibid.). Most of these deaths are caused by waterborne pathogens introduced to waterways by untreated or partially treated human and animal waste. The diarrheal symptoms they produce account for 90% of the deaths among children under the age of five and contribute to child mortality, morbidity, malnutrition, and lifelong growth stunting (Gaon Connection 2021). In all, 38 million Indian people suffer from waterborne diseases each year, adding to an estimated annual disease burden of more than 8.7 million disability-adjusted life years[1] (DALYs) from unsafe water and sanitation (Waghela 2021). The World Bank (2010) estimates that inadequate water supply and sanitation cost India about 2.4 trillion Rs per year (US$53.8 billion) in health costs, lost wages, water treatment, time spent accessing water and sanitation, and lost tourism revenue. But the impacts of water and sanitation are not limited to human health. Water pollution is a key threat to freshwater biodiversity, which impacts both ecosystem health and the productivity of fisheries (Dudgeon 2019; Khambete 2021b). Overall, water quality impacts agricultural production, industrial efficiency, and ecosystem health and potentially reduces economic growth by as much as a third (World Bank 2019).

These statistics draw attention to some pressing realities of the role of water in sustainable development in India and beyond. First, thanks in part to significant progress made under the Millennium Development Goals (MDGs), water access rates have improved in India and much of the rest of the world. India now has basic water access for at least 90% of its population (WHO/UNICEF 2021). Yet far fewer people have access to *safe* and *healthy* water, a trend that reflects growing concerns about water quality around the world. Second, progress in sanitation improvements has been far less impressive. Just over 50% of India's population had access to basic sanitation at the end of the MDG period, and while those rates have improved significantly under the Sustainable Development Goals (SDGs), progress remains slow (ibid.). Sanitation is emerging as the top priority in the water agenda for the coming decades. Finally, the impacts of poor water quality and sanitation are not limited to the water provision and management agenda alone. We can clearly see how poor sanitation, as in India where 70% of wastewater goes untreated before flowing largely into freshwater sources, leads to poor water quality by introducing pathogens into the water cycle that cause disease, limit economic productivity, and negatively impact environmental health (WaterAid 2016). Today, untreated sewage and wastewater are largest point sources of water pollution in India and in the world (CPCB 2008; Mateo-Sagasta et al. 2017). Drawing attention to the role of sanitation and water quality in the water cycle also means globalizing the "water in development" agenda, as pollution is not a problem limited to the Global South. Indeed, water quality is a global crisis and the primary challenge for water management and provision around the world, as urbanization, ageing infrastructure, and climate change push water systems to their limits (Schrecongost et al. 2020).

[1]Disability-adjusted life year (DALY) indicates the number of years of life lost due to illness, disability, or premature mortality due to disease. One DALY is equivalent to the loss of 1 year of health (Waghela 2021).

7.2 Broadening the Water-Waste Cycle: The Inclusive Goals of SDG 6

Sustainable Development Goal Six (SDG 6) acknowledges this connection by placing, for the first time, targets associated with drinking water provision, sanitation services, and ecosystem protection under a single banner (see Table 7.1). The previous Millennium Development Goals (MDGs) addressed water under Goal 7, which focused primarily on halving the number of people without access to safe drinking water and sanitation. This "taps and toilets" approach meant that work under MDG 7 was dominated by the water, sanitation, and hygiene (WASH) agenda and therefore by development experts focused almost exclusively on funding infrastructure improvements in developing countries while "largely ignor[ing]" both the environmental dimensions of water and the significance of water-sanitation interrelationships (Wiegleb and Bruns 2018, p. 1162). This focus on water and sanitation *access* meant that water quality issues and environmental concerns that impact development in countries around the world went largely ignored during the MDG era. While SDG Targets 6.1 and 6.2 continue the work on water and sanitation provision that started during MDG period, the remaining targets (6.3–6.6) emphasize water quality, efficiency, management, and ecosystem health. Goal 6.6 acknowledges, for the first time, the importance of environmental health in sustainable development by specifically calling for the protection and restoration of water-related ecosystems (Wiegleb and Bruns 2018). This makes SDG 6 both a more comprehensive plan than the MDGs and a formalized acknowledgment of the significance of the entire water cycle in sustainable development.

Table 7.1 SDG 6 targets

7.1	Universal and equitable access to safe and affordable drinking water for all
7.2	Access to adequate and equitable sanitation and hygiene for all and end open defecation, paying special attention to needs of women/girls/people in vulnerable situations
7.3	Improve water quality by reducing pollution, eliminating dumping and minimizing release of hazardous chemicals and materials, halving the proportion of untreated wastewater, and substantially increasing recycling and safe reuse globally
7.4	Substantially increase water use efficiency across all sectors and ensure sustainable withdrawals and supply of freshwater to address water scarcity and substantially reduce the number of people suffering from water scarcity
7.5	Implement integrated water resource management at all levels, including through transboundary cooperation as appropriate
7.6	Protect and restore water-related ecosystems, including mountains, forests, wetlands, rivers, aquifers, and lakes (by 2020)
7.A	Expand international cooperation and capacity-building support to developing countries in water- and sanitation-related activities and programs
7.B	Support and strengthen the participation of local communities in improving water and sanitation management

The SDGs break from the MDGs not only in scope and approach but also in process. The MDGs were crafted by development experts in a largely top-down manner that led to critiques of the MDGs having too narrow of a development agenda, being too conventional in their approach, and lacking any real enforcement mechanisms (Wiegleb and Bruns 2018). The SDGs, however, were the result of a lengthy consultative process that drew input from UN member states and civil society representatives, resulting in "millions of people… involved in the largest development dialogue the world has ever known" (Solberg 2015, p. 61). The result was a far more expansive agenda that increased the MDG's list of 8 goals with 21 targets to the SDG's list of 17 goals and 169 targets. To some, this expansion represented a "new development trajectory" and "more holistic and inclusive agenda" that resulted from an inclusive decision-making process (Wiegleb and Bruns 2018, p. 1159). For others, the participatory process created an agenda that is too expansive and even "convoluted" (Vandemoortele 2017, p. 41). They argue its ambitiousness is a weakness and that this new agenda with its numerous targets will be harder to achieve than the (still unattained) MDGs (Mukherjee 2016; Wiegleb and Bruns 2018). What seems clear is that the SDGs reflect a development agenda crafted by a greater spectrum of participants that is meant to better reflect the priorities and issues as they play out in individual countries. As a legally binding agreement, the SDGs provide at least a measure of implementation strength that lends them far more political accountability than the predecessor MDGs (Vandemoortele 2017). While the breadth of each goal, and abstract nature of many targets, may render this agreement somewhat toothless, the mechanisms for pursuing each goal are purposefully left to each individual country.

SDG 6 may be more expansive, but it is by no means comprehensive, and leaves much room for countries to pursue each of the targets individually, while all but ignoring the push for a holistic joint water agenda (Wiegleb and Bruns 2018). Some of the SDG 6 goals are also more tightly bound to specific, measurable, and enforceable targets and indicators than others. Targets 6.1, 6.2, and 6.3 include clear indicators of success. Elements of 6.4, along with 6.6, are far less specific and not tied to measurable indicators but rather reference desired changes in water use efficiency and changes in the extent of water-related ecosystems over time. Both 6.a and 6.b include specific indicators referencing increases to the amount of development assistance related to water and sanitation and the proportion of local administrative units with some policies and procedures promoting public participation in water and sanitation management. For an agenda that will be driving policy-making and international funding through 2030, these indicators leave so much room for interpretation that their implementation will certainly produce far different outcomes at the individual country level. To understand both challenges and prospects for global achievement of the SDGs, we must therefore look to individual countries, where the goals are brought to life through incorporation into planning and policy-making efforts.

As one of the largest countries in the world, India serves as a veritable linchpin for achievement of the SDGs. This is particularly true for SDG 6, as India continues to lead the world in sheer number of people (not proportion of population) without access to safe water and sanitation. Despite significant gains in recent years, India also continues to lead the world in a number of people practicing open defecation (Rubin and Kapur-Gomes 2020). This is partially due to the size of India's growing population. In late 2021 or very early 2022, India likely exceeded 1.4 billion people, representing about 18% of the total global population.[2] Population size alone is a key contributor to water use and the production of sanitation woes. These are compounded by rapid urbanization, which strains urban infrastructure. The Central Pollution Control Board (2021) estimates that only 13.5% of sewage from Class I and Class II cities is treated. Outside of cities, sewage treatment is nearly absent, and the practice of open defecation while declining continues to be common. Untreated sewage is the primary cause of surface water pollution in India and represents one of the greatest threats to both environmental and human health. India's resource-intensive agricultural sector is another key water user, contributing to the country's astounding withdrawal of 647.5 billion cubic meters of freshwater per year, far more than any other country (World Bank 2017). Population growth, economic growth, industrialization, and rising consumption rates are likely to continue after the current COVID pandemic, with stark consequences for both water use rates and water quality. India is also expected to be one of the first major world economies to suffer extreme and irreversible impacts from climate change, which is certain to have significant ramifications on the country's water cycle, including the intensification of the monsoon cycle that affects everything from agricultural practices to severe weather events (NIC 2009). India's response to these challenges is an important measure of SDG 6 achievement and will undoubtedly influence other countries in the region and across the world.

[2] Population projections from Worldometer (2021), based on extrapolations of UN population data. Despite falling fertility rates, the India's age structure should continue to produce population growth at least through the end of the century, when it is projected to house the largest population in the world.

In the next section, I review the progress that India has made during the first 5 years of the SDG agenda, examining how the country has structured its response to the targets as well as their achievements to date. I focus this examination on the SDG-related initiatives in India's largest watershed, the Ganges River Basin, which is home to a large segment of the country's population, economic and agricultural output, and water woes. Drawing on more than a decade of research on water access and water quality in the basin, I highlight the challenges associated with providing sustainable access to water and sanitation across India and identify some of the potential pathways and prospects for achieving the aims of SDG 6.

7.3 Water, Sanitation, and Sustainable Development in India: Progress by 2020

India's efforts to meet SDG 6 have largely been structured under three efforts: the National Rural Drinking Water Programme and Jal Jeevan Mission, Swachh Bharat Abhiyan, and the Namami Gange Programme. The National Rural Drinking Water Programme (NRDWP), established in 2009, predates drafting of the SDGs. In keeping with the WASH approach of the earlier MDGs, the NRDWP emphasized provision of water supply and access for drinking and other household needs. The program envisioned a gradual improvement and extension of services that started with protection and upgrades to the groundwater wells relied upon in many rural areas across India, followed by installation of communal hand pumps and finally provision of household connections to half of Indian households within the first decade of the program. Secondary efforts included improvement of local water testing capacity and investment in infrastructure to develop surface water catchment and groundwater replenishment systems. By the time the SDGs were enacted in 2016, it was already clear that the NRDWP would fall short of its fundamental goals. A 2018 audit (Comptroller and Auditor General of India (CAG)) found that the program improved basic water provision by only 8% at 40 L per capita per day (lpcd), which was far lower than both the 55 lpcd set by the program and the 70 lpcd target set by the Ministry of Drinking Water and Sanitation (Paliath 2018). The audit found that only 44% of rural households had access to safe drinking water and only 17% had piped connections (CAG 2018, p. 93). Accused of poor implementation, mismanagement of funds, and insufficient coordination between central and state authorities, the program was restructured and refunded but nonetheless subsumed by the Jal Jeevan Mission in 2020.

The Jal Jeevan Mission significantly expanded and accelerated the goals of the NRDWP, aiming to achieve universal piped water access and supply of at least 55 lpcd in households and public centers by 2024. The Mission also incorporates additional measures focused on ensuring sustainable water supply, promoting community participation in water management as well as community-level ownership, along with improved monitoring and dissemination of information about water quality. While the main mission continues to focus on rural water provision, an urban sub-scheme seeks to close the "gap" in urban household tap connections, promote water recycling and reuse, and improve public awareness of water conservation (Ministry of Housing and Urban Affairs 2021). While it is perhaps too soon to measure the success of the Jal Jeevan Mission, early reports boast rapid progress toward universal tap water provision, with about 5 crore (50 million) new household connections between 2109 and 2021 (National Jal Jeevan Mission 2021). Some of the most significant progress has been made in India's 117 "aspirational districts," identified by NITI Aayog as those with the lowest socioeconomic indicators in each state (Haq 2021). The most recent data from the Jal Jeevan Mission (2022) indicates that 47.92% of Indian households currently have a tap water connection and that 6 of India's 28 states and union territories have universal household tap water connectivity.

Swachh Bharat Abhiyan ((SBA) or Clean India Campaign) was launched by Prime Minister

Narendra Modi shortly after his election in 2014. A broad effort focused on sanitation, waste management, and hygiene, SBA represents a restructured version of the earlier, marginally successful Nirmal Bharat Abhiyan, a sanitation program best known for achieving incremental improvements in household access to toilets and piped sewage. Phase I of the SBA represents the first 5-year period (2014–2019) for which the program was initially envisioned. It included a strong focus on solid waste management, toilet installation, and eradicating open defecation. Results of these efforts were mixed and drew attention to the program's negative impacts on informal waste pickers and the disconnection between toilet availability and toilet use (Oates et al. 2018; Jain et al. 2020). Phase II (2021–2026) emphasizes sewage and solid waste management issues like waste segregation, single-use plastics, and landfills, along with an ongoing focus on making India "open defecation-free" (Mehotra 2021). It also modifies earlier approaches in response to critiques of Phase I through programs that incorporate waste pickers into formal solid waste management and promote behavioral change and public engagement for ending open defecation and monitoring waste practices (Joshi 2020). SBA has achieved impressive progress to date, helping install toilets in more than 100 million households, which has in turn produced positive economic and health effects, as well as improved water quality outcomes (UNDESA 2022; Ahluwalia 2021). In the last 2 years, SBA has played an important role in combatting the spread of COVID-19 by promoting handwashing and other hygienic behaviors meant to slow rates of disease spread and infection. PM Modi touted the importance of the program's provision of toilets in helping to check the virus' spread and questioned if social distancing would have been possible with people practicing open defecation (Hindustan Times 2020).

Finally, the Namami Gange Programme is unique among India's efforts related to SDG 6 in that it is a centrally sponsored mission focused on conservation of a regional river. Namami Gange perhaps reflects best the shift in priorities and approach from the Millennium Development Goals to the Sustainable Development Goals by acknowledging the vital connections between sewage management, surface water quality, and environmental health. It is often described as critical to India's approach to SDG 6 and sustainable surface water management (Swarajya 2019). The pillars of this integrated approach to river management include riverfront development, sewage treatment, pollution abatement, biodiversity conservation, industrial effluent monitoring, and afforestation, along with a significant emphasis on public awareness-raising efforts. But Namami Gange is not the first program of its kind to adopt a basin-wide, comprehensive approach. Since the 1980s, the Ganga River has served as a laboratory for programs promoting interstate cooperation and as a "model" for sewage and river management meant to be replicated in other river basins (see Kedzior 2018). Proponents of Namami Gange argue that it breaks from historical approaches by placing more weight on the impact of water diversion and discharge on pollution concentrations, emphasizing local management and community engagement, adopting a "life cycle" approach to river management, and incorporating biodiversity conservation principles (True Picture 2018). Its initial budget outlay of Rs.20,000 crore (about $2.6 billion) also dwarves the 4000 crore (about $525 million) spent on various Clean Ganga initiatives between 1985 and 2015 (Dutta 2017a; Chengappa 2021). This funding supported more than 300 new projects along the river, including 152 sewage infrastructure projects, 28 riverfront development projects, and a variety of conservation projects, biodiversity centers, and research and education centers. Other accomplishments include improved mapping, water quality testing, industrial inspections, and mechanisms for promoting public engagement, like events, reporting apps, and even a Gange theme song (PBNS 2021). Critics point out that increased funding and novel approaches aren't producing marked changes in water quality along the main stem of the river, where water quality consistently falls short of standards for drinking and even bathing at the vast majority of testing sites along the river (Kaur 2018).

As we mark the sixth year since adoption of the SDGs in 2016, India has achieved significant progress toward Goal 6 under each of these programs. Recent snapshot data (UN WATER 2020) indicates that 95% of India's population uses an improved drinking source, with 90% using at least basic services and only 1% relying directly upon untreated surface water. Sanitation and hygiene access rates were not as high, with 83% of the population using improved sanitation facilities and only 68% of the population having access to handwashing facilities with soap and water at home. India's economic survey of 2019 found that 93.1% of households have access to toilets, though water availability and other considerations have significant impacts on use (versus availability) rates (Swarajya 2019). Significant, double-digit percentage reductions in open defecation rates indicate that the proportion of the population using open defecation may have fallen as low as 15% (UN WATER 2020; WHO/UNICEF 2021). But data on this measure are highly variable and, at best, reflect changing individual practices over time and place. Interestingly, the GoI (Niti Aayog 2020) reported that India has achieved the target of being "open defecation free", despite strong evidence to the contrary in their own report, which started that 2.78% of urban households still do not have toilets and 3% of rural households do not use the toilets they have (67). Improvements in the treatment of domestic wastewater have been marginal, with 27% now being safely treated (UN WATER 2020). Hundreds of new sewage management projects and strengthened industrial regulation are certainly positive developments for water quality outcomes, though their impacts on specific water quality indicators remain unclear. The proportion of withdrawal from freshwater sources is down to 66% after peaking in 2009, and water use efficiency is gradually improving over time (ibid). But environmental and water management issues need addressing, recycling lags behind world norms, and the efficacy of both biodiversity conservation efforts and public awareness-raising campaigns remains unclear.

These data draw our attention to the ways in which progress toward SDG 6 is being measured and assessed. Measured at the national scale, high rates of infrastructure provision seem impressive. But the sheer size of India's population means that even a small percentage of people without access to safe water or sanitation ran represent a population of millions. If only 15% of India's total current population of about 1.4 billion people are practicing open defecation, that amounts to about 210 million people or nearly one third of the total global number of people practicing open defecation (WHO/UNICEF 2021). Assessment of progress using national-scale statistics also masks regional and demographic differences among populations. State-level data reveals incredible inequity in access to handwashing facilities of 67 percentage points between Odisha (with 29% access) and Sikkim (96% access). While differential access between rural and urban performances is often presented in SDG data, the impact of wealth on access rates is not often highlighted. A large gap of 80 percentage points separates India's poorest and richest quintile rates of access to basic sanitation (at 11% and 91%, respectively) and is larger than the access gap between states and subnational regions. Gaps in open defecation rates are large both by wealth (at 2% among the wealthiest quintile and 83% among the poorest) and by region (at 1% in the lowest subregion and 70% in the highest) (ibid.). Of course, these differentials point to the significant influence of location and socioeconomic indicators on achievement of SDG goals. Finally, we must be attuned to the use of language to shape and change meanings associated with each of these indicators. Progress begins to pale when examining rates of access to *safe and clean* water and sanitation or when considering adoption and use over availability of water and sanitation infrastructure. While the UN report cited above (UN WATER 2020) touts 95% rates of access to improved drinking sources, a closer look reveals that only 66% of the population has improved water supplies *on the premises*, only 63% of supplies are *accessible* when needed, and only 56% are *safely managed* (WHO/UNICEF 2021). Similarly, while 83% of the population has access to improved sanitation facilities, only 46% use a *safely managed* sanitation service. In addition, water from hand pumps and tube wells are counted as safe sources of piped drinking water, despite being carriers of waterborne diseases (Kelkar-Khambete 2015). While this does not mean we should discount progress out of hand, we must take a closer look into the complex contexts in which these SDG-related programs and achievements are taking place. In the next section, I examine the Ganga River Basin as microcosm of these processes in order to understand the challenges associated with providing sustainable access to water and sanitation across India and to identify some of the potential pathways and prospects for achieving the aims of SDG 6.

7.4 The Ganga River Basin: Ground Zero for SDG 6 in India

The river Ganga is incredibly significant for sustainable development of the people, environment, and economy of India. Most of its tributaries originate deep in the Himalayas near the borders between India, China, and Nepal. The main branch of the Ganga starts at the Gangotri Glacier in the Indian State of Uttarakhand and flows more than 2500 km (1550 miles) to meet the Brahmaputra River before emptying into the Bay of Bengal. It is one of the most voluminous rivers in the world, with a net discharge less than only the Amazon and Congo Rivers, though its volume is variable and dependent upon a monsoon cycle that brings dry conditions through the winter and heavy rains in the late summer months. Flood risks at this time are high and compounded by increased runoff from Himalayan snowmelt. Over millennia, sediment deposition from seasonal flooding of the Ganga-Brahmaputra and nearby Indus rivers gradually formed the Indo-Gangetic plain and delta that constitute the primary landforms of Pakistan, Bangladesh, and North India. Rich soil and abundant water created fertile conditions for early human settlement and the development of a cultural hearth in which the Ganges River valley civilization and many subsequent empires were based (Chapman 1995). This hearth birthed three of the world's major religions, including Hinduism, in which Ganga is a goddess best known for her power to cleanse the body of illness and the soul of impurity. Both the goddess and the physical river are widely perceived as nurturing, powerful, pure, emancipatory, and supremely sacred (Pandey 1984; Narayanan 2001). Ganga was declared the "National River of India" in 2008 and, today, supports the largest concentration of population, agriculture, and industry in a country that is the second largest by population and the fifth largest economy in the world.

The river Ganga has long been the focus of, and laboratory for, approaches to water management and pollution control in India. The history of river governance is one of competing interests between state and central governments and of fragmented approaches to the management of water resource, sanitation, and environmental concerns. India's constitution initially placed water resources under state, rather than central authority. This meant that interstate bodies of water, like large rivers, were subject to inefficient management by multiple legislative authorities. Later constitutional reforms moved water governance to the concurrent list, meaning that both central and state governments would hold jurisdiction over water resources. The central government established individual river boards, charged with mitigating disputes between river-sharing states, and both Central and State Pollution Control Boards, responsible for ensuring the cleanliness of water resources. But these Boards operated in a largely advisory capacity and had little legislative power beyond establishing standards and settling disputes between states. State authorities retained the power to draft and implement programs. The 1985 Ganga Action Plan (GAP) therefore represented a novel approach that placed basin management under a central government authority with the ability to enforce laws and standards. It also served as a model for later river management programs and foreshadowed future governmental reorganizations by combining pollution abatement, sanitation provision, and environmental protection under a single umbrella organization.

The desire to better unify governance of the Ganga, and to reflect the intersection of issues of water resource governance, sanitation, and environmental preservation, led to the reorganization of ministerial bodies multiple times. Before the Ganga Action Plan, regulation of river waters was divided between the Ministry of Irrigation and Power, the Department of Drinking Water Supply (under the Ministry of Rural Development), and the Department of Environment. This led to competing legislative interests during an era in which economic development and agricultural production were prioritized over both household supply and water quality. For example, policies promoting the intensive use of pesticides and fertilizers, industrial agglomeration, and disinvestment in sewage treatment increased water pollution significantly through the latter half of the twentieth century.[3] The Department of Irrigation split from that of Power and later became the Ministry of Water Resources, which was further expanded to the Ministry of Water Resources, River Development, and Ganga Rejuvenation. The Department of Drinking Water Supply added sanitation to its mission in 2010 and became its own ministry in 2011. Both ministries were combined in 2019 into the Ministry of Jal Shakti, which also took over the later National Mission for Clean Ganga from the Ministry of Environment, Forests, and Climate Change. Jal Shakti subsumes nearly all aspects of water governance in the nation and is charged with administration of the Jal Jeevan Mission, the urban arm of Swachh Bharat Abhiyan, and the Namami Gange Programme, the three main schemes under which India is structuring its drive to achieve SDG 6. Because management of the river Ganga played such a formative role in shaping efforts to manage water and water quality in India, understanding the dynamics surrounding clean water and sanitation in the Ganga River Basin (GRB) can help us to identify some of the challenges and prospects for achieving the aims of SDG 6.

7.5 Water Availability, Use, and Efficiency

Availability of water in the GRB varies significantly over the course of the year. Flow and subsequent flood risk are highest during the late summer monsoon season and can be dangerously low during the winter and early spring, posing significant problems for water quality and the health of those who live along the river. While water storage techniques were well established before the colonial period, the colonial and later independent India developed a system of dams and canals meant to improve year-round irrigation and water availability and manage flood risk. But they also contribute to the variability of water supply along the course of the river. The Upper Ganges Canal, located where the river first enters the plains near Haridwar, Uttarakhand, has a total discharge of 370 m3/s and irrigates about 20 lakh (or 2 million) hectares in the fertile *doab* region (Jal Shakti 2021). The river's reduced flow after the Upper Ganges Canal diversion compounds water quality and siltation problems in the upper central basin, especially between Kanpur and Allahabad, where the Lower Canal returns water in the form of agricultural runoff. Here, flow is strengthened by inputs from a main tributary, the Yamuna River. The Farakka Barrage, located further downstream near the border with Bangladesh, diverts another 76,500 m^3/s to improve navigability around Kolkata Port (ibid.). Other principal diversions include the Damodar Valley Project, the Tehri Dam, and the Pancheswar Dam, as well as the proposed National River Linking Project (NRLP). They are important parts of India's efforts to improve water supply for irrigation and provide "green" hydroelectric energy, but they are also subject to much debate and opposition (Prabhat 2003). These projects reduce river water levels and impact both water availability and water quality by concentrating pollutants, increasing siltation, and reducing groundwater

[3]Indeed, PM Indira Gandhi argued at the 1972 United Nations Conference on the Human Environment in Stockholm that economic development was the best approach to pollution abatement and environmental protection, as increased wealth would reduce direct dependence on natural resources (Khator 1991; Dwivedi and Khator 1995).

replenishment. They also demonstrate a fundamental prioritization of water for irrigation, which accounts for 90% of India's water consumption, over water for drinking and other household uses, which is a priority for SDG 6 (Ayyar 2018).

India is a water-stressed country, and the amount of available water per capita in the GRB seems to be declining (Ministry of Water Resources 2017). Annual rainfall has decreased over the past two decades, impacting both the river's base flow and rates of groundwater replenishment and depletion (Patel et al. 2021; Mukherjee et al. 2018). At the same time, demand is increasing. Some estimates indicate that water use rates across the country could double that of water availability by 2030, driven by agriculture, population growth and urbanization, and changing lifestyles (Water Resources Group 2022). Most households in the GRB rely on tube wells, powered by machine or hand pumps, rather than direct surface water use. Because the water table in the river basin is still relatively high and surface water pollution is a major concern, tube wells are an attractive option for household water provision that provides an additional (though arguably insufficient) filtering system for improving water quality. The use of tube wells is largely unregulated, enabling private citizens to bypass centralized municipal water systems, and their purchase is often subsidized by the government. The preferred water supply choice of wealthy residents, tube wells, has been shown to extract an average of 12.9 L per second in subtropical regions of South Asia when attached to a diesel pump (Ullah et al. 2016). Hand pumps are a far more common and represent the most frequent method of securing of household water supply in the GRB (Khambete 2021a). Tube wells, particularly those that are privately owned and located on household premises, count as a safe or "improved" piped water source under the SDGs, provided they are relatively free of fecal and other chemical contamination. They account for a large share of India's high rates of compliance with SDG 6.1. But the tube wells aren't the most reliable or safest source of drinking water. Dropping water tables, drought, and contamination from sources like nearby latrines pose risks, and even tube well water should be treated or filtered for drinking water use (Islam et al. 2016; Chandran 2017). River water is infrequently used as a primary source of drinking water, even in rural areas of the GRB, but it is often used for household chores and animal care and is consumed in religious practices associated with river worship. In informal housing settlements, where neither well nor tap water is available, direct river water use does occur. While access to centralized tap water is increasing rapidly under the Jal Jeevan Mission, central basin state Uttar Pradesh has the largest number of households still without tap water connection (Jal Jeevan Mission 2022). The program is beset by challenges and accusations of inaccurate reporting, lack of water treatment, and the so-called "dry pipes" or instances of piped tap infrastructure without a water supply (Paliath 2018; Mashal and Kumar 2021; Thakur 2021).

7.6 Open Defecation Sewage Treatment and Water Quality

Attempts to address sanitation challenges in the GRB have taken on new life under the SDGs. In rural areas, India's primary focus is the push to be open defecation-free (ODF). The GoI claims 100% toilet coverage in rural areas, but a 2018 report by the National Statistical Office (NSO) contradicts this claim, saying that only 71.3% in rural areas and 96.2% in urban areas have access to toilets. Evidence from the GRB also illustrates that *access* and *use* are not the same, as open defecation continues to be a popular practice despite increased rates of toilet provision (India Water Portal 2017; Nawaz 2019; Verma 2019b). Open defecation is often conceived of as a default option or coping mechanism for those with no other sanitation facilities and presented as an affront to human and environmental health, modernity, urban cleanliness, and women's safety (e.g., see Mathiesen 2015; McCarthy 2016; Dutta 2017b; and Sangomla 2019). But the choice between toilets and open defecation is complex and often reflects factors like time to access, cost of access, cleanliness of shared toilets, location or ownership of shared toilets, and reliance of toilets on water supply (Desai et al. 2015; Bhatt et al. 2019). Open defecation and toilet use are also not mutually exclusive, with some people opting for toilets when only when certain conditions, like crowding and cleanliness, are met. The cost of toilet use, whether privately installed or part of a shared public scheme, also nearly always exceeds that of open defecation, which is a particularly accessible option in rural areas. We do know that open defecation is a highly gendered practice that can pose risks for women using public spaces or going to the bathroom at night. For example, women are twice as likely to be raped when compared with women using a home toilet (Jadhav et al. 2016). But public toilets are not necessarily significantly safer, and many women will opt for the familiarity of open defecation (O'Reilly 2015; Reddy et al. 2019). Indeed, it seems that it is not the presence of the toilet but rather the use of public spaces that heightens insecurity. Even household toilets are frequently unused by residents and seem to have a marginal impact on open defecation rates, except where policing and coercion are used to deter open defecation (Mathiesen 2015; Verma 2019b). Where toilets are not adequately connected to sewage treatment facilities and flow directly into either surface or ground water, they are not necessarily a healthier or more sanitary option than open defecation (Sangomla 2019).

While open defecation is still practiced in urban areas of India, rates are far lower and both privacy and open space harder to secure. Here, the primary sanitation challenge is untreated sewage, which is released by most cities directly into the river. Recent calculations estimate a total of 6.07 billion liters of largely untreated wastewater are discharged into the Ganga each day, though estimates tend to range between 2 and 12 billion (ibid.). Even where sewage treatment infrastructure is available, facilities are often at least periodically inoperable due to a lack of insufficient power supply and maintenance (Subramanian 2009). Under the Namami Gange Programme, the state Government of Uttar Pradesh recently established 104 new sewage treatment plants with a combined capacity of 3 billion liters per day (Srivastava 2021). Projects in other basin states are certain to increase this volume and further improve rates of sewage interception in the central river basin. But the growing urban population is likely to continue stressing treatment facilities unless this pace of development can be maintained in the long term. While the turn of the century population of the GRB was estimated at only 100 million people, current calculations place it nearer 650 million (McNeill 2000; WWF 2022). The GRB hosts about a third of the nation's cities and another third of its urban residents (Kishore 2008). Future population growth will be concentrated in urban centers, further taxing efforts to improve sanitation (Desai 2021). Open defecation seems to be attracting more attention from both Swachh Bharat and Namami Gange Missions. This focus is politically expedient for the GoI, which faces an incredible amount of social and political pressure from both domestic and international sources to "Clean the Ganga." Toilet provision is also a far less expensive endeavor than sewage treatment plants. The truth is we have very little data on the impact of open defecation practices on health or water quality separate from that of untreated sewage. While untreated sewage continues to be the primary source of pollution in the Ganges, river management must involve the treatment of all sources of untreated sewage as well as exploration of novel solutions to both household and municipal sanitation problems.

7.7 Ecosystem Protection and Biodiversity

Plans to protect and restore water-related ecosystems in the GRB hinge mostly on water management and sanitation efforts, outlined above, that seek to improve water quality by increasing river volume and treating sewage from surface water sources (NMCG 2022). Other efforts, organized under the Namami Gange Programme, focus on riverfront development, biodiversity conservation, and afforestation. While riverfront development (include ghat beautification programs) is intended to reduce solid waste pollution and afforestation aims to enhance forest coverage in head water areas, biodiversity conservation is the key approach to ecosystems protection under SDG 6. Ganga supports an incredible range of biodiversity from the high peaks of the Himalayas, through the temperate foothills, dry plains, and finally largest river delta in the world, where mangrove forests of the Sundarbans span the border between India and Bangladesh. The basin is still home to a variety of animals, including bears, monkeys, and many bird species. The river also supports a variety of fish and other animals, including crocodiles and the blind Ganges dolphin (Sharma 2009). Human settlement in the basin has a significant impact on the health and distribution of this natural ecology, destroying habitats for native wildlife, including the Bengal tiger, Asian elephant, rhinoceros, and musk deer, which now live only in highland forests or in conservation parks (ibid., Ghosh 2015). The most fragile ecosystems and the Ganges are found in the upper and lower sections of the river, where high-mountain and tropical riparian zones support threatened biodiversity hotspots (Kishore 2008).

A species restoration plan developed by the Wildlife Institute of India focuses on biodiversity conservation with the aim of maintaining or increasing ecosystems goods and services in the basin. Along with other research groups, the Wildlife Institute established the Ganga Aqualife Conservation Monitoring Centre, charged with conducting scientific research to inform the work of both Jal Jeevan and Namami Gange, as well as planning at the village scale. A key aspect of this work thus far has involved everyday citizens participating in the collection of scientific data and acting as "guardians" of the river (*Ganga Praharis*). Interpretation and education centers are meant to provide public education and outreach, inspiring public engagement in river conservation. The first of these include *Ganga Tarini*, a floating museum on a houseboat, and *Ganga Darpan*, a physical facility in Dehra Dun, display information on "the ecological and sociocultural values of Ganga," and advocate for its conservation (Kumar 2019). This model has been scaled up and will serve as a model for conservation of other rivers throughout India.

7.8 Public Participation in Water Governance

Encouraging public participation in water governance is a key aspect of the work of SDG 6 programs in the GRB. India has a long history of participatory environmental governance, with some of the earliest measures incorporated into social forestry programs that decentralized or devolved decision-making to the level of the panchayat or local village council (Mohanty 2009). More a form of communal management than public participation, the panchayati system was nonetheless duplicated in approaches to water supply management. Participatory irrigation management (PIM) was introduced in Uttar Pradesh during the MDG era in 2009 (Kaushal et al. 2019). It led to the establishment of local water user associations (WUA) with the power to manage irrigation water access, promote efficiency, and discourage overuse. Communal management is present as well in the Ganga Gram, a Jal Shakti program in coordination with NMCG to engage 24 panchayat village councils in the management of waste, ODF practices, and other issues associated with water management (Niti Aayog 2020). The Ganga Action Plan (GAP) was the first major water quality legislation to call for public participation. Participatory approaches became enshrined across all forms of environmental governance in the 1990s, following economic restructuring and an international push for public participation as an aspect of "good governance" associated with sustainability (Cooke and Kothari 2001).

In theory, public participation is meant to empower people to shape the decision-making processes that impact their lives. In practice, participation is appealing to planners and policymakers because it promises to improve project success and sustainability by increasing public buy-in and responsibility (ibid.). This latter form of participation is most often found in Swachh Bharat Abhiyan and the Namami Gange

Programme. Two illustrative examples are the Ganga Prahari Program and the Ganga Data Collector app. The Ganga Praharis, or Guardians of the Ganga, are volunteer corps drawn from local communities and charged with monitoring water quality, promoting awareness about "the benefits of a clean and vibrant Ganga," and acting as local representatives for various agencies (WII-NMCG n.d.). The Ganga Data Collector is a mobile application developed by the Wildlife Institute of India under the National Mission for Clean Ganga (NMCG). Members of the public can download the app and use its various features to provide data to research centers responsible for monitoring aquatic species and other aspects of aquatic health. The goal of crowdsourcing participatory data collection is to allow for faster more comprehensive collection of data. Ganga Praharis will help to train and educate members of the broader public on how to use the app (Verma 2019a). Both programs exemplify "back-end" participatory approaches, where participants who are members of the public are recruited to participate in preconceived, top-down planned programs in order to improve the program's success or sustainability. "Front-end" participation, where members of the public participate in decision-making processes and are engaged in program formulation, is less common in this context.

Many participatory programs emphasize awareness-raising along with or as a precondition to public participation, and awareness-raising programs in the GRB are often structured under participation initiatives (Kedzior 2018). Awareness-raising efforts of Jal Jeevan, Swachh Bharat Abhiyan, and Namami Gange are broad and include events and workshops, seminars and conferences, rallies, exhibitions, shram daan (volunteer drives), cleanliness drives, public competitions, as well as general publicity campaigns on television and radio and in print and electronic social media. While these types of efforts are considered a fundamental part of any large-scale social program, they've also been the subject of critique. Scholars have argued that these programs attribute sanitation, hygiene, and water use practices to lack of awareness rather than to the differential structural constraints faced by individual residents (Anuradha et al. 2017; O'Reilly et al. 2017; Singh et al. 2018). Awareness-raising programs may then be more about promoting behavioral change and increasing public compliance with promoted measures than about educating the public or encouraging their genuine engagement in environmental decision-making (Gatade 2015; Kedzior 2018). Encouraging public awareness of both water and sanitation management challenges and their possible solutions is an important aspect of sustainable governance. But the broader public must be empowered to make decisions and not looked to as an expeditious method for executing preconceived programs. Efforts to elicit genuine "front-end" participation take time and are certainly challenged by the rapid pace with which these programs are being developed and rolled out and by the unlimited timeframe in which the goals of SDG are meant to be achieved. With few years left until the SDG agenda concludes, we must consider what ongoing challenges impact India's progress and what prospects remain for achievement of these goals.

7.9 Challenges and Prospects: SDG 6 by 2030

The time left to achieve SDG 6 is limited and it is unlikely that India will meet each of the component goals. Nonetheless, progress in the last few years has been impressive. An incredible amount of political, social, and financial commitments has been made in order to improve water provision, water quality, and sanitation in the Ganga River Basin and beyond. Only a few of those have been discussed here in any detail. The number of programs, institutions, and governmental organizations involved in the various aspects of this endeavor is too long to enumerate, but they are active at every scale from the smallest village to the central government. Efforts to achieve these goals are also coming from multiple angles, including top-down infrastructure provision, community mobilization, and expanded scientific research. It remains to be seen whether the programs currently being implemented under Jal Jeevan, Swachh Bharat Abhiyan, and Namami Gange will produce the results they promise in terms of providing accessible and true safe drinking water, ending open defecation, or expanding the provision of safe and healthy sanitation options to more of the population and improving water quality and environmental health. So far, open defecation certainly continues as one of the most common forms of sanitation in both rural and urban settings, residents struggle with the operability and sustainability of both tap and toilet infrastructures, and water quality in the Ganges River has not yet improved drastically. No matter how these results may come to pass, what is clear is that the achievements gained in the past 7 years are only the beginning of what will necessarily become much longer-term undertakings. In the coming years, these programs will likely need to expand their scope in order to address new and continuing challenges.

Climate change and associated impacts on water sources in India is likely to pose the greatest single challenge to water provision and water quality in medium to long term. Increased temperatures, droughts, and flooding will impact water supply and water quality, as well as biodiversity. Glacial melt may yield more runoff into snow-fed rivers like the Ganges in the short term but reduce or even disappear in the long term (NIC 2009). Likely changes in weather patterns including increases in the frequency or severity of storms, along with potential changes in the monsoon cycle, will impact infrastructure, water availability, and water quality. More communities will likely experience severe water stress, and water storage and management solutions that harvest water during the wet season and store it through the dry season are already emerging as popular mitigation responses. But increased heat and irregular supply may make conventional infrastructure solutions, like reservoir storage and piped tap water, less viable. Plans to expand canal infrastructure and interlink the rivers of northern India promise to improve reliability but are likely to exacerbate the negative impacts of water diversion programs away from the Ganges and other major rivers and for the damaged water quality and riparian ecosystem health. Improving infrastructure to expand subsurface water storage or reduce the amount of water lost to leaky infrastructure can help (Muthuwatta et al. 2017; Ayyar 2018). But the greatest prospect for improving water supply probably lies in water recycling and reuse. Water recycling and reuse programs are already part of India's SDG 6 agenda, but they seem to be playing backseat to other goals. Attention to opportunities for water recycling should be a part of ongoing infrastructure projects, particularly in the construction of new sewage treatment plants. Some creative alternatives have long been proposed by regional nongovernmental associations and underscore the need for robust "front-end" public participation in the planning process (Mishra 2005). The most sustainable path forward will likely involve taking steps to close the water cycle.

The other great challenge thrown into stark relief by emerging data on India's progress on SDG 6 is that of social inequality. As discussed above, differences in service provision, access, and risk exist not only between rural and urban populations but also according to state or region, economic status, age, and gender. Rural areas are significantly less likely to have access to safe and convenient water sources, while people living in poverty are far less likely to have access to sanitation services than their wealthier counterparts. Children continue to be far more likely than adults to suffer from death and long-term health effects of waterborne disease, while women and girls are far more likely than other family members to suffer from direct exposure to pollution and disease due to household roles, caretaking and hygiene responsibilities, and inequitable sanitation access (Gaon Connection 2021). Many of the initiatives responding to SDG 6 involve public participation that occurs, at least partially, in the form of financial buy-in. Paying for water and sanitation services can improve public investment and interest in sustaining these services, but the pay-for-service approach can lead to the exclusion of the poorest segments of the population from water and sanitation solutions and create a tiered service system where wealthy people effectively pay less for better water supply and poor people bear the brunt of service cost.

Overlapping with economic status are problems with the way most of these programs target "recognized" households for service provision. It seems clear that this category excludes some, though certainly not all, informal housing units (or "slums"). The outcome of their exclusion from service programs is likely to pose major challenges, particularly as slum populations increase in the coming decades. Providing centralized sewage and sewage treatment to municipally unrecognized communities may not be possible without a major redesign of urban spaces. Creatively rethinking solutions for urban water and sanitation provision must be pursued feverishly and made a higher priority in current research and planning within the urban Indian context (Schrecongost et al. 2020). Too often, behaviors that emanate from poor sanitation access, including littering, open defecation, and direct surface water use, are associated with individual morality or tied to ideas of modernity and backwardness (Luthra 2018). Awareness-raising and participatory programs that emphasize behavioral change perpetuate the idea that people litter or use the bathroom in the open because they have poor hygiene or they don't know better while failing to acknowledge the structural issues that limit some people's water and sanitation options. In my own research on water and sanitation in the Ganges River Basin, the structural limitations about water and sanitation provision are a common theme, with comment like "I have to shit [or drink or bathe] somewhere" being a common refrain (pers. comm., 2009). The greatest prospect for addressing structural inequity may lay outside SDG 6 measures and involve formalizing informal housing, expanding housing options, and narrowing the wealth gap. Within water and sanitation programs, progress may come from broadening definitions of "recognized" households, expanding service areas of established and new infrastructure projects, and investigating alternatives to centralized services.

Finally, a key challenge may lay in the ability of Jal Jeevan, Swachh Bharat Abhiyan, and Namami Gange to perform course correction in the middle of their short tenure. India's progress on tap water provision and toilet provision has been impressive. But the exceeding rapid rollout of these programs also contributes to their failure when they are insufficiently coordinated with other services. For example, toilets are only superior to open defecation if they are accessible, clean, usable, and functioning and if the waste they produced is treated. Otherwise, the negative health and environmental impacts of open sanitation may simply be moved indoors, displaced, or concentrated in a focused area. So too tap water must be coordinated with a sustainable supply and, often, treatment to ensure that taps do not run dry or provide unsafe water. This level of service provision and coordination for a population of 1.4 billion people inside a 15-year timeframe is no small task. An approach marked by laudable goals and short timelines reflects the incredible politicization of water and waste in India. It is politically expeditious for politicians to promise to provide taps and toilets, to modernize the country by ending open defecation, or to "save the Ganges." Many elections are won and fought on these platforms. In interviews, state-level politicians have indicated that it is better to promise and fail to deliver once elected than it is to display pragmatism. Voters can be forgiving of programmatic failures, to which many have become accustomed, particularly in the development sector. It's not that cynicism pervades but that program failure needs to be understood within a complex political context that encourages lofty promises without demanding complete delivery. Understanding this context also helps to explain why rapid program development and delivery have led to problems like poor quality latrine construction, supply-driven toilet and tap distribution, and long term reluctance to adopt behavioral change. If the various organizations and institutions administering these programs can learn from these shortcomings and course correct before the SDG agenda expires in 2030, India will certainly improve its chances of successfully meeting the outcomes of SDG 6. The shift from Millennium Development Goals to Sustainable Development Goals has expanded the water agenda by recognizing that water supply, water quality, sanitation, and environmental health are not disparate links on the water-waste chain but coupled together and central to the processes associated with sustainable development.

References

Ahluwalia P (2021) Seven years of Swachh Bharat Mission [Press release], October 2. PRS Legislative Research. https://prsindia.org/theprsblog/seven-years-of-swachh-bharat-mission

Anuradha R, Datta R, Raja JD, Lawrence D, Timsi J, Sivaprakasam P (2017) Role of community in Swachh Bharat Mission. Their knowledge, attitude and practices of sanitary latrine usage in rural areas, Tamil Nadu. Indian J Community Med 42(2):107–110

Ayyar K (2018) Global water crisis: what are India's taps running dry? Time Magazine, June 6. https://time.com/5302661/water-crisis-drinking-india-drought-dry/

Bhatt N, Budhathoki SS, Lucero-Prisno DE, Shrestha G, Bhattachan M, Thapa J, Sunny AK, Upadhyaya P, Ghimire A, Pokharel PK (2019) What motivates open defecation? A qualitative study from a rural setting in Nepal. PLoS One 14(7):e0219246

Central Pollution Control Board (2008) Evaluation of operation and maintenance of sewage treatment plants in India-2007, Control of Urban Pollution Series: CUPS/2007. http://admin.indiaenvironmentportal.org.in/files/sewage.pdf

Central Pollution Control Board (2021) Status of STPs, March 8. https://cpcb.nic.in/status-of-stps/

Chandran R (2017) India must regulate tubewells as water becomes scarce with urbanization. Reuters, September 4. https://www.reuters.com/article/us-india-landrights-water/india-must-regulate-tubewells-as-water-becomes-scarce-with-urbanization-idUSKCN1BF27W

Chapman GP (1995) Introduction: the Ganges-Brahmaputra basin. In: Chapman GP, Thompson M (eds) Water and the quest for sustainable development in the Ganges Valley. Mansell Publishing, London, pp 3–24

Chengappa R (2021) Ode to a sacred river: why the Namami Gange Programme is different from previous efforts to clean Ganga. India Today, March 21. https://www.indiatoday.in/magazine/nation/story/20210329-ode-to-a-sacred-river-1781262-2021-03-21

Comptroller and Auditor General of India (CAG) (2018) Performance audit of the National Rural Drinking Water Programme. Ministry of Drinking Water and Sanitation. https://cag.gov.in/cag_old/content/report-no15-2018-performance-audit-national-rural-drinking-water-programme-ministry-drinking

Cooke B, Kothari U (2001) Participation: the new tyranny? Zed Books, London

Desai S (2021) India's population data and a tale of two projections. The Hindu, July 11. https://www.thehindu.com/opinion/lead/indias-population-data-and-a-tale-of-two-projections/article32329243.ece

Desai R, McFarlane C, Graham S (2015) The politics of open defecation: informality, body, and infrastructure in Mumbai. Antipode 47(1):98–120

Dudgeon D (2019) Multiple threats imperil freshwater biodiversity in the Anthropocene. Curr Biol 29(19):R960–R967

Dutta S (2017a) From open defecation to indiscriminate waste disposal, why cleaning-up Ganga is an arduous task. NDTV, May 2. https://swachhindia.ndtv.com/from-open-defecation-to-indiscriminate-waste-disposal-why-cleaning-up-ganga-is-an-arduous-task-7011/

Dutta S (2017b) What Namami Gange can do that earlier Ganga clean-up plans couldn't. NDTV, June 23. https://swachhindia.ndtv.com/what-namami-gange-can-do-that-earlier-ganga-clean-up-plans-couldnt-9139/

Dwivedi OP, Khator R (1995) India's environmental policy, programs, and politics. In: Dwivedi OP, Vajpeyi DK (eds) Environmental policies in the third world: a comparative analysis. Greenwood Press, Westport, pp 47–69

Gaon Connection (2021) 90% diarrhoea deaths in India due to lack of safe drinking water and basic sanitation facilities, September 3. https://en.gaonconnection.com/diarrhoea-children-death-india-drinking-water-sanitation-unicef-wash-world-health-organization-43029/

Gatade S (2015, October 31) Silencing caste, sanitizing oppression: understanding Swachh Bharat Abhiyan. Econ Polit Wkly 50(44):29–35

Ghosh P (2015) Conservation and conflicts in the Sundarban Biosphere Reserve, India. Geogr Rev 105(4):429–440

Global Alliance on Health and Pollution (2019) Pollution and health metrics. Global, regional, and country analysis. https://gahp.net/wp-content/uploads/2019/12/PollutionandHealthMetrics-final-12_18_2019.pdf

Haq Z (2021) Jal Jeevan Mission's progress in 117 poorest districts outpaces national average. Hindustan Times, July 1. https://www.hindustantimes.com/india-news/jal-jeevan-mission-s-progress-in-117-poorest-districts-outpaces-national-average-101625130026146.html

Hindustan Times (2020) Swachh Bharat has played key role in Covid fight: PM Modi. Hindustan Times. https://www.hindustantimes.com/india-news/swachh-bharat-has-played-key-role-in-covid-fight-pm/story-yDKgXlJhUt8OUQmIwK2JlI.html

India Water Portal (2017) Open defecation continues in rural India despite increased toilet construction, April 24. https://www.indiawaterportal.org/articles/

open-defecation-continues-rural-india-despite-increased-toilet-construction
Islam MS, Mahmud ZH, Islam MS, Saha GC, Zahid A, Ali AZ, Massan MQ, Islam K, Jahan H, Hossain Y, Hasan MM, Cairncross S, Carter R, Luby SP, Cravioto A, Endtz HP, Faruque SM, Clemens D (2016) Safe distances between groundwater-based water wells and pit latrines different hydrogeological conditions in the Ganges Atrai floodplains of Bangladesh. J Health Popul Nutr 35(26):1–10
Jadhav A, Weitzman A, Smith-Greenaway E (2016) Household sanitation facilities and women's risk of non-partner sexual violence in India. BMC Public Health 16:1139
Jain A, Wagner A, Snell-Rood C, Ray I (2020) Understanding open defecation in the age of Swachh Bharat Abhiyan: agency, accountability, and anger in rural Bihar. Int J Environ Res Public Health 17(4):1384
Jal Jeevan Mission (2022) Functional household tap connection (FHTC) in every home. Jal Jeevan Mission Dashboard. https://ejalshakti.gov.in/jjmreport/JJMIndia.aspx
Jal Shakti (2021) Historical background of Upper Ganga Canal. https://idup.gov.in/en/article/ganga
Joshi P (2020) Informal waste pickers to be part of Swachh Bharat. The Times of India, 9 January. https://timesofindia.indiatimes.com/city/aurangabad/informal-waste-pickers-to-be-part-of-swachh-bharat/articleshow/73162361.cms
Kaur B (2018) Namami Gange: 5 reasons why Ganga will not be clean by 2020. Down to Earth, October 15. https://www.downtoearth.org.in/coverage/pollution/namami-gange-5-reasons-why-ganga-will-not-be-clean-by-2020-61891
Kaushal N, Babu S, Mishra A, Ghosh N, Tare V, Kumar R, Sinha PK, Verma RU (2019, June 25) Towards a healthy Ganga-Improving river flow through understanding trade-offs. Front Environ Sci 25:83
Kedzior SB (2018) "Preemptive participation" and Environmental awareness across Indian Water Quality Policy. J Environ Dev 26(3):272–296. https://doi.org/10.1177/1070496517697768
Kelkar-Khambete A (2015) Verbal jugglery lets 86% Indian households have 'safe drinking water'. India Spend, October 7. https://www.indiaspend.com/verbal-jugglery-lets-86-indian-households-have-safe-drinking-water-62297
Khambete AK (2021a) Everything you want to know about drinking water. India Water Portal, January 9. https://www.indiawaterportal.org/faqs/everything-you-want-know-about-drinking-water
Khambete AK (2021b) Water can kill! India Water Portal, November 12. https://www.indiawaterportal.org/articles/water-can-kill
Khator R (1991) Environment, development and politics in India. University Press of America, Lanham
Kishore K (2008) The Holy Ganga. Rupa & Co., New Delhi
Kumar S (2019) Floating museum creating awareness for a clean Ganga in Varanasi. Hindustan Times. https://www.hindustantimes.com/lucknow/floating-museum-creating-awareness-for-a-clean-ganga-in-varanasi/story-bdPCPtVcPSL7clQ9YFCx5O.html
Luthra A (2018) 'Old habits die hard': discourses of urban filth in Swachh Bharat Mission and The Ugly Indian. J Multicult Discourses 13(2):120–138
Mashal M, Kumar H (2021) In India's water stressed villages, Modi seeks a tap for every home. The New York Times, December 21. https://www.nytimes.com/2021/12/21/world/asia/india-water-modi.html
Mateo-Sagasta J, Zadeh SM, Turral H (2017) Water pollution from agriculture: a global review. Food and Agriculture Organization of the United Nations. https://www.fao.org/3/i7754e/i7754e.pdf
Mathiesen K (2015) Can $15m worth of toilets finally clean up the Ganges? The Guardian, September 11. https://www.theguardian.com/global-development-professionals-network/2015/sep/11/india-amma-15-million-dollars-toilets-open-defecation-clean-ganges
McCarthy J (2016) Open defecation is a problem in India's cities & rural areas. Global Citizen, November 16. https://www.globalcitizen.org/en/content/urban-rural-two-sides-of-indias-sanitation-problem/
McNeill JR (2000) Something New Under the Sun: An Environmental History of the Twentieth-Century World. W.W. Norton and Co, New York
Mehotra K (2021) Phase 2 of Swachh Bharat Mission to focus on waste segregation at source. The Indian Express, March 3 https://indianexpress.com/article/india/phase-2-of-swachh-bharat-mission-to-focus-on-waste-segregation-at-source-7210207/
Ministry of Housing and Urban Affairs (2021) Union Budget 2021–2022, Ministry of Housing & Urban Affairs [Press release]. Press Information Bureau. https://pib.gov.in/PressReleasePage.aspx?PRID=1694420
Ministry of Water Resources (2017) Shortage of water [Press release]. Press Information Bureau. https://pib.gov.in/newsite/PrintRelease.aspx?relid=168727
Mishra VB (2005) The Ganga at Varanasi and a travail to stop her abuse. Curr Sci 89(5):755–763
Mohanty R (2009) Institutional dynamics and participatory spaces: the making and unmaking of participation in local forest management in India. IDS Bull 35(2):26–32
Mukherjee K (2016) MDGs to SDGs: Lessons for UHC for India. Glob J Med Public Health 5(4):1–3
Mukherjee A, Bhanja SN, Wada Y (2018) Groundwater depletion causing reduction of baseflow triggering Ganges river summer drying. Sci Rep 8:12049
Muthuwatta L, Amarasinghe UA, Sood A, Surinaidu L (2017) Reviving the "Ganges Water Machine": where and how much? Hydrol Earth Syst Sci 21:2545–2557
Narayanan V (2001) Water, wood, and wisdom: ecological perspectives from the Hindu traditions. Daedalus 130(40):179–206

National Jal Jeevan Mission (2021) Two years of Jal Jeevan Mission. Department of Drinking Water and Sanitation, Ministry of Jal Shakti, Government of India. https://jalshakti-ddws.gov.in/sites/default/files/two-years-of-jal-jeevan-mission_1.pdf

National Mission Clean Ganga (NMCG) (2022) Aquatic biodiversity conservation in Ganga River Basin. https://nmcg.nic.in/AboutBio.aspx

National Statistical Office (NSO) (2018) Drinking water, sanitation, hygiene and housing condition in India [NSS 76th Round report]. Government of India. https://www.thehinducentre.com/resources/article30979980.ece/binary/Report_584_final_0_compressed.pdf

Nawaz MS (2019) Open defecation continues along Ganga in Haridwar. The Times of India, October 19. https://timesofindia.indiatimes.com/city/dehradun/open-defecation-continues-along-ganga-in-haridwar/articleshow/71653269.cms

NIC (2009) India: impact of climate change to 2030, a commissioned research report, NIC 2009-03D

Niti Aayog (2020) India voluntary national review. Government of India. https://sustainabledevelopment.un.org/content/documents/26279VNR_2020_India_Report.pdf

O'Reilly K (2015) From toilet insecurity to toilet security: creating safe sanitation for women and girls. Wiley Interdiscip Rev Water 3(1):19

O'Reilly K, Dhanju R, Goel A (2017) Exploring "the remote" and "the rural": open defecation and latrine use in Uttarakhand, India. World Dev 93:193–205

Oates L, Sudmant A, Gouldson A, Gillard R (2018) Reduced waste and improved livelihoods for all: lessons on waste management from Ahmedabad, India. Coalition for Urban Transitions. https://newclimateeconomy.net/content/cities-working-papers

Paliath S (2018) National Rural Drinking Water Programme 'failed' to achieve targets: government auditor. Here's why. IndiaSpend, November 25. https://www.indiaspend.com/national-rural-drinking-water-programme-failed-to-achieve-targets-government-auditor-heres-why/

Pandey S (1984) Ganga and Yamuna in Indian art and literature. Prakashan, Chandigarh

Patel A, Goswami A, Dharpure JK, Thamban M (2021) Rainfall variability over the Indus, Ganga, and Brahmaputra river basins: a spatio-temporal characterization. Quat Int 575–576:280–294

PBNS (Prasar Bharati News Services) (2021) Namami Gange: 6 years of transforming and rejuvenating river Ganga. News on Air, July 12. https://newsonair.com/2021/07/12/namami-gange-6-years-of-transforming-and-rejuvenating-river-ganga/

Prabhat A (2003) Drive to link Indian rivers. BBC News, May 23. http://news.bbc.co.uk/2/hi/south_asia/3050485.stm

Reddy YM, Raghavan S, Vedala SC (2019) A narrative exposition on public toilet use by women: a study from Warangal. Indian J Gend Stud 26(2):108–137

Rubin BF, Kapur-Gomes S (2020) India spent $30 billion to fix its broken sanitation. It ended up with more problems. CNET. https://www.cnet.com/culture/india-spent-30-billion-to-fix-its-broken-sanitation-it-ended-up-with-more-problems/

Sangomla A (2019) Pollution 'time bomb' ticking for Ganga despite ODF. Down to Earth News, April 2. https://www.downtoearth.org.in/news/waste/pollution-time-bomb-ticking-for-ganga-despite-odf-63790

Schrecongost A, Pedi D, Rosenboom JW, Shrestha R, Ban R (2020) Citywide inclusive sanitation: a public service approach for reaching the urban sanitation SDGs. Front Environ Sci 8(19). https://doi.org/10.3389/fenvs.2020.00019

Sharma M (2009) Passages from nature to nationalism: Sunderlal Bahuguna and Tehri Dam opposition in Garhwal. Econ Polit Wkly 21(February 21):35–42

Singh SL, Kunwar N, Sharma A (2018) Impact of Swachh Bharat Abhiyan in Indian society. Int J Home Sci 4(1):215–219

Solberg E (2015) From MDGs to SDGs: the political value of common global goals. Harv Int Rev, Fall 2015

Srivastava S (2021) UP govt sets up 104 sewage treatment plants, water quality in Ganga, other rivers improves. India Today, October 2. https://www.indiatoday.in/india/story/uttar-pradesh-government-sewage-treatment-plants-water-quality-ganga-rivers-1859825-2021-10-02

Subramanian S (2009) Ganga reels as Varanasi spews sewage. Mint, September 2. http://www.livemint.com/2009/09/02000924/Ganga-reels-as-Varanasi-spews.html

Swarajya (2019) Economic Survey 2018–19 focuses on resource efficiency for SDG 2030; Namami Gange key to sustainable water management. Swarajya Magazine, July 4. https://swarajyamag.com/insta/economic-survey-2018-19-focuses-on-resource-efficiency-for-sdg-2030-namami-gange-key-to-sustainable-water-management

Thakur A (2021) Homes with taps but no water, village water tanks lying empty, a story of yet another official pipe dream. Times of India, September 20. https://timesofindia.indiatimes.com/india/homes-with-taps-but-no-water-village-water-tanks-lying-empty-a-story-of-yet-another-official-pipe-dream

True Picture (2018) Why Namami Gange will succeed, unlike Ganga Action Plan, December 13. https://www.thetruepicture.org/namami-gange-will-succeed-unlike-ganga-action-plan/

Ullah Z, Ullah R, Ali M, Junaid M (2016) Performance evaluation of diesel and electric operated tube wells irrigation system in sub-tropical conditions. Pure Appl Biol 5(1):142–148

UN WATER (2020) India. https://www.sdg6data.org/country-or-area/India

United Nations Department of Economic and Social Affairs (UNDESA). (2022). Swachh Bharat Abhiyan (Clean India Mission). UNDESA Sustainable Development. https://sdgs.un.org/partnerships/swachh-bharat-abhiyan-clean-india-mission

Vandemoortele J (2017) From MDGs to SDGs: critical reflections on global targets and their measurement. In: van Bergeijk PAG, van der Hoeven R (eds) Sustainable Development Goals and income inequality. Edward Elgar, Cheltenham, pp 32–50

Verma L (2019a) An app to map Ganga water, aquatic life. The Indian Express. https://indianexpress.com/article/india/an-app-to-map-ganga-water-aquatic-life/

Verma R (2019b) Despite toilets in place, a quarter of rural population stuck to open defecation. Down to Earth News, January 10. https://www.downtoearth.org.in/news/waste/despite-toilets-in-place-a-quarter-of-rural-population-stuck-to-open-defecation-62770

Waghela D (2021) WaSH: creating a health future for underserved communities. TATA Trusts. https://www.tatatrusts.org/our-work/water-sanitation-and-hygiene

Water Resources Group (2022) Background of our work in India (National and Uttar Pradesh). World Bank Group. https://www.2030wrg.org/india/background/

WaterAid (2016) An assessment of faecal sludge management policies and programmes at the national and select state levels. https://smartnet.niua.org/content/ee28f99a-6b85-4205-a398-c3b425e4054b

WHO/UNICEF (2021) Progress on household drinking water, sanitation and hygiene, 2000–2020. https://www.who.int/publications/i/item/9789240030848

Wiegleb V, Bruns A (2018) Hydro-social arrangements and paradigmatic change in water governance: an analysis of the sustainable development goals (SDGs). Sustain Sci 13:1155–1166

WII-NMCG (n.d.) Ganga Prahari. https://nmcg.nic.in/wii/prgbggp.aspx

World Bank (2010) The economic impacts of inadequate sanitation in India. https://www.wsp.org/sites/wsp/files/publications/wsp-esi-india.pdf

World Bank (2017) Annual freshwater withdrawals, total [data]. Retrieved from https://data.worldbank.org/indicator/ER.H2O.FWTL.K3

World Bank (2019) Worsening water quality reducing economic growth by a third in some countries: World Bank [Press release]. https://www.worldbank.org/en/news/press-release/2019/08/20/worsening-water-quality-reducing-economic-growth-by-a-third-in-some-countries

Worldometer (2021) India Population (LIVE). Available: https://www.worldometers.info/world-population/indiapopulation/ (accessed 30 Dec 2021).

World Wildlife Foundation (WWF) (2022) The Ganges. https://www.wwf.org.uk/where-we-work/ganges

Sya Buryn Kedzior is an Associate Professor of Geography and Environmental Planning and Assistant Director of Environmental Science and Studies at Towson University in Towson, Maryland. She is a human geographer and political ecologist with interests in environmental issues and politics in India and the United States. Her published work examines relationships between public environmental knowledge and resource use, the role of civil society in environmental politics, and participatory environmental governance.

India's Journey Toward Cleaner and Affordable Energy for Achieving SDG 7: Progress and Prospects

8

Anurag Piyamrao Wasnik

Abstract

The Sustainable Development Goals (SDGs) are 17 intertwined international sets of goals introduced by the UN in 2015. SDG 7, which talks about affordable and clean energy, is essential for sustainable human development and for powering economic growth. It is a prerequisite for achieving other SDGs, including health, education, gender equality, and poverty eradication. Hence, this research chapter strives to understand the progress made by India in achieving SDG 7 goals and recognize the opportunities from a national and subnational point of view. India is no exception to the clean energy transition, which has become a global phenomenon, and this study shows how India has used the international SDG principles and has applied them in its local context. Despite the significant challenges India has faced in bringing shared responsibility to its population, it has embraced a sustainable energy future and clean technologies and remains the only G-20 nation on track to meet the 2 °C global warming targets.

This chapter is developed by qualitative methods drawn out by resourcing secondary research tools. The study has been segregated into different areas, intending to cover government policies, corporate efforts, and international collaborative efforts toward achieving SDG 7. The chapter highlights critical policy initiatives such as Pradhan Mantri Sahaj Bijli Har Ghar Yojana and Pradhan Mantri Ujjwala Yojana and initiatives undertaken to strengthen the EV ecosystem in the country. Similarly, the research helps one understand India's present and future paths of energy mix and energy efficiency. The study, through SWOT analysis, intends to examine inevitable challenges restricting the efforts and potential institutive actions to overcome or avoid them. The consolidated information illustrated in this chapter will benefit current researchers and motivate future investigators to take up SDG 7 in India as part of their proposed research.

Keywords

Sustainable development goals · Affordable and clean energy · SWOT · Sustainability · India

8.1 Introduction

SDG 7 aims to ensure all access to affordable, reliable, and sustainable energy by 2030. SDG 7 achievements highly correlate with other SDGs'

A. P. Wasnik (✉)
Indian Institute of Management Ahmedabad (IIMA), Ahmedabad, Gujarat, India
e-mail: p17anuragw@iima.ac.in

V. Dutta, P. Ghosh (eds.), *Sustainability: Science, Policy, and Practice in India*, Sustainable Development Goals Series, https://doi.org/10.1007/978-3-031-50132-6_8

progress, including health, education, gender equality, and poverty eradication. The goal further emphasizes substantially increasing the share of renewable energy in the country's total energy mix and improving its energy efficiency. Enhanced international cooperation, investments in innovation, capacity building, and clean energy infrastructure suffice to realize the more sterile energy goals. India's progress on the goals of SDG 7 stands crucial as it has a population of around 1.3 billion and is the fifth economy by GDP.

India has undertaken several initiatives and policy shifts and strengthened its international collaborations in line with its commitments toward accomplishing SDG 7 by 2030. The focus has been reducing carbon footprints by expanding its power generation mix with renewable energy sources, implementing cleaner cooking and transportation measures, and promoting economic growth. India has successfully reached 99.9% home electrification by implementing projects such as Bijli Har Ghar Yojana – Saubhagya, Pradhan Mantri Ujjwala Yojana, Suryashakti Kisan Yojana (SKY), Solar Urja, and solar PV electrification. With subsidy placed on adopting clean fuel usage, the coverage has reached 2842 lakh households with LPG and 72 lakh with PNG connection (India VNR 2020, 2020). The emphasis is on promoting a cleaner transportation system owning to its higher consumption of fossil fuels and carbon emissions. With policies such as FAME I, FAME II, PLI Scheme for National Programme on Advanced Chemistry Cell Battery Storage, battery swapping, and inclusion of energy storage in the harmonized list of infrastructure in place, India accelerates in the path of bringing in EV and clean energy-powered vehicles into the market. The volatility in fossil fuel prices during the recent global unrest and its effects on our daily lives and economic activities show how important it is for India to have its independent power supply. Increasing investment in renewable energy plants and existing capacity building through public-private partnerships and multilateral cooperation will help us become self-sufficient in our energy demands and achieve a cleaner and more efficient energy mix.

India is localizing the SDGs with a collaborative effort from the State, civil society organizations, businesses, entrepreneurs, and the country's citizens. It makes an effort to be more inclusive and participative, where no one is left behind in access to clean and affordable energy. The Voluntary National Review (VNR) by NITI Aayog has helped reflect on how each State is performing under SDGs, promoting transparency in the efforts made, and facilitating the exchange of best practices across states to achieve the goal efficiently. The report presents an overview of the SDG 7 scenario in India, an analysis of some of its flagship initiatives, current standing, challenges, and way forward to accomplishing its commitments.

8.2 Literature Review

Energy sources that are both affordable and reliable are essential for modern civilization to function and thrive. A developed energy system underpins every industry, from agriculture and education to infrastructure, communications, and high technology. Historically, intensive development patterns have been based on inexpensive fossil fuels, which have a high energy density, and have been the primary source of greenhouse gas emissions driving climate change. On the other hand, the availability of new, clean technology can reroute development more sustainably (Smith et al. 2019). The governments are accelerating the transition to an affordable, dependable, and sustainable energy system by emphasizing energy-efficient practices, investing in renewable energy resources, and deploying clean energy technology and infrastructures. They are willing to innovate and develop cutting-edge new technologies, upending the status quo of the global energy system with investment in R&D, positioning them at the forefront of initiatives to combat climate change with sustainable energy sources.

While electrification plans must achieve the upper tiers of energy access, current policies emphasize ensuring equitable, high-quality energy services. Special attention has been paid to off-grid communities where the standardization of technologies and business models could support more sustainable and affordable operations (Ramchandran et al. 2016). To advance the clean cooking initiative, reliable and affordable energy distribution networks should be strengthened. In contrast, consideration should be given to integrating renewable sources with electrification, power grid expansion, and upgrading initiatives for clean fuel cooking (Suyash Jolly et al. 2012). Though it's inevitable that the national access to clean cooking in the region largely correlates to economic income levels, countries with difficult geographic conditions and significant rural populations have frequently seen severe share discrepancies (IRENA 2022). As the window for access to affordable, reliable, and sustainable energy is narrowing, the adoption rate must accelerate.

With the growing demand for energy supplies and stricter environmental quality standards, access to affordable and clean energy has been the primary goal of SDG 7. According to REN21, the proportion of renewable energy in global final energy consumption was 17.9% in 2018, marginally lower than the previous year's level of 18.1% (REN21 2020). This suggests that the expansion of renewable energy is struggling to keep up with the rising global demand for energy. To achieve the necessary increases in the proportion of the world's energy consumption covered by renewables, there must be apparent worldwide increases in the rate of expansion. Modern renewables contributed to over 60% of all new capacity for electricity generation in 2014, whereas developing countries, notably Eastern Asia, accounted for nearly 72% of the total growth in energy consumption from modern renewable sources between 2010 and 2012 (United Nations Statistics Division (UNSD) 2022). With a focus on growing the significant contribution of renewable energy in the global energy mix and improving energy efficiency, the proportion of renewable energy sources in overall energy output is rapidly increasing.

Access to clean fuels and cooking technologies is becoming more crucial to policymakers in the Asian region. Though progress is modest and the area is still far from achieving universal access by 2030, this is due to the complexity and diversity of the challenges and low levels of investment pledges by the world powers (IRENA 2020). The difficulties have only worsened with the most recent increases in fuel costs. Households have faced financial and logistical barriers under lockdown as supply chains were disrupted and prices rose, causing many to opt for cheaper and conventional options. With the lowest renewable energy tariff in the Asia-Pacific region, India can concentrate on raising the proportion of non-renewable energy sources in its energy mix.

India's socioeconomic growth is significantly contingent on people having universal and equitable access to affordable, reliable, and clean energy. The core elements of SDG 7 are supported by India's national energy policy, which includes universalizing access to electricity and clean cooking fuel, prioritizing renewable energy in the energy mix, and increasing energy efficiency. The country's energy mix is rapidly diversifying through the expansion of renewable energy options such as solar, wind, hydro, and waste-to-energy conversion, as well as increased installed capacity and access for all (Kumar et al. 2010).

India, which has the world's highest clean cooking deficit (548 million people), is also the world's fastest-improving country and may thus provide some powerful lessons for the entire region. Access rose by an average of 4.4 percentage points in India between 2016 and 2020, partly due to the large-scale Pradhan Mantri Ujjwala Yojana (PMUY) initiative and 99.9% household electrification under the PM Sahaj Bijli Har Ghar Yojana (Saubhagya Dashboard 2019). Energy conservation, commercial building regulations, energy consumption requirements for energy-intensive firms, and modernizing regulatory institutions and infrastructure will suffice the energy efficiency of India (Arumugam 2019).

Building the capacity and physical assets by 2030 to serve growing industrial, residential, and infrastructure demand is continuous. With the active participation of the corporate sector and tech-driven entrepreneurs, India is dedicated to universalizing sustainable energy. Like in most economies, India is determined to develop a modern electricity system focused on increasing the number of renewable energy sources in its energy mix. It was previously primarily restricted to a centralized power system based on fossil fuels, particularly coal. With several policies and structural changes in place, India is well on increasing renewable energy capacity to 450 GW by 2030, up from 132 GW in 2020 (India VNR 2020, 2020).

8.3 SDG 7 in the Indian Context

India launched its Voluntary National Review (VNR) to transition SDGs from global to local in 2017. The VNR reflects the country's overall development, identifies the main challenges to reaching SDG targets, and is based on feedback and collaboration between subnational and local governments, civil society, and markets. India attempts to make the SDGs more approachable and participative at all levels. It is based on significant participation and consultation with over a thousand civil society organizations working with vulnerable groups, business sector leaders, and subnational governments (India VNR 2020, 2020). India has undertaken multiple initiatives such as Bijli Har Ghar Yojana – Saubhagya, Pradhan Mantri Ujjwala Yojana, Suryashakti Kisan Yojana (SKY), Solar Urja, solar PV electrification, National Biogas and Manure Management Programme, etc. and collaborated with the international- and local-level institutions to synergize in the realization of the affordable and clean energy goals by 2030 (Jain et al. 2018).

8.3.1 Monitoring Progress at the National and State/Union Territory Level

Two national-level indicators have been identified to assess India's progress toward the affordable and clean energy goal, each of which captures one of the five SDG targets for 2030 outlined under this goal. These indicators were chosen based on data available at the subnational level to ensure comparability across states and UTs. There are two indicators for mapping the progress: the percentage of households electrified and the percentage of LPG + PNG connections against the number of households. Both goals are consistent with the global SDG 7, which calls for everyone to access affordable, reliable, and modern energy services (NITI Aayog 2021).

As per the SDG India Index and Dashboard 2020–2021, the SDG Index Score for Goal 7 ranges between 50 and 100 for states and 71 and 100 for UTs. Fifteen states and 5 UTs have entered the category of achievers (with an index score of 100), and 12 states and 3 UTs bagged a position in the category of front runners (score range between 65 and 99, including both).

Household electrification: India has made significant progress on household electrification and is on track to meet its target of providing every household access to electricity. About 99.99% of households had electricity by the end of March 2019 (Saubhagya Dashboard 2019). Except for Chhattisgarh, all states have achieved 100% electrification status. Among the UTs, Delhi, Ladakh, Jammu, Kashmir, and Puducherry have gained universal electricity access.

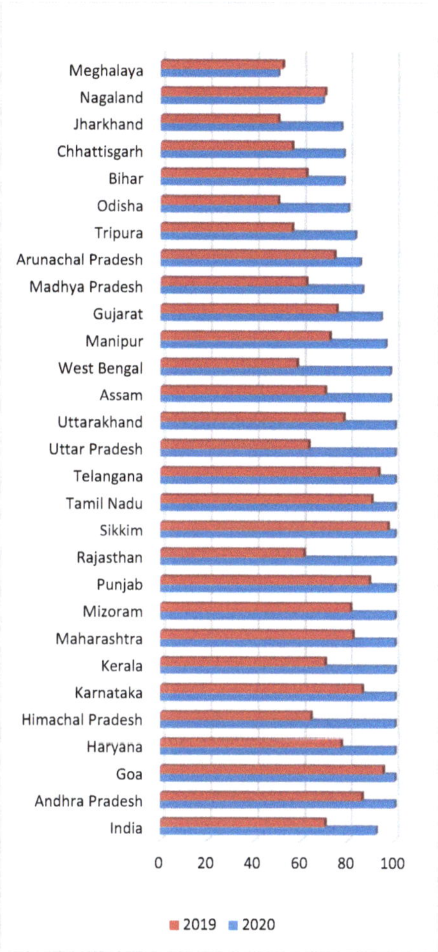

Region wise score (out of 100)		
State	**Year**	
	2019	**2020**
India	70	92
Andhra Pradesh	86	100
Goa	95	100
Haryana	77	100
Himachal Pradesh	64	100
Karnataka	86	100
Kerala	70	100
Maharashtra	82	100
Mizoram	81	100
Punjab	89	100
Rajasthan	61	100
Sikkim	97	100
Tamil Nadu	90	100
Telangana	93	100
Uttar Pradesh	63	100
Uttarakhand	78	100
Assam	70	98
West Bengal	58	98
Manipur	72	96
Gujarat	75	94
Madhya Pradesh	62	86
Arunachal Pradesh	74	85
Tripura	56	83
Odisha	50	80
Bihar	62	78
Chhattisgarh	56	78
Jharkhand	50	77
Nagaland	70	69
Meghalaya	52	50

(SDG India Index and Dashboard | iTech Mission, 2020)

Fig. 8.1 Affordable and clean energy (SDG 7) dashboard of India

Clean cooking fuel (LPG + PNG coverage): As of July 2020, there are 2824 lakh connections of liquefied petroleum gas (LPG) and 72 lakh connections of piped natural gas (PNG) in the country. Gujarat leads in PNG connections (26 lakh), while West Bengal has no PNG connection. Among the UTs, Delhi leads in both LPG (49.8 lakh) and PNG (9 lakh) connections (NITI Aayog 2021) (Fig. 8.1 and Table 8.1).

Table 8.1 State-wise performance on affordable and clean energy (SDG 7)

Index	
Achiever	100
Front runner	65–99
Performer	50–64
Aspirant	0–49

8.3.2 India's Current Standing and Efforts in Ensuring Universal Access to Affordable, Reliable, and Modern Energy Services

The Indian Government has launched a plethora of initiatives to ensure that all Indians have access to affordable, dependable, and modern energy services. To meet its commitment to affordable and clean energy by 2030, it has launched ambitious initiatives such as Pradhan Mantri Sahaj Bijli Har Ghar Yojana – Saubhagya, Pradhan Mantri Ujjwala Yojana, Suryashakti Kisan Yojana (SKY), solar PV electrification, and Solar Urja, to name a few. The emphasis has been on increasing the proportion of renewable energy in the country's total energy mix through a series of measures, policy shifts, and collaborations. Since the transportation sector consumes roughly 33% of the total energy in the country, the priority has shifted to a green transportation ecosystem through the promotion of EVs, clean fuel, and several sustainable measures (NITI Aayog 2022b).

The policy initiatives undertaken to strengthen the EV ecosystem in the country are the FAME I and II Scheme, PLI Scheme for National Programme on Advanced Chemistry Cell Battery Storage, draft Battery Swapping Policy, and inclusion of energy storage in the harmonized list of infrastructure, which will promote the EV adoption and strive toward the self-reliance on the clean energy sources.

8.3.3 Pradhan Mantri Sahaj Bijli Har Ghar Yojana – Saubhagya

The PM Sahaj Bijli Har Ghar Yojana – Saubhagya was introduced in October 2017 to achieve universal household electrification by connecting all underprivileged households in rural and urban parts of the nation to electricity (Pradhan Mantri Sahaj Bijli Har Ghar Yojana – Saubhagya 2022). With the cooperation and concentrated efforts of the center and the states, it is one of the most extensive global programs for universal electrification. The nodal organization for implementing the program nationwide is the Rural Electrification Corporation Limited (REC). The program's objective is to provide power to all unconnected rural families and low-income urban households to achieve universal household electrification across the nation. For the scheme's implementation, the Socio Economic and Caste Census (SECC) 2011 data were used to identify the potential beneficiary families for the free connection. According to SECC data, ineligible households would be given electrical connections for a fee of Rs 500, which would be paid over ten electricity bills. Additionally, financial assistance was available to all DISCOMs, including private sector DISCOMs, state power departments, and RE cooperative societies.

To track the information and development of household electrification, a unique SAUBHAGYA online portal has been created. The scheme aims to reduce the dependency on kerosene fuel; improve education, health services, communication, and public safety; and create job opportunities through universal access to electricity for all willing households.

As of March 2019, all households were reported as electrified by the State under the program, except 18,734 households in Chhattisgarh regions affected by left-wing extremists. Following this, seven states – Assam, Chhattisgarh, Jharkhand, Karnataka, Manipur, Rajasthan, and Uttar Pradesh – reported that nearly 19.09 lakh unelectrified households, identified before March 2019, that were previously unwilling to receive electricity connections have now expressed a willingness to do so (Saubhagya Dashboard 2019). All union territories (UTs) and states have signed memorandums of understanding (MoUs) with the Central Government to guarantee enough power supply to agricultural consumers starting on April 1, 2019, as well as a 24×7 power supply to all households, businesses, and industries. All states have been asked to create specialized initiatives to identify families that have not been electrified yet and then link those with the power network.

8.3.4 Pradhan Mantri Ujjwala Yojana

The PM Ujjwala Yojana (PMUY) was launched by the Ministry of Petroleum and Natural Gas (MOPNG) in May 2016 as a flagship program to assist rural and low-income families with access to clean cooking fuels such as LPG. Previously, these households relied on traditional cooking fuels such as firewood, coal, cow-dung cakes, and conventional fuels. Using traditional cooking fuels negatively influences the environment and rural women's health (Ministry of Petroleum and Natural Gas 2021). The campaign aimed to empower women and protect their health while decreasing the severe health risks connected with cooking with fossil fuels. It sought to reduce the number of deaths in India caused by unclean cooking fuel and to protect young children from acute severe respiratory infections caused by indoor air pollution produced by the combustion of fossil fuels.

The scheme gained traction as it expanded to include 80 million low-income households, up from 50 million previously, with an extra Rs. 4800 crore infusion from the Government. In September 2019, the Government met its goal of 8 crores of LPG connections under the Pradhan Mantri Ujjwala Yojana; however, only three states have gone kerosene-free: Haryana, Punjab, and Andhra Pradesh are among them. Delhi, Chandigarh, Daman and Diu, Dadra and Nagar Haveli, Andaman and Nicobar Island, and Puducherry are among the union territories that have gone kerosene-free.

A major section of the Indian population still uses solid fuels despite the Government's support for developing clean cooking fuels. A few of the approaches that policymakers can use to achieve the exclusive use of clean energy in rural India include raising awareness of the risks of solid fuels and the benefits of cleaner fuels, lowering the cost of LPG cylinder refills in rural areas, and promoting gender equality within households, particularly in cooking and related tasks. The Pradhan Mantri Ujjwala Yojana 2.0 was launched to distribute eight million LPG connections to low-income families by March 2020 as an expansion of the previous program. The scheme's delivery of eight million LPG connections has also helped to boost LPG coverage from 62% on May 1, 2016, to 99.8% on April 1, 2021 (NIC 2022).

8.3.5 Policy Initiatives Undertaken to Strengthen the EV Ecosystem in the Country

The Indian electric vehicle market is projected to grow at a CAGR of 47.09% to USD 15,397.19 billion by 2027 (Mordor Intelligence 2022). The road transport sector contributes significantly to carbon emissions and accounts for nearly 33% of particulate matter emissions (NITI Aayog 2022b). The shift to clean mobility, spearheaded by electric vehicles (EVs), is vital to creating a low-carbon transportation ecosystem.

India has pledged to achieve the net-zero emission target by 2070 at the 26th Conference of the Parties (COP26) in November 2021. Comprehensive strategies for decarbonizing high-GHG-intensive sectors such as transportation and energy are at the forefront of any climate-related policy initiatives of the Government. To realize the targets set, the Government has formulated several policies emphasizing the decarbonization of the transport ecosystem by transitioning to clean mobility, led by electric vehicles (EVs). Several policies and institutional-level initiatives have been implemented, such as the Faster Adoption and Manufacturing of (Hybrid) Electric Vehicles in India (FAME) I and II and the Production-Linked Incentive (PLI) program for the "National Programme on Advanced Chemistry Cell (ACC) Battery Storage" to mark the journey of creating the green transport ecosystem with the synergy of Government, stakeholders, and the consumer at all level.

8.3.5.1 FAME II

Sustainable transportation solutions are critical for the future, and enhanced incentives for electric two-wheelers will boost uptake and promote more domestic investments in future technologies. FAME II is an initiative formulated by the

Government of India to encourage the development of electric vehicles. The policy is significant in light of global efforts to address climate change. The project was created to achieve the goals of the National Electric Mobility Mission Plan (NEMMP), which aims to make hybrid and electric vehicles the first choice for buyers, so that these vehicles can replace conventional vehicles and thus reduce liquid fuel consumption in the country from the automobile sector (Ministry of Heavy Industries & Public Enterprises, 2020). FAME's Phase I spanned from 2015 to 2019, while Phase II began in 2019 and is projected to be completed by 2022. The GoI initially introduced the FAME II Scheme in 2019. It offers 50% more subsidy at Rs. 15,000 per kWh on electric scooters and bikes in India. The recent FAME II subsidy has encouraged faster adoption and manufacturing of electric vehicles in India.

Under FAME II, the Government has allocated Rs 10,000 crores for 3 years, from 2019 to 2022. Phase II of FAME will give incentives for public transportation in ten cities. 10,000 crores have been allocated for the FAME II project over 3 years till 2022. The Government has sanctioned 8596 crores for incentives, with 1000 crores set aside for installing electric vehicle charging stations in India (PMY Team 2022). The Government would provide incentives for commercial electric buses, three-wheelers, and four-wheelers. Plug-in hybrid cars and those with a large lithium-ion battery and electric motor will also be included in the plan, with financial assistance available based on the size of the battery.

Furthermore, the Central Government would provide incentives for purchasing 5 lakh three-wheelers, 7000 electric buses, and 35,000 four-wheelers. Additionally, the maximum subsidy for an electric two-wheeler would be 40% of its cost, up from 20% (Ministry of Heavy Industries & Public Enterprises 2019). A revised set of criteria for Phase II of the Faster Adoption and Manufacturing of (Hybrid) Electric Vehicles Scheme would encourage manufacturers of electric and hybrid vehicles to domesticate a significant portion of their components. The Government's continuing support for electronic vehicles (EVs) will make electric two-wheelers cheaper, benefiting electric motorbike and scooter buyers tremendously.

8.3.5.2 PLI Scheme for National Programme on Advanced Chemistry Cell Battery Storage

With a budgetary outlay of Rs. 18,100 crores, the Government has launched the Production-Linked Incentive (PLI) Scheme as the "National Programme on Advanced Chemistry Cell (ACC) Battery Storage" to boost India's manufacturing capabilities by achieving a manufacturing capacity of 50 gigawatt hours (GWh) of ACC. By rolling out this initiative, the Government is focusing on increasing local value addition while ensuring that India's levelized battery production cost is competitive globally. The program is intended to be technology-neutral, concentrating exclusively on the required output of the batteries. As a result, the beneficiary business will be free to choose relevant advanced technology, associated equipment and machinery, raw materials, and other intermediate products for establishing a cell manufacturing facility to cater to any demand.

The program anticipates attracting domestic and foreign investment that will support local manufacturing while facilitating the building of battery storage demand for electric cars and stationary storage and establishing a complete domestic supply chain in the country. The ACC PLI plan is intended to significantly affect the nation's savings due to a significant reduction in crude oil imports and an increase in the proportion of renewable energy on the national grid.

Together with the previously unveiled PLI Scheme for the automotive sector (25,938 crores) and the Faster Adoption of Manufacturing of Electric Vehicles (FAME) (10,000 crores), this PLI Scheme for Advanced Chemistry Cell (ACC) (18,100 crores) will enable India to switch from conventional fossil fuel-based automobile transportation systems to environmentally cleaner, sustainable, cutting-edge, and more effective electric vehicle (EV)-based transportation system (Ministry of Heavy Industries 2022).

8.3.5.3 Draft Battery Swapping Policy

Battery swapping falls within the purview of battery as a service (BaaS) business models, which entail consumers purchasing an EV without the battery, which drastically cuts the upfront costs, and paying a monthly subscription fee to service providers for battery services throughout the vehicle's lifetime (NITI Aayog 2022a). BaaS is suitable for fixed and removable batteries and is the conduit for implementing swapping solutions. Battery swapping offers three significant advantages over charging, i.e., saving time, space, and cost since each swappable battery is optimized.

The proposed Battery Swapping Policy for domestic electric vehicles (EVs) was ideated by the NITI Aayog. The policy aims to improve the effectiveness of the battery-swapping ecosystem for three-wheeled electric rickshaws and electric scooters, which will promote EV adoption. According to the proposed strategy, developing a battery-swapping network would be prioritized in the first phase for all major cities with a population greater than 40 lakh. The policy outlines the minimal technical and operational requirements that battery-swapping ecosystems must meet in order to implement battery-swapping infrastructure that is effective, efficient, reliable, safe, and customer-friendly while also providing financial support to the battery providers (for the cost of batteries) and EV users.

The draft policy suggests that the Goods and Services Tax Council consider reducing the tax rate differential on lithium-ion batteries and electric vehicle supply equipment, which will boost its manufacturing and consumption. The tax rate on lithium-ion batteries is 18% and 5% on electric vehicle supply equipment. The policy also proposes assigning a unique identification number (UIN) to swappable batteries throughout the manufacturing process to aid tracking and monitoring. The Bureau of Energy Efficiency (BEE) is the Central Nodal Agency responsible for rolling out EV public charging infrastructure and implementing battery-swapping networks across the country. An approach includes switching out exhausted batteries for charged ones. The initial cost of the vehicles is reduced due to battery swapping, which unlinks the vehicle from fuel sources (in this case, the battery). Smaller vehicles, such as two- and three-wheelers, will be primary beneficiaries of the battery swapping since their batteries are smaller and more accessible to switch than those in other automobile categories, which will eventually happen over time.

8.3.5.4 Inclusion of Energy Storage in the Harmonized List of Infrastructure

The Rakesh Mohan Committee (1996) classified infrastructure as power, gas, water supply, telecommunications, highways, industrial parks, railways, ports, airports, urban infrastructure, and storage, while the Reserve Bank of India (RBI) identifies infrastructure for external commercial borrowing funds as power, telecommunications, railway, roads including bridges, seaports, airports, industrial parks, urban infrastructure (water supply, sanitation, and sewage projects), mining, exploration, and refining (Planning Commission (Secretariat for Infrastructure) 2008).

Due to the definition of infrastructure discrepancy, the Cabinet Committee on Infrastructure created a harmonized master list of five key infrastructure sectors and several infrastructure subsectors to assist all agencies responsible for supporting infrastructure in various ways. Harmonized Master List of Infrastructure notified by the DEA and the Ministry of Finance has five sectors with multiple subsectors: transport and logistics, energy, water and sanitation, communication, and social and commercial infrastructure.

The inclusion of transport and logistics energy opens the window of opportunity to strengthen the EV ecosystem in the country and promote clean energy sources, as the sector accorded with the infrastructure status is entitled to avail multiple benefits and concessions. In the Union Budget of 2022–2023, Union Finance Minister Nirmala Sitharaman laid down sets of opportunities that the sector enlisted in the list of infrastructure can avail. With longer tenure and more accessible conditions, the industry can raise capital from insurance companies, pension funds, and foreign lenders. The status allows enterprises to access cheaper foreign currency finance through external commercial borrowing. Furthermore, it will be eligible to borrow through the India Infrastructure Finance Company (IIFCL) and benefit from GST exemptions, make in India exemptions, and so forth (Government of India 2022). A great benefit of this process is that it ultimately leads to local job creation and promotes research and development in eco-friendly mobility.

8.4 India's Current Standing and Efforts in the Share of Renewable Energy in the India's Energy Mix

India ranks third in the globe for overall energy consumption. At the same time, it was positioned third in the July 2021 EY's Renewable Energy Country Attractiveness Index (RECAI), with the parameters focusing on the country's total installed capacity and renewable energy investments (Ernst and Young 2021).

Renewable energy is rapidly replacing traditional energy sources in the country, evident from various projections on the country's renewable energy growth. India is striving to expand its renewable energy mix under its commitment to improve energy efficiency and reduce overall emission intensity while protecting the vulnerable sectors of the economy. The National Action Plan on Climate Change (NAPCC) aims to achieve 15% renewable energy in its total energy mix by 2020. The NITI Aayog has set an ambitious objective of producing 410 GW of wind and 420 GW of solar power by 2047 (Asian Development Bank Institute 2018).

India's global climate commitments, made through intended nationally determined contributions (NDC) at UNFCCC, reinforce India's commitment to a cleaner energy mix by 2030. India's global climate pledge to source 40% of its energy from renewable sources by 2040 reveals its policy objective of stiff focus on promoting renewable energy (Ministry of Environment, Forest and Climate Change 2015). India's energy mix trends further reflect shifting governmental emphasis on clean energy production. The detailed projections by IEA under India's new policy scenario are presented below (Fig. 8.2 and Table 8.2):

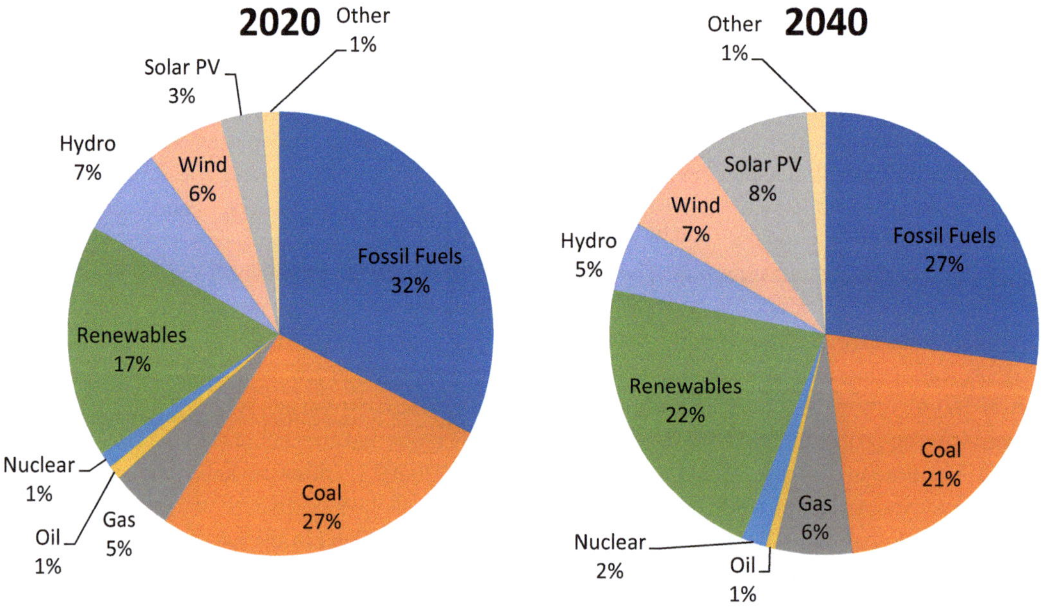

Fig. 8.2 Energy mix of India (2020–2040)

Table 8.2 Power generation capacity by type in India

	2014	2020	2030	2040	Shares (%) 2014	Shares (%) 2040	CAAGR[a] 2014–2040
Fossil fuels	204	280	419	576	71	53	4.1
Coal	174	230	329	438	60	41	3.6
Gas	23	41	76	122	8	11	6.6
Oil	7	9	13	15	3	1	2.9
Nuclear	6	10	24	39	2	4	7.6
Renewables	79	147	304	462	27	43	7
Hydro	45	58	83	108	15	10	3.5
Wind	23	50	102	142	8	13	7.2
Solar PV	3	28	100	182	1	17	16.4
Other	7	11	18	30	3	3	5.5
Total	289	436	746	1076	100	100	5.5

[a]Compound average annual growth rate (IEA 2015)

Through its policy thrust to transition to a greener energy mix by 2040, India has committed to increasing its power generation capability from nonfossil fuel sources from 41% in 2022 to 52% by 2040 (IEA 2015) (Central Electricity Authority (CEA) 2022).

8.5 India's Current Standing and Efforts in Improving Energy Efficiency

The Environmental Conservation Act of 2001 offered a concrete shape to India's long-held energy efficiency concerns. The launch of the Bureau of Energy Efficiency (BEE) and the National Mission for Enhanced Energy Efficiency provided additional institutional support to focus on lowering energy intensity and CO_2 emissions. Enforcing standards and labeling have effectively targeted household appliances in residential and commercial sectors, and this has raised consumer awareness and resulted in a successful reduction in household electricity consumption.

With the country's rising urbanization and the development of smart cities, the potential for energy efficiency in the construction sector is being realized. The Energy Conservation Building Code (ECBC) and BEE's voluntary star grading scheme for buildings aid in the shift to green buildings (India VNR 2020, 2020). Other efforts, such as encouraging energy efficiency in micro-, small-, and medium-sized businesses, increasing distribution companies' ability, and developing state institutional capacity, broaden the scope of energy efficiency.

BEE's Super Efficient Equipment Programme (SEEP) focuses on appliances that achieve a five-star certification. This endeavor resulted in the manufacturing and popularization of 35-W ceiling fans over 70-W ceiling fans. The distribution of 362 million LED appliances across the country under the UJALA Scheme has resulted in energy savings of 47 billion kWh and a reduction of CO2 emissions of 38 million tons per year (Ministry of Power 2015). The substantial distribution of LPG to low-income households under the Pradhan Mantri Ujjwala Yojana has also resulted in considerable CO2 emission reductions. Other initiatives, such as encouraging energy efficiency in micro-, small-, and medium-sized businesses, building distribution companies' capacity, and developing states' institutional capacity, broaden the scope of energy efficiency in India.

8.6 India's Current Standing and Efforts in Strengthening International Collaboration

With international and multilateral collaboration, India is bringing innovation, research, and learning together to realize the goals of clean and affordable energy by 2030 and contribute to its commitment toward the country's economic growth with a cleaner and more efficient energy mix (Government of India, Ministry of Power 2022) (Table 8.3).

Table 8.3 India's international collaborations

Sector	Country/institution/agency	Year of collaboration	Brief
Energy security	ASEAN-India Energy Cooperation	1993	Ensure access to affordable, reliable, sustainable, and modern energy for all by 2030 and moving toward low-carbon growth path through promoting different means of renewable energy source
Energy conservation	Indo-German Energy Programme (IGEN)	1995	Energy-efficient cooling has been identified as an area of collaboration. Supports policies and programs of the Energy Conservation Act
Fossil fuel-based power plants, renewable energy, and demand-side energy efficiency and low carbon growth strategies	Indo-German Energy Forum (IGEF)	2006	Promote dialogue and cooperation with involvement of public and private sector in the areas of energy security, energy efficiency, renewable energy, investment in energy projects and collaborative R&D
Energy efficiency and awareness	India-France	2006	Energy information centers for creating awareness on energy efficiency have been established at state designated agencies of Haryana and Punjab. Sharing of energy efficiency best practices and best available technologies in MSMEs and benchmarking and mapping Indian MSME energy consumption
Energy conservation	Indo-Japan Energy Dialogue	2006	Cooperation project on energy conservation guidelines and energy management manual, capacity building of energy manager and energy auditor, waste heat recovery, heat pump in buildings and best practice guide in the foundries sector
Knowledge sharing	Clean Energy Ministerial (CEM)	2009	Global forum to share lessons learned and best practices and to encourage the transition to a global clean energy economy
Energy efficiency	International Partnership for Energy Efficiency Cooperation (IPEEC)	2009	Promotes energy efficiency worldwide by exchanging information related to energy efficiency, developing partnerships between energy efficiency sectors, and supporting energy-efficient initiatives
Energy-efficient residential and public buildings	Indo-Switzerland	2011	Technical assistance to builders and developers for designing energy-efficient buildings through integrated design process
Energy efficiency	India-China	2012	Cooperation in enhancing energy efficiency in industries (cement, paper, and steel) Considered as the major energy-intensive industries with a very large bandwidth of energy consumption within the sectors
Energy efficiency	India-Russia	2013	Exchange of experience in the field of energy management, energy audits, and energy services

(continued)

Table 8.3 (continued)

Sector	Country/institution/agency	Year of collaboration	Brief
Industrial efficiency, building energy efficiency, energy efficiency financing, and institutional strengthening	India-USA collaboration	2015	Deploying and transferring innovative clean energy technologies
Renewable energy, energy efficiency	India-UK	2015	Market reforms, regulatory structures, and the role of competition in the supply and distribution of electricity including regulations and incentives for renewable energy deployment
Green energy	India-Germany	2015	Indo-German Energy Programme-Green Energy Corridors (IGEN-GEC). The objective of this program is to improve and accelerate the sector framework and conditions for gridintegration of renewable energy projects with conventional power grids
Solar energy	India-International Solar Alliance	2015	To undertake innovative and concerted efforts to reduce the cost of finance and technology for deployment of solar generation assets
Energy security and clean energy	India-International Energy Agency	2017	Collaboration across a range of vital areas including energy security and clean energy transitions
Renewable and hydrogen push	India-International Renewable Energy Agency	2022	Will work closely to assess the potential role green hydrogen can play both as an enabler of the transition in India and as a new source of national energy exports
Zero emission vehicle	India-California	2022	Addressing policy, technology and investment strategies for a zero emission vehicle (ZEV) transition, it would identify opportunities for India to emerge as a strategic leader in the global ZEV transition
Biofuel, ethanol blending, and hybrid electric vehicles	India-Maruti Suzuki	2022	Collaboration with National Dairy Development Board (NDDB) to study business model for the future commercialization of biogas and its potential for widespread use, in view of establishing a joint venture. Work upon the development of biomethane fuel-based vehicle
Battery swapping	India-International Council on Clean Transportation (ICCT)	2022	Explored the landscape of battery swapping for E2Ws in India and undertook quantitative analysis to ascertain the impact of various parameters on the total cost of ownership (TCO)
Battery supply chain	India Energy Storage Alliance (IESA)-Future Battery Industries Cooperative Research Centre (FBICRC), Australia	2022	The pact will focus on critical minerals and supply chain for advanced battery manufacturing in India. Will cater to supply chain development and integration of the battery supply chain between India and Australia

Government of India, Ministry of Power (2022)

8.7 SWOT Analysis

SWOT analysis of India's journey toward cleaner and affordable energy:

Strengths	**Weaknesses**
Increasing public-private partnerships in the renewable energy generation sector Policy support of subsidies, tax waivers, and incentives on local manufacturing and the adoption of cleaner energy options Robust collaborations for R&D in the clean energy segment High-level awareness and adaptability of the clean energy sources in rural households with successful implementation of several governmental schemes Less volatility and influence of external factors in the price of energy in comparison to fossil fuels	Large population size thus long time for implementation and monitoring of the policies The high upfront cost of setting up power generation plants Rely on heavy imports of raw materials for manufacturing of PV cells and panels, EV batteries High speculation among people for EV vehicles with regard to safety, mileage, and availability of charging infrastructure Reliance on foreign technology and innovations for the power generation
Opportunities	**Threats**
Job creation with the establishment of research centers, manufacturing or assembly hub Increase in the R&D centers/hub leading to innovation in the cleaner energy segment Increase in production and procurement from local raw material manufacturers Improved income of the farmers with promotion of the ethanol blending in the petrol Less expenditure on the imports of fossil fuels-strong forex reserve, controlled inflation	Risk of monopoly with private players entering into the renewable power generation space; price of the energy may remain high Fossil fuel still being a cheaper and more reliable option over the cleaner energy sources Lack of procurement, distribution, regulatory, and compliance-related regulations Connectivity of renewable energy with the the national grid is still at a nascent stage

8.8 Challenges and Way Forward in the Implementation of SDG 7 in India

India's total primary energy demand is expected to increase by 63% by 2030, while its contribution to global energy-related CO2 emissions is expected to grow from 6.7% in 2017 to 10.6% (International Energy Agency 2021); thus, achieving low-carbon energy security is critical for the country's sustainability journey. Building the capability and physical resources required by 2030 to meet the rising industrial, residential, and infrastructure demand is an ongoing effort. Significant barriers include a lack of enough human capital, technical know-how, financial resources, and funding to support domestic renewable energy generation and the power industry expansion. Another issue the nation must address is how to increase power supply more quickly to fulfill current unmet industrial, commerce, home, and future demand.

Grid parity has essentially been achieved in the renewable energy industry, owing to competitive auctions that have significantly reduced prices for solar and wind power installations. However, in the case of solar panels and equipment, there is a significant reliance on imports to the tune of more than 85% (India VNR 2020, 2020). Other resources, such as battery storage equipment and permanent magnets for electric vehicles, are in short supply. As a result, indigenous technological development and industrial capacity expansion are being intensively pushed.

Continuity should be given to initiatives such as the FAME Scheme, PLI Scheme for National Programme on Advanced Chemistry Cell Battery Storage, Battery Swapping Policy, and inclusion of the energy sector in the harmonized list of infrastructure to accelerate the adoption of sustainable mobility options in the country while enhancing the domestic production capability. The Government of India can bring in the concept of a model state that is actively taking measures toward adopting cleaner and affordable energy sources and sharing the knowledge and practices with other states.

The Ethanol Blending Programme will be crucial to achieving the nation's energy security amid rising urbanization, population, and climate change. Ethanol Roadmap 2021, presented by NITI Aayog, has set the target of ethanol blending of 20% by 2030, which will reduce our carbon footprint while boosting the income of our farmers (NITI Aayog, Ministry of Petroleum and Natural Gas 2021). Nuclear energy can be another thrust source for cleaner energy. Currently, nuclear energy contributes a total of 3.11% of the total energy generation in India. Atomic energy can be crucial for energy efficiency, given its smaller land footprints and waste. The Government has amended its Atomic Energy Act to allow public sector companies to form joint ventures to build nuclear power plants (Department of Atomic Energy 2015), increasing the share of nuclear energy in its total energy mix and providing a sustainable alternative to conventional energy sources.

Finally, adopting cleaner and more efficient energy options should come from the people themselves. Government should promote subsidies and waivers for cleaner energy sources, primarily in the energy-intensive sector like households, transportation, and manufacturing.

8.9 Conclusion

Intending to achieve universal clean and affordable energy by 2030, the Government of India is building its capacity to fulfill the rising energy demand with urbanization and economic growth. It includes policy reforms, international collaborations, engagement with the private sector, research, and innovation in the clean energy segment. The increase of the renewable energy share to our overall energy mix has been prioritized with increased governmental expenditure, public-private partnerships, and several policy related incentives. Initiatives such as the Pradhan Mantri Sahaj Bijli Har Ghar Yojana – Saubhagya, the Pradhan Mantri Ujjwala Yojana, and the Suryashakti Kisan Yojana (SKY) were critical in increasing the adaptability of cleaner energy sources in rural households in India and increasing the mission's inclusivity.

Voluntary National Review by NITI Aayog has helped track and compare the progress made under each goal at the national and subnational level, which brings responsibility to the state and related stakeholders to develop innovative solutions/policies to help achieve the targets. It has further facilitated sharing of the best practices and initiatives within the states and the capacity building of the performing states. With several policies in place, the constant efforts from the Government, the business sector, and the end consumer of India will help realize cleaner and more affordable energy goals by 2030.

8.10 Future Directions and the Limitations of the Study

The impact of recent policy undertakings by the Government, such as the Battery Swapping Policy, Ethanol Blending Programme, etc., has not been incorporated in the paper since the policies are still in the implementation phase. The dataset extracted and observations drawn from them are from secondary sources; thus, the inferences may differ from those made over time. With continuous efforts being made to realize the goals of SDG 7, future researchers can depend on the recent policy reports and publications from the governmental agencies to map India's journey toward ensuring sustainable, affordable, and modern energy for all.

References

Arumugam DA (2019) National Institute for Transforming India (NITI) Aayog and achieving gender equality in the sustainable development framework by the year 2030. Think India 22(3):904–911. https://doi.org/10.26643/think-india.v22i3.8427

Asian Development Bank Institute (2018) Green energy finance in India: challenges and solutions. https://www.adb.org/sites/default/files/publication/446536/adbi-wp863.pdf

Central Electricity Authority (CEA) (2022) Power sector at a glance ALL India. Ministry of Power, Government

of India. https://powermin.gov.in/en/content/power-sector-glance-all-india
Department of Atomic Energy (2015) The atomic energy (amendment) bill, 2015. Prsindia. https://prsindia.org/billtrack/the-atomic-energy-amendment-bill-2015
Ernst and Young (2021) Renewable energy country attractiveness index. https://assets.ey.com/content/dam/ey-sites/ey-com/en_gl/topics/power-and-utilities/power-and-utilities-pdf/ey-recai-57-top-40-ladder.pdf
Government of India (2022) Union budget 2022–23. Press Information Bureau https://pib.gov.in/newsite/union-budget2022.aspx
Government of India, Ministry of Power (2022) International cooperation. Bureau of Energy Efficiency. https://beeindia.gov.in/content/international-cooperation
Hajikhani A, Suominen A (2022) Mapping the sustainable development goals (SDGs) in science, technology and innovation: application of machine learning in SDG-oriented artefact detection. Scientometrics 127:6661. https://doi.org/10.1007/s11192-022-04358-x
IEA (2015) World energy outlook 2015. https://www.iea.org/reports/world-energy-outlook-2015
India VNR 2020 (2020) NITI Aayog. https://www.niti.gov.in/node/1242
International Energy Agency (2021) India energy outlook 2021. IEA https://iea.blob.core.windows.net/assets/1de6d91e-e23f-4e02-b1fb-51fdd6283b22/India_Energy_Outlook_2021.pdf
IRENA (2020) Tracking SDG 7: The Energy Progress Report (2020). https://www.irena.org/publications/2020/May/Tracking-SDG7-The-Energy-Progress-Report-2020
IRENA (2022) Tracking SDG 7: the energy progress report (2022). https://www.irena.org/publications/2022/Jun/Tracking-SDG-7-2022
iTech Mission. (2020). SDGs Dashboard: Track, Monitor and Report Data on Global Goals. www.sdgsdashboard.org. https://www.sdgsdashboard.org/
Jain A, Tripathi S, Mani S (2018) Access to clean cooking energy and electricity. CEEW https://www.ceew.in/sites/default/files/CEEW-Access-to-Clean-Cooking-Energy-and-Electricity-11Jan19_0.pdf
Jolly S, Raven R, Robjin H (2012) Upscaling of business model experiments in off-grid PV solar energy in India. Springer Link. https://link.springer.com/article/10.1007/s11625-012-0163-7
Kumar A, Kumar K, Kaushik N, Sharma S, Mishra S (2010) Renewable energy in India: current status and future potentials. Renew Sust Energ Rev 14(8):2434–2442. https://doi.org/10.1016/j.rser.2010.04.003
Ministry of Environment, Forest and Climate Change (2015) India's intended nationally determined contributions. https://pib.gov.in/newsite/PrintRelease.aspx?relid=128403
Ministry of Heavy Industries (2022) (PLI) scheme for advanced chemistry cell (ACC) battery storage. Press Information Bureau. https://heavyindustries.gov.in/sites/default/files/2023-09/MHI_Programme%20Agreement.pdf
Ministry of Heavy Industries & Public Enterprises (2019) Implementation of National Electric Mobility Mission Plan. Press Information Bureau. https://pib.gov.in/newsite/PrintRelease.aspx?relid=191337
Ministry of Heavy Industries & Public Enterprises, Government of India (2020) National Electric Mobility Mission Plan 2020 Vision Statement. https://heavyindustries.gov.in/sites/default/files/2023-07/NEMMP-2020.pdf
Ministry of Petroleum and Natural Gas (2021) Pradhan Mantri Ujjwala Yojana 2.0. PMUY. https://www.pmuy.gov.in/about.html
Ministry of Power (2015) Energy Efficiency. Ministry of Power, Government of India. https://powermin.gov.in/en/content/energy-efficiency
Mordor Intelligence (2022) India electric vehicle market – growth, trends, covid-19 impact, and forecasts (2022–2027). https://www.mordorintelligence.com/industry-reports/india-electric-vehicle-market
NIC (2022) Pradhan Mantri Ujjwala Yojana (PMUY). Presentations. https://presentations.gov.in/logos/pradhan-mantri-ujjwala-yojana-pmuy/
NITI Aayog (2020) SDG India Index. Sdgindiaindex https://sdgindiaindex.niti.gov.in/#/ranking
NITI Aayog (2021) SDG\India Index & Dashboard 2020–2021. https://www.niti.gov.in/writereaddata/files/SDG_3.0_Final_04.03.2021_Web_Spreads.pdf
NITI Aayog (2022a) Battery swapping policy. https://www.niti.gov.in/sites/default/files/2022-04/20220420_Battery_Swapping_Policy_Draft.pdf
NITI Aayog (2022b) Draft battery swapping policy.https://pib.gov.in/PressReleasePage.aspx?PRID=1818569
NITI Aayog, Ministry of Petroleum and Natural Gas (2021) Roadmap for Ethanol Blending in India 2020–2025. NITI Aayog. https://www.niti.gov.in/sites/default/files/2021-06/EthanolBlendingInIndia_compressed.pdf
Planning Commission (Secretariat for Infrastructure) (2008) Definition of infrastructure. NITI Aayog. https://niti.gov.in/planningcommission.gov.in/docs/sectors/ppp_report/reports_guidelines/Definition%20of%20Infrastructure.pdf
PMY Team (2022) Fame India scheme. PM Modi Yojana. https://pmmodiyojana.in/fame-india-scheme/
Pradhan Mantri Sahaj Bijli Har Ghar Yojana – Saubhagya (2022) Pradhan Mantri Sahaj Bijli Har Ghar Yojana – Saubhagya. National Portal of India. https://www.india.gov.in/spotlight/pradhan-mantri-sahaj-bijli-har-ghar-yojana-saubhagya
Ramchandran N, Pai R, Parihar AKS (2016) Feasibility assessment of Anchor-Business-Community model for off-grid rural electrification in India. Renew Energy 97:197–209. https://doi.org/10.1016/j.renene.2016.05.036

REN21 (2020) Renewables 2020 global status report. https://www.ren21.net/wp-content/uploads/2019/05/gsr_2020_full_report_en.pdf

Saubhagya Dashboard (2019) Ministry of Power. https://saubhagya.gov.in/

Smith K, Liu Y, Liu S (2019) Urban drinking water challenges and solutions: energy nexus. In: Köster S, Reese M, Zuo J (eds) Urban water management for future cities, Future city, vol 12. Springer, Cham. https://doi.org/10.1007/978-3-030-01488-9_4

United Nations Statistics Division (UNSD) (2022) Affordable and clean energy. Unstats. https://unstats.un.org/sdgs/report/2016/goal-07/

Anurag Piyamrao Wasnik is interested in working on the intersections of sustainability, innovation, and entrepreneurship with a focus on public policy issues. He is currently working with NITI Aayog, Atal Innovation Mission (Public Policy Think-Tank of Govt. of India, erstwhile Planning Commission), leading multiple strategic initiatives vital to the Innovation & Entrepreneurship Ecosystem building in India. He has completed his M.Sc. in International Management from Bocconi Universita, Italy and MBA from IIM Ahmedabad.

Fuelling Women Lives with LPG: A Sociological Study of Fuel Use Pattern Amidst Policy Intervention in Peri-Urban Areas of Uttar Pradesh

9

Manish K. Verma and Moni Chandra

Abstract

Access to clean cooking fuels is essential as cooking food with biomass not only accelerates air pollution but has negative consequences for public health and the environment. Its importance can be ascertained from the fact that it is one of the SDGs that pertains to clean and affordable energy to all by 2030. This article attempts to study the fuel use pattern in peri-urban areas of selected cities of Uttar Pradesh, where despite availability of clean fuel, majority of households still use biomass. Though with the launch of Ujjwala Yojana, the use of LPG has increased but the persistence of solid biomass fuel is still there. Using descriptive statistics, the paper examines the underlying factors that hinder usage of LPG on a sustained basis. The findings prove that socio-economic and cultural factors play an important role in the adoption of clean fuel i.e., LPG. The study suggests that apart from expanding access and improving economic conditions, future research and policy should also consider the socio-cultural norms and attitudes that discourage the use of LPG and work on changing the perception of people towards the ill-effects of biomass.

Keywords

Fuel · Biomass · LPG · Women · Peri-urban · SDG

M. K. Verma (✉)
Department of Sociology, Ambedkar School for Social Sciences, Babasaheb Bhimrao Ambedkar University, Lucknow, Uttar Pradesh, India
e-mail: mkvbbau@gmail.com

M. Chandra
Ambedkar School for Social Sciences, Babasaheb Bhimrao Ambedkar University, Lucknow, Uttar Pradesh, India
e-mail: chandramonee@gmail.com

9.1 Introduction

Access to clean fuels for cooking is unevenly distributed across the globe. Although the SDG7 ensures access to affordable, reliable, sustainable and modern energy, still, one third of the global population or 2.4 billion people worldwide remain without access to clean cooking. Moreover, cooking with polluting fuels is a major contributor to diseases and deaths, particularly for women and children land and low quality of life in most of the developing countries (WHO 2022). Among developing countries, India is a major contributor to global biomass use. Though India's population is about one-seventh of the world's population but there are about one-third of global biomass users (WHO 2016), particularly in rural areas. It doesn't need to be reiterated that biomass use for cooking adds to air pollution and illnesses in developing countries (Smith

V. Dutta, P. Ghosh (eds.), *Sustainability: Science, Policy, and Practice in India*, Sustainable Development Goals Series, https://doi.org/10.1007/978-3-031-50132-6_9

2000; Salvi and Barnes 2009). The burning of biomass also releases carbon and other greenhouse gases in the environment that contributes to air pollution and is a significant factor of climate change (Wilkinson et al. 2009), forest degradation and depletion of natural resources (Agarwalla and Saha 2021). Besides, cooking on biomass not only exposes women to smoke-filled environment but is a potent cause of direct and indirect health issues (Karimu 2015; Desai et al. 2004). Even simultaneous use of traditional biomass with clean fuels hinders in controlling household air pollution. Due to the harmful effects that biomass has on human health and the environment, the Sustainable Development Goals call for universal access to clean cooking fuels by 2030. Notably, despite policy interventions, biomass use continues in many developing countries in the world (Masera et al. 2000; Gould and Urpelainen 2018; Mani et al. 2020).

Though the launch of Ujjwala Yojana, which aimed at providing 8 Crore LPG connections in the name of women in BPL (Below Poverty Line) households across the country, has considerably expanded access to clean fuels, the transition to clean cooking fuels is still at a slow pace. Still, many households use alternative cooking fuels like wood, cow dung and agri-residue. Undoubtedly, LPG connections have increased, but a large number of *Pradhan Mantri Ujjwala Yojana* (PMUY) beneficiaries are not refilling cylinders on a sustained basis. This questions the sustainability of PMUY. Jain and other scholars (2018) argued that though a hike in LPG ownership occurred between 2015 and 2018, rise in fuel stacking and the usage of multiple cooking fuels was also noticed. According to Kar et al. (2019), a study of LPG sales data from one district in Karnataka showed that the rapid growth in connections as a result of Ujjwala was not complemented by an increase in LPG use. This irregularity in the use of LPG (clean fuel) is due to the multiple factors operating simultaneously. Heltberg et al. (2000) pointed out that the use of clean fuels is determined by the household's socio-economic and demographic attributes. Besides, factors involved in clean fuel transition like accessibility and affordability of the households (Pandey and Chaubal 2011; Behera et al. 2015; Karimu et al. 2016) and awareness regarding clean fuels (Kumar et al. 2016) and household (Leach 1992) also play an important role in clean fuel transition.

In rural north India, women are primarily responsible for cooking and preparing biomass (Sharma 2018; Hirway and Jose 2011), and the decision of refilling the cylinders lies on the men who are the main breadwinners and financial decision-makers in the household (Palriwala 1993). Gould and Urpelainen (2019) and Choudhuri and Desai (2020) argued that households where women equally participate in decision-making are more likely to adopt LPG. Women adding income to the household and are earning is also associated with the use of clean fuels (Choudhuri and Desai 2020). Other studies mentioned that in South Asia where women are head of the households are more likely to use LPG (Das et al. 2014; Behera et al. 2015; Mottaleb et al. 2017). The household size also governs fuel requirement through the actual demand and available persons for collection. Similarly, education is also a strong influencer in fuel choices (Islam 2014; Heltberg 2004; Behera et al. 2015; Rahut et al. 2016; and Dash et al. 2018) as the household income and wealth are dependent on it. According to Lewis and Pattanayak (2012), clean cooking fuels are positively associated with education level. Also, higher education of the head of the household aids in greater adoption of LPG (Lewis and Pattanayak 2012; Swain and Mishra 2019). Pandey and Chaubal (2011) pointed out household size, size of the land and possession of BPL ration card as the governing factors of clean fuel choice. In other words, better education aids in better income and better income aids in usage of better fuel and better kitchen appliances that check household dependency on the forest. Reddy and Srinivas (2009) also reiterated that there is a direct correlation between level of income and standard of living and education. The poverty-stricken households or economically poor households are expected to rely more on the basic fuels like firewood, crop residues and dung (Gupta and Köhlin 2006; Gundimeda and Köhlin

2008). Further, the use of clean fuels is more in urban households in comparison to rural households due to its better accessibility (Lewis and Pattanayak 2012; Rahut et al. 2014). Unlike the urban areas with less access to biomass, fuel switching uptake in the peri-urban areas, especially in less developed districts, is challenging due to the high cost (price) of refilling and availability of biomass (Lewis and Pattanayak 2012). Wambua (2011) is of the view that low per capita income and the price of fuel also play an essential role in clean fuel use. As such, the poor households depend more on the forest and cow dung for their cooking needs than better off households. Moreover, households with farm holdings tend to rely more on crop residue produced from their farms. This echoes that income level is an important factor that is responsible for the dependence of biomass and a barrier in the use of clean fuel, but not the only determining factor for the type and level of fuel consumption. Kapur (2020) and Dr'eze and Sen (2013) are of the view that gender norms, attitudes and beliefs too affect the penetration of public policy. At present, majority of the Indian households have the infrastructure to cook with LPG, but simultaneous use of biomass unveils the importance of gender norms and attitudes in cooking practices. The norms and attitudes of our patriarchal society encourage the use of biomass and preserve LPG in the region. The traditional families restrain women from working outside their home. The prevalent dogmas encourage women to preserve LPG, support women's work that facilitates the use of biomass and impede the decision-making regarding LPG refills. Besides, the cooking practices of households, the taste preferences, the cooking location, food habits and cooking appliances also determine fuel choice pattern, and they are all culturally determined. Further, in some households where LPG is used predominantly, it is done to facilitate the adherence to norms of seclusion that prevent women from leaving the home to collect biomass (Vyas et al. 2020). While in many households, spending money to refill the cylinder was not seen as a priority as expenses related to education of children, medicines and replenishing household food stocks always take precedence over refilling the cylinder. Also, the households practicing animal husbandry and farming and residing near the agricultural fields and jungle use firewood, cow dung and crop residue as alternate fuels to save LPG. Therefore, animal husbandry and farming facilitate the continued use of the chulha, because they produce waste that can be used as biomass. Animal husbandry is that economic activity in which women may acceptably engage. According to Mandelbaum (1988) and Jeffery et al. (1989), in rural north India, a daughter-in-law, who generally cooks in the household, is expected to behave with respect before her father-in-law and other elder members in the family, who generally decide when to refill the cylinder. This household hierarchy hinders communication between a daughter-in-law and the decision-maker and slows the process of getting an LPG refill. A cook who is the wife of the household head, on the other hand, does not face the same restrictions. The patriarchal norms and attitudes promote women's work that facilitates the use of biomass and impedes communication between the person especially women and the decision-maker (mainly the male members or the breadwinner of the family) regarding LPG refills. This sheds light on the fact that in addition to low income, the sociocultural norms and attitudes related to women's status also play a key role as to why the use of biomass still persists despite the availability of clean fuel. Thus, the transition, from biomass to clean fuel, depends upon location (Damette et al. 2018) like urban, rural, peri-urban and household attributes (size of family, consumption level, education and income of the family). Notably, Ujjwala Yojana contributed to a rapid change in the cooking fuel of Indian households. However, most of the literature on usage of cooking fuel post-Ujjwala focuses documented the continued use of biomass among LPG owners in rural and urban areas, and there is a dearth of studies on peri-urban areas. Hence, the present study is undertaken to find out why biomass is still used in peri-urban areas despite having an LPG connection in the household; as without sustained usage of clean cooking fuels, the SDG 7 (clean and affordable energy to all by 2030 is impossible to achieve) is also essential for achieving the SDGs, particularly in developing countries. The peri-urban approach aids in differentiating the diversity in fuel use practices but also the rate and direction of change in fuel use between the traditional rural places and the more modern rural places as well as between the modern urban places and the less modern urban places (Saksena et al. 2014).

9.2 Methods

The present study is descriptive in nature and uses both primary and secondary data. The secondary data was compiled from the Census of India (2011). Necessary primary data and information were collected with the help of survey method from the beneficiary households of PMUY. Overall, 300 beneficiary's households were surveyed from peri-urban areas of 2 selected districts of UP, namely, Hardoi and Lucknow. The districts were selected on the basis of highest and lowest LPG users in the central region of UP. Lucknow has the highest LPG connections in the central region of UP, while Hardoi occupies the last spot. In the second stage, two blocks have been chosen on geographical location criteria from each district due to the unavailability of data of Ujjwala beneficiaries at block level. Pihani and Sandila blocks were selected from Hardoi district, while Bakshi ka Talab and Mohanlalganj were selected from Lucknow district. From each block of peri-urban area, 75 households were selected. The interview questions were analysed using descriptive statistical techniques with the help of SPSS.

9.3 Results

9.3.1 Socio-economic Background of the Beneficiary Households

Table 9.1 shows the socio-economic characteristics of beneficiary households of Hardoi and Lucknow districts. Age is one of the basic demographic features that play an important role in ascertaining the views of different age groups about particular problems in social research. In both the districts under study, majority of the respondents fall in the age group of 31–50 years. The average size of the household stands at five persons. If district-wise data is checked, it comes to around five and four in Hardoi and Lucknow, respectively. It is obvious from the table that out of the total sample households in both the districts, 65% of the households were found residing in 'pucca' houses, while only 2% were residing in 'katcha' houses. However, out of the total sample households from Lucknow district, only 0.7% was 'katcha' as compared to Hardoi's 3.3%. None of the respondents is found residing in any other type of house structure/residence. It shows the economic progress of the people of the study area. Semi-pucca has the second-highest percentage (33%), while 'katcha' house structure has lowest per cent (only 2%). The importance of pucca house can be ascertained with the fact that without access to the pucca house, it is difficult to ensure the availability and accessibility of safe drinking water, sanitation facility and cooking fuel within the premise. Mere providing access to basic amenities is not sufficient unless the sustainability of the services is ensured. Caste is one of the most important social categories in Indian society. Dash et al. (2018) considered caste category as an important determinant of energy choice. More than half of the respondents belonged to OBC (57%), followed by the SC (31.7%) and general caste (11%). The respondents from ST category are just 1(0.3%) in a sample of 300, i.e. almost negligible. In Hardoi district, around 51% of the respondents were from OBC, while from Lucknow, 63.3% of the respondents belonged to this category. The table illustrates that around 41% of the respondents belonged to the middle level (class 8th) and 25% were having education of secondary level (class 10th). Notably, only 5% of the total respondents under study were graduates, and 11% of the total sampled households were found illiterate. The maximum qualification of the majority of respondents from both the blocks of the two districts under study is up to middle level (class 8th). Rahut et al. (2016) considered occupation an important determinant of fuel choice. It deter-

Table 9.1 Socio-economic background of the study sampled households

S. no.	Socio-economic variables		Hardoi	Lucknow	Total
1.	**Age group**	**20–30 years**	64	60	124
			(42.3%)	(40.4%)	(41.3%)
		31–50 years	84	86	170
			(56.4)	(57%)	(56.7%)
		Above 50 years	2	4	6
			(1.3)	(2.6)	(2.0)
2.	**Average household size**		5.05	4.25	4.65
3.	**Type of house structure**	**Kutcha**	5	1	6
			(3.3)	(0.7)	(2.0)
		Semi-Pucca	59	40	99
			(39.3)	(26.7)	(33.0)
		Pucca	86	109	195
			(57.3)	(72.7)	(65.0)
4.	**Caste**	**SC**	52	43	95
			(34.7)	(28.7)	(31.7)
		ST	1	–	1
			(0.7)	–	(0.3)
		OBC	76	95	171
			(50.7)	(63.3)	(57.0)
		GEN	21	12	33
			(14.0)	(8.0)	(11.0)
5.	**Education**	**Illiterate**	14	19	33
			(9.33)	(12.67)	(11.00)
		Literate	13	11	24
			(8.67)	(7.33)	(8.00)
		Primary	10	21	31
			(6.67)	(14.00)	(10.33)
		Middle (class 8th)	55	67	122
			(36.67)	(44.67)	(40.67)
		Secondary (class 10th)	48	27	75
			(32.00)	(18.00)	(25.00)
		Graduate	10	5	15
			(6.67)	(3.33)	(5.00)
6.	**Occupation**	**Private job**	10	26	36
			(6.71)	(17.22)	(12.00)
		Business/shop	37	30	67
			(24.83)	(19.87)	(22.33)
		Labour work	71	86	157
			(47.65)	(56.95)	(52.33)
		Farming	4	2	6
			(2.68)	(1.32)	(2.00)
		Street vendors	27	7	34
			(18.12)	(4.64)	(11.33)
7.	**Income**	**Average household income (in Rs.) (monthly)**	7067	9253	8160
		Per day household income (in Rs.)	236	308	272
		Per person Household income (in Rs.) (monthly)	1398	2176	1754

Source: Based on primary data collected from the field

mines the work pattern of a person. The inclination of households towards traditional fuels is more when farming is their primary occupation as farming provides crop residue and other biomass (Swain and Mishra 2019). In both the districts under study, it was found that half of the total respondents (52.3%) were involved in labour work (construction labourers, mechanics, carpenters, washer men and sweepers). Only 12% of the total respondents were doing private job. The private job here includes manager, accountant, guard, peon, teacher, compounder and technician. It is interesting to note that though living near the farms, only 2% of the total respondents from both the districts were involved in farming activity, and it's almost negligible. A little more than 11% depicts street vendors (fast food, fruits and vegetable sellers). Income of a person plays an important role in shaping the economic condition of an individual. Looking district wise, it is found that the average income of the beneficiary households in Hardoi stands at Rs. 7067, while that of Lucknow is Rs. 9253. However, the average income of the sampled households is Rs. 8160. This reflects that it is poverty that prompts them to use biomass along with LPG.

9.4 Status and Pattern of Usage of Cooking Fuel

Table 9.2 gives information regarding the pattern of usage of wood for different purposes in the total sampled households in both the districts under study. It is found that 9 and 12.7% of the households still use wood for cooking and boiling water, respectively, in both the districts (combined). Looking district wise, it is noticed that 34% of the sampled households in Hardoi district still cook food on wood (18% on a daily basis) despite having LPG, while 22.7 use it for boiling water. While in Lucknow, 5.3% often use wood for cooking but not on a daily basis. Thus, the picture is far much better in the capital city of Lucknow where more than 90% of the sampled households don't use wood, both for cooking and boiling purposes.

Table 9.3 gives information regarding the pattern of usage of agri/crop residue for different purposes in the total sampled households in both the districts under study. It is found that 12.6% and 11% of the households still use agri/crop residue for cooking and boiling water in both the districts. Looking district wise, it is noticed that 20.7% of the sampled households in Hardoi district still cook food on agri/crop residue on a daily basis despite having LPG, while 18% use it for boiling water. While in Lucknow, 4.7% use agri/crop residue for cooking but not on daily basis. The picture is far much better in the capital city of Lucknow where more than 95% of the sampled households don't use agri/crop residue, both for cooking and boiling purposes.

Table 9.4 gives statistics regarding the pattern of usage of cow dung for different purposes in the total sampled households in both the districts under study. It is found that 24% and 4.7% of the households still use cow dung for cooking and boiling water in both the districts. Looking district wise, it is noticed that 42% of the sampled households in Hardoi district still cook food on cow dung despite having LPG, while 6.7% use it for boiling water. While in Lucknow, 6% use cow

Table 9.2 Pattern of wood usage

	Cooking food			Boiling water			
District	Daily	Often	Never	Daily	Often	Never	No. of HH
Hardoi	27	24	99	34	4	112	150
	(18.0)	(16.0)	(66.0)	(22.7)	(2.7)	(74.7)	(100.0)
Lucknow	–	8	142	4	1	145	150
		(5.3)	(94.7)	(2.7)	(0.7)	(96.7)	(100.0)
Total	27	32	241	38	5	257	300
	(9.0)	(10.7)	(80.3)	(12.7)	(1.7)	(85.7)	(100.0)

Source: Based on primary data collected from the field

Table 9.3 Pattern of crop residue usage

	Cooking food			Boiling water			
District	Daily	Often	Never	Daily	Often	Never	No of HH
Hardoi	28	3	119	26	1	123	150
	(18.7)	(2.0)	(79.3)	(17.3)	(0.7)	(82.0)	(100.0)
Lucknow		7	143		6	144	150
		(4.7)	(95.3)		(4.0)	(96.0)	(100.0)
Total	28	10	262	26	7	267	300
	(9.3)	(3.3)	(87.3)	(8.7)	(2.3)	(89.0)	(100.0)

Source: Based on primary data collected from the field

Table 9.4 Pattern of cow dung usage

	Cooking food			Boiling water			
District	Daily	Often	Never	Daily	Often	Never	No. of HH
Hardoi	27	24	99	34	4	112	150
	(18.0)	(16.0)	(66.0)	(22.7)	(2.7)	(74.7)	(100.0)
Lucknow		8	142	4	1	145	150
		(5.3)	(94.7)	(2.7)	(0.7)	(96.7)	(100.0)
Total	27	32	241	38	5	257	300
	(9.0)	(10.7)	(80.3)	(12.7)	(1.7)	(85.7)	(100.0)

Source: Based on primary data collected from the field

Table 9.5 Pattern of LPG usage

	Cooking food		Boiling water				
District	Daily	Often	Daily	Often	Occasionally	Never	No. of HH
Hardoi	133	17	75	1	1	73	150
	(88.7)	(11.3)	(50.0)	(0.7)	(0.7)	(48.7)	(100.0)
Lucknow	139	11	138	1		11	150
	(92.7)	(7.3)	(92.0)	(0.7)		(7.3)	(100.0)
Total	272	28	213	2	1	84	300
	(90.7)	(9.3)	(71.0)	(0.7)	(0.3)	(28.0)	(100.0)

Source: Estimated from primary data

dung for cooking but not on daily basis. Thus, as compared to Hardoi, the usage of clean fuel is far much better in the capital city of Lucknow, where more than 90% of the sampled households don't use cow dung, both for cooking and boiling purposes.

Table 9.5 gives statistics regarding the pattern of usage of **LPG** for different purposes in the total sampled households in both the districts under study. It is found that around 90% of the households use LPG for cooking on a daily basis in both the districts. Looking district wise, it is noticed that in Hardoi district, around 11% still don't use LPG for cooking on a daily basis, while in Lucknow, only 7.3% don't use it daily. As far as boiling of water is concerned, 71% of sampled households use LPG on a daily basis. It is interesting to note that 48.7% never boil water on LPG, while in Lucknow, 72% use it for boiling also apart from cooking.

9.5 Discussion

The present study examines the use of LPG for cooking under the PMUY and identifies the underlying factors behind the simultaneous use of biomass despite having LPG connection. The findings make it clear that biomass use has not stopped even after the launch of Ujjwala Yojana, and beneficiary households continue to use biomass on regular basis. No doubt, there is considerable increase in number of LPG connection under PMUY, but the households continue to use wood, cow dung and crop residue because of their easy access and cheaper price. Both economic and non-economic factors play a role in explaining irregular use of LPG and use of biomass that exists. The results show that the usage of clean fuel is above 90% in peri-urban blocks of the capital city as compared to Hardoi district. For, e.g. in peri-urban households in Pihani block of Hardoi district, people living next to the forest (Tarai belt) use firewood as the prime alternative biomass fuel with LPG. Also, fuel stacking is present, and the households have not totally abandoned the use of firewood as they adopt other fuels, especially LPG. The finding is supported by Swain and Mishra (2019) that the probability of decline in the use of firewood is less for the households that are benefitted by the PMUY. This may be largely due to the lack of affordability to refill LPG cylinder by the PMUY beneficiaries as they are poor and get LPG connection by spending less. Due to this reason, the level of LPG use is found low on average than it would have been if LPG were used all the time. The respondents stated that they used both LPG and chulha and that they use LPG only in exceptional circumstances like when it was raining or if the biomass is wet. Besides, most cooking takes place outdoors in the warm season and in enclosed or covered kitchen areas in the wet season. One of the respondents told that for cooking small things like chai, they use LPG. The reliance on biomass such as firewood is also found in the households that can afford clean fuel; however, it is prevalent more among the poor. Moreover, as the beneficiaries of Ujjwala are from the BPL category, the poverty level (that had increased due to COVID-19) decreases the probability of households to meet clean energy use (LPG) regularly. Therefore, even after getting an LPG connection from Ujjwala, the households depend on biomass like cow dung and firewood. In Hardoi district, it is found that PMUY beneficiaries are not found refilling cylinders on a regular basis. Those households, who have some agricultural land, rely on their own produced biomass (crop residue); as such, their reliance on LPG is less. Traditional cooking practices with firewood, cow dung and crop residue still persist simultaneously along with LPG, and its use is seen more in Hardoi as compared to Lucknow. Those households, who did not have animals that produced dung and farmland land that produced agricultural residue, had to rely on buying dung cakes from the neighbourhood. This shows that majority of the households especially in Hardoi district still depend upon biomass and that modern fuels thus have failed to displace traditional fuelwood completely. Notably, the pattern of LPG usage is good in peri-urban blocks of Lucknow. The data on Ujjwala users of both the districts signify that the probability of LPG (clean fuels) usage increases with increasing household income (Lewis and Pattanayak 2012; Behera et al. 2015), and low household income is positively related to the purchase and use of biomass fuel. However, this finding is contradictory to Swain and Mishra (2019) finding that there is no significant influence of household income on the adoption of LPG. The per capita income of Hardoi is low as compared to Lucknow; as such, the poor households depend more on the forest and cow dung for their cooking needs than better off households. As such, complete transition from biomass to LPG is still slow there.

The study also found that the educational level of the head of the household and of females also is one of the potent reasons beside household income that prompts beneficiaries for sustained usage of traditional fuel in spite of having an LPG connection. Education is a potent driver of growth and development that aids in reducing poverty and better health. It changes the attitudes and the perceptions of the people regarding any particular social phenomena. The way a person responds is mostly determined by his educational status. It is hypothesized that educated household heads are more concerned about the ill effects of traditional fuels and the benefits of clean fuels. Such households would prefer lesser use of firewood (Swain and Mishra 2019). According to Mekonnen and Köhlin (2009), the education level of the household head increases working opportunities and household income, leading to less dependence on biomass fuels. As such, it is imperative to know the educational background of the sampled households. Also, it is the second most important variable after household income in fuel choice in peri-urban areas under study, since non-beneficiaries with better income are using LPG despite low educational level as compared to beneficiaries. It is found that more than 40.7% of the total sampled respondents have education up to class eighth. With this level of education, one can't expect better awareness regarding the ill effects of biomass. Swain and Mishra (2019) also echoed that the education level of the household head emerges as an important factor that can motivate the households towards less use of firewood and more of LPG for cooking. It is argued that the educated household head would be more concerned about environmental- and health-related adverse impact of the traditional fuels and benefits of cleaner fuel like LPG. Heltberg (2004), Rahut et al. (2014), Behera et al. (2015) and Dash et al. (2018) pointed out that educated household heads prefer to use cleaner cooking fuels for checking health hazards and reducing opportunity costs of time associated with collection of firewood. The household fuel consumption patterns depict that in peri-urban households, LPG is used, but its use is not regular and on a sustained basis. This provides valuable insight in distinguishing between different conceptual models of the energy transition process. The data from the study area confirmed that energy ladder model is a more accurate description of household energy pattern.

The empirical results show that household head's age, educational attainment, household size, income, type of dwelling unit, the duration of food cooked and price of fuelwood are factors influencing households' choice of cooking fuel. Besides, the sociocultural norms, attitude of the people and certain myths that food cooked on LPG are not tasty attempt to scuttle the programme. However, the low level of income is a major hurdle in the regular use of LPG thereby impeding the effective implementation of the programme. Lastly, many respondents also opined that LPG connection, availed through PMUY, has benefitted the women in terms of cooking food on LPG and that they are engaged in part-time income-generating activities like carrying out sewing works, *papad* and *achaar* making and helping in their husband's fast-food work. This is so because, with LPG connection, women are not regularly required to cook food using biomass. These activities are gradually making them socially and economically empowered.

9.6 Conclusion and Policy Implications

The paper delves upon factors such as income, education and occupation affecting fuel usage in peri-urban areas of UP. As the household's income rise, there is a higher likelihood of modern (LPG) fuel adoption; however, households have not stopped using traditional fuels rather they are following a multiple fuel use approach. The study highlights that economic reasons are not the only determining factors for fuel choice and that non-economic factors (taste, cultural norms, behaviour pattern, etc.) equally affect fuel choice. For that, understanding households' perception is essential as collection (availability), and the use of biomass is not restricted. This limits the use of other cleaner fuels like LPG. The findings reveal that PMUY had led to an increase in usage of clean fuel, i.e. LPG, but, for sustained use of clean fuels (LPG), some long-term strategies are required to make it accessible economically like creating opportunities of job for women that reduce their financial dependence. Besides, greater subsidy on LPG makes its relatively cheaper and induces households to use it more regularly. So, greater subsidy and not simply raising household income also incentivize the beneficiaries to use LPG regularly. Apart from improving economic conditions, the trust of future policy studies should be on enhancing the level of education, as only education can change the stereotyped approach of the people. The households having educated heads, the probability of using biomass is less and vice versa. Though it is difficult to modify the household cooking practices, raising the level of education and better income opportunities will be worthwhile to change the people's perception regarding biofuels. As the use of biomass have multiple social, health and environmental concerns, it's necessary that people should be apprised with the ill effects of biomass, so that they may move towards the use of more sustainable energy alternatives. Hence, it is recommended that more studies on peri-urban areas should be done separately as their socio-economic and cultural practices are different from rural and urban areas. This will aid in understanding the attitudes of the people residing in peri-urban areas, and the policies and institutions can be fine-tuned accordingly. Last but not the least, simply formulating policy and leaving it in the hands of gas agencies to execute does not end the responsibility, rather the government needs to evaluate and re-evaluate the existing problems. Media reporting, surprise inspection of the homes using wood and timely supply of LPG should be strictly monitored.

References

Agarwalla S, Saha D (2021) An empirical study on attitudes towards willingness to pay for forest ecosystem services at Simlipal National park. Indian J Ecol 48(3):634-640

Behera B, Rahut D, Jeetendra A, Ali A (2015) Household collection and use of biomass energy sources in South Asia. Energy 85:468e480

Census (2011) Primary Census Abstract 2011, Registrar General & Census Commissioner, India. Ministry of Home Affairs, Government of India, accessed through www.censusindia.gov.in

Choudhuri P, Desai S (2020) Gender inequalities and household Fuel choice in India. J Clean Prod 265:121487. https://doi.org/10.1016/j.jclepro.2020.121487

Damette O, Delacote P, Del Lo G (2018) Households energy consumption and transition toward cleaner energy sources. Energy Policy 113:751e764

Das S, De Groote H, Behera B et al (2014) Determinants of household energy use in Bhutan. Energy 69:661–672

Dash M, Behera B, Rahut DB (2018) Understanding the factors that influence household use of clean energy in the Similipal Tiger Reserve, India. Nat Res Forum 42(1):3e18

Desai MA, Mehta S, Smith KR, World Health Organization (2004) Indoor smoke from solid fuels: assessing the environmental Burden of disease at national and local levels. World Health Organization, Geneva

Dr'eze J, Sen A (2013) An uncertain glory: India and its contradictions. Princeton University Press, Princeton

Gould CF, Urpelainen J (2018) LPG as a clean cooking fuel: adoption, use, and impact in rural India. Energy Policy 122:395–408

Gould CF, Urpelainen J (2019) The gendered nature of liquefied petroleum LPG stove adoption and use in rural India. J Dev Stud:1–21 https://doi.org/10.1080/00220388.2019.1657571

Gundimeda H, Köhlin G (2008) Fuel demand elasticity for Energy and environmental policies: Indian sample survey evidence. Energy Econ 30(2):517–546

Gupta G, Köhlin G (2006) Preferences for domestic Fuel: analysis with socio-economic factors and rankings in Kolkatta, India. Ecol Econ 57:107–121

Heltberg R (2004) Fuel switching: evidence from eight developing countries. Energy Econ 26(5):869e887

Heltberg R, Arndt TC, Sekhar NU (2000) Fuelwood consumption and forest degradation: a household model for domestic energy substitution in rural India. Land Econ 76(2):213–232

Hirway I, Jose S (2011) Understanding women's work using time-use statistics: the case of India. Fem Econ 17(4):67–92

Islam T (2014) Household-level innovation diffusion model of photo-voltaic (PV) solar cells from stated preference data. Energy Policy 65:340e350

Jain A, Tripathi S, Mani S, Patnaik S, Shahidi T, Ganesan K (2018) Access to clean cooking energy and electricity: survey of states 2018, CEEW Report, Council on Energy, Environment and Water (CEEW), New Delhi, India

Jeffery P, Jeffery R, Lyon A (1989) Labour pains and labour power: women and childbearing in India. Zed Books, London

Kapur D (2020) Why does the Indian state both fail and succeed? J Econ Perspect 34(1):31–54

Kar A, Pachauri S, Bailis R, Zerriffi H (2019) Using sales data to assess cooking LPG adoption and the impact of India's Ujjwala programme in rural Karnataka. Nat Energy 4(9):806–814

Karimu A (2015) Cooking fuel preferences among Ghanaian households: an empirical analysis. Energy Sustain Dev 27:10e17

Karimu A, Mensah JT, Adu G (2016) Who adopts LPG as the main cooking fuel and why? Empirical evidence on Ghana based on national survey. World Dev 85:43e57

Kumar P, Rao RK, Reddy NH (2016) Sustained uptake of LPG as cleaner cooking fuel in rural India: role of affordability, accessibility, and awareness. World Dev Perspect 4:33e37

Leach G (1992) The energy transition. Energ Policy 20(2):116–123. https://doi.org/10.1016/0301-4215(92)90105-B

Lewis JJ, Pattanayak SK (2012) Who adopts improved fuels and cook stoves? A systematic review. Environ Health Perspect 120(5):637e645

Mandelbaum DG (1988) Women's seclusion and men's honor: sex roles in North India, Bangladesh, and Pakistan. University of Arizona Press

Mani S, Jain A, Tripathi S, Gould CF (2020) The drivers of sustained use of liquified petroleum LPG in India. Nat Energy 5(6):450–457

Masera OR, Saatkamp BD, Kammen DM (2000) From linear fuel switching to multiple cooking strategies: a critique and alternative to the energy ladder model. World Dev 28(12):2083–2103

Mekonnen A, Köhlin G (2009) Determinants of household fuel choice in major cities in Ethiopia, Working Papers in Economics No 399. University of Gothenberg, Sweden. Accessed through http://www.rff.org/files/sharepoint/WorkImages/Download/EfD-DP-08-18.pdf

Mottaleb KA, Ali A et al (2017) An exploration into the household energy choice and expenditure in Bangladesh. Energy 135:767–776

Palriwala R (1993) Economics and patriliny: consumption and authority within the household. Soc Sci:47–73

Pandey VL, Chaubal A (2011) Comprehending household cooking energy choice in rural India. Biomass Bioenergy 35(11):4724e4731

Rahut DB, Behera B, Ali A (2016) Patterns and determinants of household use of fuels for cooking: empirical evidence from sub-Saharan Africa. Energy 117:93e104

Rahut DB, Das S, De Groote H, Behera B (2014) Determinants of household energy use in Bhutan. Energy 69:661e672

Reddy BS, Srinivas T (2009) Energy use in Indian household sector-an actor-oriented approach. Energy 34(8):992–1002. https://doi.org/10.1016/j.energy.2009.01.004

Saksena S, Fox J, Spencer J, Castrence M, diGregorio M, Epprecht M, Sultana N, Finucane M, Nguyen L, Vien TD (2014) Classifying and mapping the urban transition in Vietnam. Appl Geogr 50:80–89

Salvi SS, Barnes PJ (2009) Chronic obstructive pulmonary disease in non-smokers. Lancet 374(9691):733–743

Sharma K (2018) Living with pain: women's everyday lives and health in rural Bihar, The Hindu Center for Public Policy, Policy Report No. 23

Smith KR (2000) National burden of disease in India from indoor air pollution. Proc Natl Acad Sci 97(24):13286–13293

Swain SS, Mishra P (2019) Determinants of adoption of cleaner cooking energy: experience of the Pradhan Mantri Ujjwala Yojana in rural Odisha, India. J Clean Prod. https://doi.org/10.1016/j.jclepro.2019.119223

Vyas S, Gupta A, Khalid N (2020) Gender and LPG use after government intervention in rural North India. SocArXiv. https://doi.org/10.31235/osf.io/3v4cj

Wambua SM (2011). Household energy consumption and dependency on common pool forest resources: the case of Kakamega Forest, Western Kenya, Dissertation to obtain the Ph.D. Degree in the international Ph.D. Program for agricultural sciences in Goettingen (IPAG) at the faculty of agricultural sciences, Georg-August-University Göttingen, Germany

Wilkinson P, Smith KR, Davies M, Adair H, Armstrong BG, Barrett M, Bruce N, Haines A, Hamilton I, Oreszczyn T (2009) Public health benefits of strategies to reduce greenhouse-LPG emissions: household energy. Lancet 374(9705):1917–1929

World Health Organization (2016) Burning opportunity: clean household Energy for health, sustainable development, and wellbeing of women and children

WHO (2022) WHO publishes new global data on the use of clean and polluting fuels for cooking by fuel type

Manish K. Verma is a Professor in the Department of Sociology at Babasaheb Bhimrao Ambedkar (Central) University, Lucknow. Earlier, he served as Head and Deputy Coordinator of UGC SAP DRS-I in the Department of Sociology and Coordinator Sports of the University. Having obtained a doctorate degree from Jawaharlal Nehru University, New Delhi, he has more than 22 years of teaching and research experience. During his two decades long experience in academic arena at various positions, he has published many books which includes *Globalisation, Environment and Social Justice: Perspectives, Issues and Concerns* (2018; Routledge, Taylor & Francis Group, UK), *Globalization, Social Justice and Sustainable Development in India* (2017; Gyan Publishing House, New Delhi), *Peri-Urban Environment* (2017; Winshield Press, Delhi), *Globalization and Environment: Discourse, Policies and Practices* (2015; Rawat Publications, Jaipur and New Delhi), and *Development, Displacement and Resettlement* (2004; Rawat Publications, Jaipur and New Delhi). His most recent book, *Environment and Sustainable Development: Perspectives and Issues*, is under publication from Routledge, Taylor and Francis Group, UK. He has published several research papers and chapters in reputed journals and edited volumes apart from conducting research projects independently. His main research interests include environment and development with micro-level studies in the areas of involuntary displacement, urban ecology, social justice, and globalization.

Moni Chandra is a Senior Research Associate at Giri Institute of Developmental Studies, Lucknow. She has worked in various research projects specifically in the areas of urban labour market, Swachh Bharat Mission, human security, and vulnerability. She has a doctorate in Environmental Sociology from Babasaheb Bhimrao Ambedkar University, Lucknow. An awardee of "MGNCRE" Doctoral Fellowship in 2021 by Mahatma Gandhi National Council of Rural Education, Ministry of Human Resource Development, Government of India, her main research interests span energy, environment, gender, sustainability, and health.

Decent Work and Inclusion: Migrant Women Domestic Worker's Search for a Better Life

10

Anindya Basu and Diotima Chattoraj

Abstract

Work plays a vital role in everyone's lives, ensuring psychological well-being, social connections, and autonomy. The term "decent" work is referred to as work that complies with standards of equity, security, and respect for human rights. A huge number of populations, known as migrants (both voluntary and forced), migrate to escape the conditions of poverty, deprivation, political conditions, demographic shifts, and shrinking resources in their native places. Irrespective of whether they are voluntarily or forcedly migrating, they all seek a better life in the places they move to. Therefore, this chapter examines the challenges that these migrants face in their search for a better life through decent work and their inclusion in society.

This chapter explores the challenges faced by female migrant workers, particularly domestic helpers/workers, in their search for a better life and resultant inclusion in society. In addition, the chapter highlights the presence of women in the work force, as well as how the feminization of domestic work affects migration. Additionally, it demonstrates the role of female domestic workers in contributing to economic growth and sustainable development in places of origin and destinations. Through qualitative and quantitative data analysis of both secondary and primary nature involving profiling, correlation, and SWOT analysis, the chapter examines the nature of female domestic workers migrating from West Bengal to other parts of the country, the challenges they face, and how their striving for decent working conditions, and inclusion is promising to lead to sustainability as envisaged through Sustainable Development Goals (Goal 8). Despite the fact that migration seems to be a symbiotic exchange, it is observed that female domestic workers contribute significantly in sociocultural, political, and economic spheres through remittances and entrepreneurial activities and can act as agents of development if channeled properly.

Keywords

Decent work · Inclusion · Migration · Feminization · SWOT · Sustainability

A. Basu
Diamond Harbour Women's University, Kolkata, West Bengal, India
e-mail: basu_anindya2004@yahoo.com

D. Chattoraj (✉)
Nanyang Technological University, Singapore, Singapore
e-mail: chattorajdiotima@gmail.com

V. Dutta, P. Ghosh (eds.), *Sustainability: Science, Policy, and Practice in India*, Sustainable Development Goals Series, https://doi.org/10.1007/978-3-031-50132-6_10

10.1 Introduction

Work plays a vital role in everyone's lives, ensuring psychological well-being, social connections, and autonomy. The term "decent" work is referred to as work that complies with standards of equity, security, and respect for human rights. Currently, migration is a major issue that deserves attention because of its ubiquity, the impact it has, and the associated issues. It is centered on people and brought about by their aspirations, fears, triumphs, and tragedies, which makes this topic not only an intriguing subject but also compelling and pertinent (Chattoraj et al. 2020). They leave behind their roots and migrate to escape the conditions of poverty, deprivation, political conditions, demographic shifts, and shrinking resources in their native countries. Irrespective of whether they are voluntarily or forcedly migrating, they all seek a better life in the destination places they move to. Labor migrants/migrant workers are the ones who predominantly indulge in rural–urban migration and take up menial jobs of domestic helpers, chauffeurs, daily laborers, and vendors (Jha and Lahiri 2020). People tend to migrate in search of a better livelihood but often face difficult work conditions. The United Nations, in this regard, have been instrumental in ensuring decent work conditions for all. Accordingly, we find the goal of Sustainable Development Goal 8, which states "Promote inclusive and sustainable economic growth, employment and decent work for all. Sustained and inclusive economic growth can drive progress, create decent jobs for all and improve living standards" (United Nations 2015).

Ravenstien's law of migration (1885) made his observations based on migration to the United Kingdom and acted as the foundation for subsequent research on migration. The majority of his 11 principles hold true even today. However, the migration of women is becoming an important component in present times; they are no longer traveling as a dependent family member only; they are becoming part of the flow of migrant workers and are often the principal wage earner for their families. Women's migration, both international and national, is in response to gender-specific labor demand reflecting existing values, norms, stereotypes, and hierarchies based on gender, and recruitment intermediaries (private or official) also reinforce such gender segregation in the labor market. As in most cases, the low-skilled job market accommodates the majority of female migrants, and thus, occupational mobility or the ability to switch between different occupations is limited for them.

Domestic work as an unregulated economic activity is quite vast, providing employment to unskilled workers. At least 60 million domestic workers across the globe are engaged in domestic services, so that others can do their various jobs outside the home arena regularly. With the changing demography and social trends more educated women are taking up jobs, extended family support is missing too for which the requirement for domestic help is increasing by leaps and bounds. Approximately 80% of domestic service providers are women and come from the disadvantaged section of society, and out of every 25 working women, 1 is engaged in domestic services (WIEGO 2021). The ILO (2015) through Global Estimates on Migrant Workers (2015) documented that worldwide, more than 150 million workers were international migrants, out of which approximately 67 million are girls and women. The presence of women is conspicuously high, as works such as cleaning, cooking, and caring are regarded as "feminine."

Domestic workers face severe exploitation not only in terms of lower wages or fewer employment benefits (such as maternity leave) but also lack provisions regarding social security (such as health coverage and pension provisions) in this informal sector. As the women who are engaged in domestic work are low-skilled from disadvantaged communities have very little employment opportunity, they face such discrimination tacitly.

10.2 Feminization of Migration: Female Domestic Workers

In several Asian countries, such as the Philippines, Indonesia, Thailand, and Sri Lanka, female migrants have outnumbered their male counterparts, and this notion has come to be known as the "feminization of migration" (Ullah 2015). Furthermore, Kanji and Menon (2001) showed the use of the term "feminization of labor" in two ways. First, it is used to refer to the rapid and substantial increase in the proportions of women in paid work since the 1980s. Globally, 70% of the 20–54 age group is in the paid workforce. In developing countries, the figure is 60% (United Nations 1999 cited in Kanji and Menon 2001). These figures do not reflect women's participation in rural and urban informal sectors in developing countries, which is typically less visible and therefore undercounted. However, these low-wage informal sectors employ a high number of women in developing and transitional countries (Mehra and Gammage 1999). With the feminization of labor, jobs have shifted from manufacturing to services in developed countries and from agriculture to manufacturing and services in developing countries. Since 1980, women's employment has grown faster than men's employment except in Africa. The differences between male and female participation rates have shrunk significantly in many regions, because the male labor force participation rate has stagnated (or slightly decreased) (United Nations 1999).

From the early 1970s and late 1980s, the migration flow in general and female migration in particular had been predominantly to the "newly" oil-rich countries in the Middle East (Kanji and Menon 2001). The booming economies in some Southeast and East Asian countries have contributed to a partial shift of Asian [female] migrants' direction. By 1997, countries such as Malaysia, Taiwan, Singapore, Japan, Hong Kong, Macau, and South Korea became popular destinations (Ullah 2015). This redirection from oil-rich countries has been marked due partly to the reported violence against them, abuses, and declining salary base (Ullah et al. 2015).

Domestic workers, according to the definition of the International Labor Organization (ILO), are those workers who perform work in or for a private household or households. "They provide direct and indirect care services which include tasks such as cleaning the house, cooking, washing and ironing clothes, taking care of children, or elderly or sick members of a family, gardening, guarding the house, driving for the family, and even taking care of household pets" (ILO 2021). A domestic worker works on a full-time or part-time basis; may work for a single household, through a service provider, or both; may be residing with the employer's family (live-in worker); or may be living in his or her own home (live-out). It is also possible for a domestic worker to work outside his/her own country as a nonnational or foreigner, thus the term migrant domestic worker. There are 75.6 million domestic workers in the world, and 76.2% of them are women, indicating that approximately a quarter of domestic workers are male (ILO 2021). Domestic work is a more important source of employment among female employees than among male employees.

Although they provide essential services, domestic workers rarely have access to rights and protection. Approximately 81% of them are in informal employment, which is double the share of informal employment among other employees. Additionally, they work in some of the most difficult conditions. Compared to other workers, their average monthly salary is 56% less, and they are also more likely to work either extremely long or extremely short hours. Moreover, they face violence and harassment as well as restrictions on their freedom to move around. Particularly vulnerable are informal domestic workers. Part of the informality in domestic work can be attributed to gaps in national labor and social security legislation and part to implementation problems.

The Asia and Pacific regions remain the world's largest employers of domestic workers (ILO 2021). It is home to 38.3 million domestic workers or 50.6% of all domestic workers worldwide. China is the world's single largest employer of domestic workers, accounting for 22 million

workers: India (4.8 million), the Philippines (2 million), Bangladesh (1.5 million), and Indonesia (1.2 million) (ibid).

Women in domestic work face many difficulties in various aspects. Some of the general problems faced by domestic workers are a lack of decent wages and working conditions, violence, abuse, and sexual harassment at the workplace, i.e., at the home of the employer. The domestic workers who come through placement agencies suffer exploitation in many forms. Human trafficking is also an issue. Since they are not recognized as "workers," they do not enjoy access to social security benefits that other workers do, such as health insurance, maternity protection, and old age security. Thus, domestic workers have the opportunity to move horizontally (i.e., change of place of domestic work) but are unable to have upward mobility because of lack of access to education and skills training. Most domestic workers are not organized into trade unions and have no representative voice.

Due to the "hidden" character of domestic work, abuse is less visible, and migrant women depend heavily on employers. To restrict the free movement of the maid, employers often confiscate passports and other related travel documents. As they are less educated, most migrants have little to no knowledge about the situation at the destination and are unaware of their rights and obligations. It is often language that acts as the main barrier preventing them from effectively communicating their problems. Thus, they do not lodge formal complaints but prefer to suffer harassment and violence (Reshmi 2005).

Nonetheless, the situation is much worse for women who migrate illegally. Women are often reluctant to report their problems because of their illegal status. Working conditions and living conditions for women migrants are also gender specific. Being women migrants, without proper overseas documentation, makes them more of a soft target to abuse (Reshmi 2005). Compared to men who work in groups such as construction or plantations, women are more likely to work individually in domestic services, where there is less chance of support from others. Domestic workers are, therefore, one of the most vulnerable groups of women workers.

10.3 Objectives and Methodology

The main objectives of this chapter are to look at the evolution of domestic workers, especially women in India in general; trace the attributes of the national level female migrant domestic workers from West Bengal; to bring out the real-life issues challenging the dignity of work; to assess how actions have been taken to ensure decent working conditions for them; to gauge the level of empowerment derived from work; and to have an idea about the future trajectory and sustainability of this sector.

For our study, we reviewed online reports, documents, and newspaper articles. Secondary data are mainly derived from document analysis and reviewing the literature on the concepts of decent work and inclusion that lead to sustainability. Analysis of secondary and primary data (n = 345) collected directly or over telephonic conversations was also carried out to perform qualitative and quantitative analysis.

First, a general profiling of the respondents was performed. Statistically, a relationship between salary and age, caste, marital status, educational level, and years of experience was calculated specifically for respondents working in major metropolitan cities (n = 80). A SWOT analysis has been carried out to ascertain how the working condition of this sector can be improved while keeping pace with the principle of the SDG's concept of decent work and inclusion.

10.4 Migration in India

Migration in India is primarily influenced by social structures and patterns of development (Madhumathi 2013). In India, since independence, the government has pursued development policies that have accelerated migration. The Economic Survey 2016 reports that among 1.2 billion citizens, there are more than 500 million migrant citizens (approximately 42%) in India (Choudhary and Priyadarshini 2021).

The main causes of migration in India, as noted by Bala (2017: 55), are as follows: "The push factors being poverty, lack of work opportunities, unemployment and underdevelopment, poor economic condition, lack of opportunities, exhaustion of natural resources and natural calamities, scarcity of cultivated land, inequitable land distribution, low agricultural productivity, etc. Pull factors, include, attracting migrants to an area (places of destination), like, employment and higher education opportunities, higher wages facilities, better working condition." Millions of people migrate from their rural homes to the big urban cities, namely, Delhi, Mumbai, and Kolkata. The result of this population shift can be seen in the appalling living conditions such as slums, unsafe water, no sanitation, overcrowding with the attendant increase in crime, insecurity for women, sexual abuse, and frequent epidemics and AIDs.

Among the different categories of migrants, migrant laborers are one of the many factors that propel India's economic growth. "e-SHRAM," the national portal for registering unorganized workers, and the National Database of Unorganized Workers (NDUW), an initiative by the central government, recently (Choudhary and Priyadarshini 2021), reached the level of 12 crores as reported by the Ministry of Labor and Employment. This prompt registration of the workers reflects the intent of the government to safeguard the unorganized workers, who represent the majority (93%) of the country's workforce (ibid).

Migration is becoming increasingly feminized in India, with marriage cited as the leading cause behind women's internal migration. In general, women migrate independently to improve their economic prospects, education, and household responsibilities. States such as Bihar and Odisha have more male migrants, while Jharkhand and Chhattisgarh have a high rate of female migration (Choudhary and Priyadarshini 2021). However, feminization of migration has also led to an increase in the informalization and exploitation of migrant workers. As migrants, women are often subjected to gender-based and migration-related discrimination.

Statistical evidence shows that quite a large number of migrant workers from rural India are domestic workers, which is the third largest sector in India after agriculture and construction. Their working hours often stretch from 14 to 16 h, combining house work and working in other houses, and they are paid very low wages. The workplace is unsafe, and they have no social security, compensation for injuries, access to drinking water, or health care (Madhumathi 2013). Stricter and frequent eviction drives the displacement of the shanties of domestic workers, destabilizing their family and work lives. Furthermore, they have to live in makeshift tents with plastic covers and are forced to bathe and defecate out in the open (ibid.). Due to their migrant worker status, they are forced to buy food grains and kerosene at higher market prices. Despite the hardships domestic workers face, there are continuous streams of new migrants eager to join in as domestic workers.

10.5 Female Domestic Workers in India

Women play a crucial role in the development of society and the economy in India and comprise almost 50% of the total population (Beri 2020). Since 1971, they have outnumbered men migrants in India (Choudhary and Priyadarshini 2021). In the past, women had only been offered low profile jobs, but now they are considered capable of high-profile jobs, such as engineering, computer, management, administration, police, etc. This has led to an enormous increase in demand for domestic help. In addition to washing utensils and clothes, getting water, cleaning, and sweeping, domestic helpers are also engaged in irregular marketing, grocery, shopping, and ration drawing. Although they engage in various activities as domestic servants, they receive the lowest wages and face the greatest difficulties. The common problems faced by domestic workers include inadequate wages, poor working conditions, and inconsistencies in receiving monetary and nonmonetary benefits. The caste system also exacerbates the stigma associated with domestic work in India, as work such as cleaning and sweeping is typically associated with people belonging to the lower castes (National Domestic Workers' Movement, NDWM 2021). They are mostly illiterate. Tasks such as cooking, washing, and cleaning, which are traditionally seen as female chores and considered subservient to women, are performed by them. On the basis of employment arrangements, domestic workers can be classified into the following categories (National Domestic Workers' Movement, NDWM 2021):

Resident or migrant (interstate/overseas)
Part-time service or full-time service
Serving single employer or serving multiple employers
Live-in or live-out
Hired directly by households or hired by third-party agency
Tied or nontied

Most of the migrant women receiving domestic help are live-in workers, residing at the place of employment. Interstate migrant domestic workers are now on the rise as overseas migration is more cumbersome, as it delivers more remuneration than being in the native area serving.

10.6 Present Status

The concept of domestic workers or "servants" was first recognized in the Census of 1931, and almost 2.7 million people fell under the category, but by 1971, it came down to 67,000 as the colonial hangover started to fade away and the middle-class population doing their own chores was on the rise (Lahiri 2017). Strikingly, Indian homes have seen a 120% increase in the number of domestic workers in the decade post-liberalization, indicating higher prosperity – in 1991, it was approximately 7.5 lakhs and reached 16.6 lakh in 2001 (The Wire 2018). As per the NSSO (2011); the number of domestic workers was 3.9 million, where nearly 90% of domestic workers in India are women or children (especially girls), while the trade unions estimated that the number was more than 10 million. This kind of underestimation was highlighted by ILO (2021), which gauged the number to be between 20 and 80 million. The census data (Census of India 2011) corroborated this observation, as the number of female workers (15–59 years) rose 17% between 2001 and 2011. For the cities, the rise was even sharper, clocking 70% from approximately 14.7 million in 2001 to 25 million in 2011. From earlier studies, it can be gathered overwhelmingly that migrant domestic workers are women, most of them coming from backward communities from rural interiors. Those with almost no specific skill, due to the "pull factor" of the large urban centers by default, join the menial work ignoring the exploitative working conditions in the unorganized sector. It is very abysmal to find such a large group to be left out from any legal provisions or social security net.

Long back in 1985, concerted efforts to ensure networking, social security, and empowerment were made through NDWM, which was initiated through multiple groups in Mumbai and Chennai and gradually spread to 16 states and is affiliated with the International Domestic Workers Federation (IDWF) (National Domestic Workers' Movement, NDWM 2021). The Self-Employed Women's Association (SEWA) is a trade union based in Ahmedabad founded in 1972 that has been able to organize over 70,000 domestic workers across the country (SEWA 2021) and helped them raise collective voice to meet their justified demands in ensuring exploitation-free employment. Similar organizational endeavors were also taken in West Bengal, Kolkata, where Paschimbanga Griha Paricharika Samity (PGPS-West Bengal Domestic Workers Society) was granted trade union status by the state government in 2018, who fights for a safe working environment and weekly paid-off for domestic help (The Wire 2018).

Efforts have been made through organizations and even start-ups to formalize the process of hiring domestic help at national and international levels for the affluent section of society. For that, higher amount has to be shelled out and a percentage needs to be paid by the workers as handling fee; so, expectations are also higher from such recruiters from both sides like trained maids or healthy working conditions. However, as there is already a huge influx of migrant labor through informal channels, the impact of such organizations is still limited.

To ensure the basic rights of domestic help, India became a signatory to the International Labor Organization's (ILO) 189th convention, known as the Convention on Domestic Workers in 2011, but it has yet to ratify. Not much progress has been made in the enactment of the Domestic Workers Welfare and Social Security Act, 2010 Bill, drafted by the National Commission for Women (NCW), which also explicitly condemns child labor. In such a situation, the Domestic Workers Regulation of Work and Social Security Bill, 2016, was framed by the National Platform for Domestic Workers. It had provisions for compulsory registration of the employer and the employee, maintenance of a social security fund, and decent working terms and conditions.

The Case Study of West Bengal

Since West Bengal is one of the leading states supplying domestic help all over the country, it has been selected as the study area. All the respondents (n = 345) are currently working as migrated domestic help in various parts of India and come from different parts of West Bengal, India. From the selected sample based on snowball sampling, the districts supplying the lion's share appear to be South 24 Parganas (25%), Murshidabad (16%), Maldah (14%), North 24 Parganas (12%), and other rural pockets of Howrah, Hooghly, and Purba Barddhaman. Interestingly, Kolkata (12%), although a busy urban center, supplies a healthy number of migratory domestic help mostly to other megalopolises, such as Delhi-NCR and Mumbai, which offer lucrative wages. Through a questionnaire survey, the study tried to bring out major reasons for migration of the respondents; 38% cited lack of job opportunities, 50% mentioned poverty, 8% faced family issues, and the remaining 4% did not reveal the actual reason. Overwhelmingly, 98% of the respondents are Hindu, and they stated that being from another minority community is not very favorable to working as domestic help due to cultural reservations prevalent even today. Sixty-two percent of the sample population belongs to the age group 25–45 years, who are agile and have experience under their belt to shoulder responsibility and chores, and 79% of them are from backward classes. The marital status of the respondents is varied – single, divorced/abandoned or widow, but married ones are most common (58%) as they have greater family responsibility and to earn more are forced to leave their native place of residence. Educational attainment is highly skewed; 52% of them are illiterate, and 25% have completed only the primary level. As educational exposure and skill attainment are low, they have very little scope to select other avenues for livelihood. In most cases, they are engaged in multitasking with a main thrust in primary works such as cooking (50%), taking care of children (20%), or looking after the elderly (15%).

Statistically, a relationship between salary and age, caste, marital status, educational level, and years of experience has been calculated specifically for the respondents working in major metropolitan cities (Delhi, Mumbai and Chennai clusters) for at least over 1 year (n = 80). H_0: Salary is not affected by age, caste, marital status, educational level, or years of experience. H_a: Salary is affected by age, caste, marital status, educational level, and years of experience. Since the observed value of F[1] in 5 degrees of freedom in the numerator and 74 degrees of freedom in the denominator is 13.58, it is more than the 5% tabulated value (2.21). Thus, the null hypothesis is rejected at the 5% level of significance. The conclusion is that salary is dependent on age, caste, marital status, educational level, and years of experience. It is known that if $p \leq \alpha$ (where $\alpha = 0.05$), the null hypothesis will be rejected. Here, the p value is 0.0210. Therefore, the null

[1] F statistic is used in a statistical F Test to compare two variances, and then based on that value, null hypothesis is supported and rejected accordingly.

hypothesis is rejected, because there is a significant relationship among salary offered, age, caste, marital status, educational level, and years of experience. Here, R^2 is 4213 means 42.13% of the variability of the response variable (salary given) is explained by the predictor variables (age, caste, marital status, educational level, and years of experience).

$$\text{Coefficient Equation: } y = \beta o + \beta 1X1 + \beta 2X2 + \beta 3X3 + \beta 4X4 + \beta 5X5$$

The coefficients of salary offered on educational level (−0.05432) and caste (−0.01010) are negative, which means there would be an expected decrease in salary, corresponding to a 1-unit increase in educational level and marital status when other predictors are constant. There is a preference for less educated domestic help, as there is little chance for them to shift. The coefficients among salary and other factors are positive, meaning that when the other two predictors are constant, salary will increase with a 1-unit increase in the level of perception of age (0.65641), marital status (0.12218), and years of experience (0.66721). Middle-aged, married women with considerable work experience are most coveted in the job market, and the standard error between the predictors and response is less.

Although the women are earning for themselves and their family and are often the main breadwinners, the final decision regarding their work-related migration is either made by their parents/guardians (15%) if they are unmarried or by their spouse (43%) if they are married; only 30% make their own decisions. Even the ones who are widowed or separated in their cases are either a senior male member of her family or her son. Therefore, with economic self-sufficiency, empowerment seems not to be intrinsically linked.

For recruitment, the respondents are dependent mostly on their peer group, who are already in the trade, such as neighbors (44%), relatives (28%), and now some registered agencies (23%), making roadways into the prior fully personalized sector, which is a step toward the formalization of this service sector.

10.7 Migrant Workers and COVID-19 Lockdowns

Every sector across the globe is struggling in its own way to fight out COVID-19-related challenges, and migrant workers have also been badly hit. Travel restrictions and lockdowns made them panic-stricken. They are more vulnerable due to inadequate health care, worse economic conditions, and overcrowded living conditions, which put them at greater risk of infection (Chattoraj 2021, 2022). Gulf countries, which are home to approximately 35 million international workers (Karasapan 2020), struggled to protect these workers, as millions of migrants from India, Bangladesh, Nepal, and Pakistan were in forced lockdown, unpaid, abandoned, and under threat of deportation (Karasapan 2020; Weeraratne 2020).

In India, according to the projections made by Rajan and Bhagat (2022), as of the 2021 Census, there is an estimated 600 million migrant population. Based on the 2011 Census, approximately one-third of all internal migrants are interstate and interdistrict migrants, which makes them a population of almost 200 million (Rajan and Bhagat 2022). The majority of internal migrants are migrant workers, who make up two-thirds or approximately 134 million people as of today. Although official estimates suggest that there are approximately 5 million domestic workers in India, according to the International Labor Organization, the numbers are somewhere between 20 million and 80 million (Saluja 2022). However, there is a lack of comprehensive data on migrant workers at the national, regional, and global levels. While these workers contribute 10% to India's economy, they are socially and politically vulnerable.

To stop the spread of the virus, India in March 2020 enforced the world's largest lockdown, putting a ban on all sorts of travel, which immediately affected 90 million domestic workers, tens of millions of migrant construction workers, and approximately 10 million street vendors (Ullah et al. 2021; Ullah et al. 2022) in terms of scarcity of food, shelter, loss of wages, fear of getting infected, and anxiety (Ahmed and Deshingkar

2020). As a result, they started fleeing to their native places on foot. Several thousand lost their lives on their way. As they reached their villages, they were seen as potential carriers of the infection and were mistreated by the police and locals (Ahmed and Deshingkar 2020). Even their families were harassed, discriminated against, and singled out (Jha and Lahiri 2020). A large number of people returned to the country after the lockdown. The very effort to stave off the pandemic turned into one of greatest human tragedies in India's recent history (Chattoraj 2021; Chattoraj et al. 2021).

The situation of the migrant domestic workers who were laid off during the COVID-19 crisis was worse, as they were mostly kept out of government relief packages. Not only was there denial of full wage, many had to go back to their native place by arranging their own transport at sky high rates. Even after the lifting up of the lockdown, many of them were not reinstated as being unnecessarily stigmatized as spreaders of infection. The state of affairs was sadder for the migrant domestic workers whose husbands worked as daily wage earners in the destination city, as going back to their native place was not a viable option, and they had no means to stay back in the city either. An estimated 50 million domestic workers are currently in lurch in terms of their employability and livelihood (Action Aid Association India 2021). Those who have managed to obtain some sort of employment are being paid less, and many unnecessary restrictions are also put on them.

10.8 Impact of the Pandemic on Health and Social Development

The returning migrants view their homes in the village as ensuring food and comfort. However, in reality, MWs are faced with many social, health, and economic challenges. The lockdown worsened the scenario, turning it into a humanitarian crisis. Limited access to health care, overcrowding, poor hygiene, and scarcity of food increase community transmission of the virus.

In June 2020, 56% of the COVID-19 cases, an increasing proportion of positive cases in rural areas, were attributed to returnee migrants from West Bengal, with a 10% positivity rate among them (Jha and Lahiri 2020). More than 70% of new cases in Uttar Pradesh were linked to migrant workers in May (ibid.). In addition to a temporal trend, epidemiological evidence also links these increasingly underprivileged migrants to the growth of this pandemic. Often, it has been seen that migrants are not eligible for various health benefits. Therefore, in the present community-widespread of COVID-19, early detection, testing, diagnosis, contact tracing, and care seeking are difficult, if not impossible, among migrants, increasing the risk of outbreaks in these populations, and symptoms may go unnoticed or even actively concealed, posing further health risks. People in rural areas rely predominantly on public healthcare systems with urban resource mobilization. In addition to the increase in infection, returning migrants to rural areas will also bring about supply problems due to the lockdown and its economic impact on the infrastructure.

10.9 Concerns and Challenges Associated

According to the Indian Constitution, women and men are entitled to equal rights and are prohibited from discriminating based on their genders, castes, or creeds. However, women in economically affluent classes, as well as those in working classes, do not enjoy the same status as men. Domestic work has been defined as a contemporary form of slavery (Beri 2020).

As domestic workers mostly work within the confines of home, it will be a mistake to feel that the working conditions are "safe heaven." Often, they are exposed to a range of unhealthy and hazardous working conditions. In particular, the migrated domestic helps who are unfamiliar with their surroundings feel isolated there. Lack of privacy is another issue that affects the mental health of workers.

Often, domestic workers are vulnerable to employers' verbal, physical, or sexual abuse, as the "elites" try to maintain the master-servant divide through acts of intimidation, such as serving stale foods, frequent frisking at security checks, and deducting payments for absence due to emergencies. Intimidation, bullying, physical abuse, and other forms of harassment are among the forms of harassment encountered by domestic workers. Incidents of molestation, rape, and even murder are not unheard of.

The live-in women migrant workers are worse off, as they do not have the fallback option, such as seeking advice from their family members directly, or if the situations are truly complex, they even leave the job at short notice or without any notice to return home.

The lack of any written contract, i.e., legally binding obligation (such as the Minimum Wages Act), robs the chance for the migrant domestic helps to even lodge complaints even if they are not paid the promised monthly wages or are fired without any notice. The provisions for regular hikes in salary, bonus, and appraisals are not prevalent, and when asked for a hike, they are often scolded and ignore ripping of their self-dignity (Nigam 2021). There might be some weak pressure from the locally formed nonregistered organization in very few localities, but their role in enforcement is still extremely weak or nonexistent.

Lack of labor laws and social protection schemes hurt most helpless child workers, the ones who mask their true age and join the domestic workforce in cities to escape from poverty or are trafficked; since they are on the wrong side of the law too, they cannot raise their voice again any injustice or torture.

Last but not least, as most domestic workers are illiterate or barely literate, they are completely unaware of the provisions available or about the struggles to achieve protection; to pressurize the government to claim their rights, there is a need to unite them within the realm of an umbrella organization having the bargaining power to ensure dignity in the workplace.

10.10 Need for Protection and Resulting Regulatory Frameworks

The need for protective instruments to ensure decent work conditions for domestic workers was first officially recognized by the International Labor Conference (ILC) in 1948. However, for ages, no convention was introduced. In 2007, a prolonged global effort, the Convention Concerning Decent Work for Domestic Workers, was adopted, and recommendations were passed in Geneva, 2011, which mandated a minimum age for employment, a minimum wage, stipulated rest hours, and protective measures against acts of violence.

The regulatory frameworks that help domestic workers in roundabout ways are the Minimum Wages Act, 1948; the Child labor Act (1986); the Unorganized Workers' Social Security Act, 2008, (UWSSA); and the Sexual Harassment of Women at Workplace (Prevention, Prohibition and Redressal) Act, 2013. However, these have never truly been applied in any real-life situation and are grossly inadequate. In addition, to take care of the health needs of domestic workers, Rashtriya Swasthya Bima Yojana (RSBY) was taken up, which is a smart card-based cashless health insurance scheme.

There are few other social security schemes especially targeting unorganized workers, such as the Indira Gandhi National Old Age Pension Scheme renamed the National Social Assistance Scheme (1995) with the National Family Benefit Scheme as one of its kinds, Janani Suraksha Yojana (2005), Atal Pension Yojana (2015), Pradhan Mantri Jeevan Jyoti Bima Yojana (2015), and Pradhan Mantri Suraksha Bima Yojana (2015), to name a few.

The salient features of the proposed draft National Policy on Domestic Workers are as follows:

- Registration of domestic workers as unorganized workers would facilitate their rights and benefits.
- Providing them with the right to form associations/unions.

- They should be entitled to a minimum wage and should have the opportunity to enhance their skills.
- To safeguard them against abuse and exploitation. Domestic workers must be provided with a grievance redressal system.
- To give them access to courts and tribunals for redress of grievances.
- It is necessary to establish private placement agencies for them

Undoubtedly, this Act shows a positive intent to safeguard the rights of workers, but it lacks an enforcement mechanism. Seven states, namely, Andhra Pradesh, Bihar, Karnataka, Kerala, Rajasthan, Jharkhand, and Odisha, have taken a step forward by trying to ensure minimum wages as per the Minimum Wages Act, 1948, although quite meagre in today's context. To improve the situation, the drafted policy suggested setting up skill building centers and grievance redressal centers (Beri 2020). However, the need for a national law is genuinely felt which might get addressed if the draft of National Policy for Domestic Workers which address the issue holistically to ensure "legitimate labor market activity" gets enacted. There has been a feeble argument that domestic workers are within the boundary of a private household and that the enactment of laws by the state would disturb privacy, but this argument has been rejected from many quarters, as the private household is actually the "workplace" in the strictest terms. However, the problems encountered by domestic workers are largely behavioral, so formulating law will not bring in sea change overnight, but regular monitoring can ensure a better situation.

10.11 SWOT Analysis

From the literature review and direct interviews of the migrated female domestic help, an analysis of the ground reality has been made on the basis of the Strength-Weakness-Opportunity-Threat (SWOT) model. Although there are several unaddressed issues and grim challenges in the way to these hardworking individuals, the road to sustainability and empowerment is not that far-fetched. To ensure a decent working environment and inclusiveness, legal frameworks are being placed, although at a very slow pace. However, to get the best out of the employees, most employers provide a healthy environment and just remunerations. The upcoming organizations by the workers themselves and the presence of organized agencies are making the sector more competitive and formal (Table 10.1).

The vulnerability of migrant domestic workers involves both physical and psychological exploitation. There are inhuman living and working conditions with low monetary return and almost no healthcare facilities.

Table 10.1 SWOT analysis

Strength	*Opportunities*
One can work from a secured home environment Steady yearlong remuneration is available There is a loose confederacy among the migrated female domestic helps which has been stronger with the availability telecom connectivity increasing their collective power of coercion	Important avenue of livelihood for low-skilled individuals The unregulated job market can be organized with certain degree of vigilance from the government or local administration Very recently, the domestic workers have come within the ambit of Unorganized Sector Social Security Act, 2008 which provides them some legal standing
Weakness	***Threats***
Tiresome job with long hours, less rest The minimum wage is often not paid as legal bindings are frail and social security measures are almost absent There is limited scope for the workers to shift to other sectors due to lack of vocational training even if they are not satisfied	Often exploited by the placement agencies as many do not have individual networking at far-off places There are chances of the "servants" being ill-treated by the "masters" (physical and psychological abuse and sexual harassment) As the sector is purely demand driven, during general economic crisis periods, there is oversupply and results in repeated decrease in salary or even job retrenchment

10.12 Conclusion

This chapter observes that female domestic workers contribute significantly in the sociocultural, political, and economic spheres through remittances and entrepreneurial activities and are agents of development if channeled appropriately. They are nevertheless the most vulnerable: they work long hours for little pay; they have no maternity leave, no health coverage, or pensions. Furthermore, the working and living conditions are poor.

The problem of migrant labor, especially domestic workers, is enormous and requires intervention in a systematic way. Trade unions have always been proactive in securing the rights of workers, and here, they also have a huge role to play. These workers in their rank and file equally engage them in sustainable development and social justice. Several NGOs with skilled manpower and resources provide trade unions with research and background. Through their organizational structures and organizing skills, trade unions can deal with employers, policymakers, and governments for better terms for migrant workers and contribute to social harmony and sustainable growth.

References

Action Aid Association India (2021) Support domestic workers. Retrieved December 1, 2021 from https://www.actionaidindia.org/campaign/support-domestic-workers/

Ahmed N, Deshingkar P (2020, April 1) Locked out under coronavirus lockdown–continuing exclusion of India's migrant workforce. Retrieved December 1, 2021 from https://archive.discoversociety.org/2020/04/01/locked-out-under-coronavirus-lockdown-continuing-exclusion-of-indias-migrant-workforce/

Bala A (2017) Migration in India: causes and consequences. Int J Adv Educ Res 2(4):54–56

Beri Y (2020) A study on female domestic workers in India. J Interdiscip Cycle Res 12(6):1394–1403

Census of India (2011) Population totals. Registrar General and Census Commissioner of India. Ministry of Home Affairs, Government of India, New Delhi

Chattoraj D (2021) Migrant workers in Singapore: experiences in times of Covid-19. Acad Lett 2

Chattoraj D (2022) "We are all migrant workers": commonality of Bangladeshi migrants' experiences in Singapore amidst COVID-19. Int J Asia Pacific Stud 18(2):9–36

Chattoraj D, Ahsan Ullah AKM, Shahidul Haque M (2020) The migration myth in policy and practice: dreams, development and despair. Migration Lett 17(3):473–475

Chattoraj D, Ullah AA, Hossain MA (2021) The COVID-19 pandemic and the travails of Rohingya refugees in the largest Bangladeshi refugee camp. Volume 1: community and society, vol 1, p 165

Choudhary S, Priyadarshini A (2021, December 23) Big strides on path for women migrant workers in eastern India. Retrieved November 12, 2021 from https://www.financialexpress.com/india-news/big-strides-on-path-for-women-migrant-workers-in-eastern-india/2388013/

ILO (2015) ILO global estimates on migrant workers: results and methodology. Labor Migration Branch, Geneva

ILO (2021) Who are domestic workers? Retrieved October 5, 2021 from Who are domestic workers (Domestic workers) (ilo.org)

Jha SS, Lahiri A (2020) Domestic migrant workers in India returning to their homes: emerging socioeconomic and health challenges during the COVID-19 pandemic. Rural Remote Health 20(4):6186

Kanji N, Menon K (2001) What does the feminization of labor mean for sustainable livelihoods? International Institute for Environment and Development.

Karasapan O (2020, September 17) Pandemic highlights the vulnerability of migrant workers in the Middle East. Retrieved December 14, 2021 from https://www.brookings.edu/blog/future-development/2020/09/17/pandemic-highlights-the-vulnerability-of-migrant-workers-in-the-middle-east/

Lahiri T (2017) Maid in India: stories of inequality and opportunity inside our homes. Aleph Book Company, New Delhi

Madhumathi M (2013) Migration for domestic work – a case of female domestic workers in Bangalore. Int J Social Sci Interdiscip Res 2(1):44–61

Mehra R, Gammage S (1999) Trends, countertrends, and gaps in women's employment. World Dev 27(3):533–550

National Domestic Workers' Movement, NDWM 2021. Retrieved December 23, 2023, from https://ndwm.org/

Nigam R (2021) Dignity is all we need: domestic help in India. Media India Group. Retrieved December 2, 2021 from https://mediaindia.eu/society/domestic-help-in-india/

NSSO (2011) NSSO Statsitics-2011-2012, 68th round. Ministry of Statistics and Programme Implementation, Government of India

Rajan SI, Bhagat RB (2022) Internal migration and the Covid-19 pandemic in India. In: Triandafyllidou A (ed) Migration and pandemics, IMISCOE research series. Springer, Cham. https://doi.org/10.1007/978-3-030-81210-2_12

Ravenstein EG (1885) The laws of migration. J Stat Soc Lond 48(2):167–235

Reshmi RS (2005) Gendered vulnerabilities, discrimination and abuse among women migrants – a special reference to return domestic workers in Kerala, India
Saluja R (2022, May 9) Pandemic brings new hardships to India's domestic workers. Retrieved on September 4, 2022 from https://undark.org/2022/05/09/pandemic-brings-new-hardships-to-india-domestic-workers/
SEWA (2021) Domestic workers. Retrieved December 17, 2021 from https://sewadelhi.org/advocacy-campaigns/domestic-workers/
The Wire (2018) In a first, Kolkata Domestic Workers' Organization gets Trade Union status. Retrieved December 13, 2021 from https://thewire.in/rights/in-a-first-kolkata-domestic-workers-organisation-gets-trade-union-status
Ullah AA (2015) Abuse and violence against foreign domestic workers. A case from Hong Kong. Reg Stud 10(2):221–238
Ullah A, Hossain M, Islam K (2015) Migration and worker fatalities abroad. Springer
Ullah AA, Nawaz F, Chattoraj D (2021) Locked up under lockdown: the COVID-19 pandemic and the migrant population. Social Sci Human Open 3(1):100126
Ullah AA, Chattoraj D, Ibrahim WZ (2022) The COVID-19 pandemic and the 'stranded' migrant population: an unequal pain. Suvannabhumi 14(2):61–88
United Nations (1999) World survey on the role of women in development: globalization, gender and work. United Nations, DESA, New York. Retrieved from https://www.un.org/press/en/1999/19991020.dev2220.doc.html
United Nations (2015) Sustainable development, transforming our world: the 2030 agenda. New York
Weeraratne B (2020) Return and reintegration without assimilation: South Asian migrant workers in the gulf during COVID-19, Working paper no. 327. Institute of South Asian Studies
WIEGO (2021) Empowering informal workers, securing informal livelihoods. Retrieved December 13, 2021 from https://www.wiego.org/informal-economy/occupational-groups/domestic-workers

Anindya Basu is working as an Assistant Professor in the Department of Geography, Diamond Harbour Women's University since 2017. Previously, she served as an Assistant Professor in Women's Christian College, Kolkata. She has a keen interest in socio-political geography, urban-environmental issues, and tourism geography. Dr. Basu undertook two sponsored research projects, presented papers in national and international conferences and seminars, and published several research papers and book chapters.

Diotima Chattoraj is a Research Fellow at Wee Kim Wee School of Communication and Information in Nanyang Technological University, Singapore. She is also an adjunct research fellow at the Department of Social and Health Sciences in James Cook University, Singapore. She completed her PhD at Ruhr University Bochum, Germany, in 2016. Her research interests include migration, development, ethnicity, international relations, and boundary making. She was a former researcher at the Saw Swee Hock School of Public Health at NUS, in collaboration with NTU and SMU. Prior to that, she worked as a researcher at the Faculty of Social Sciences (FASS) in Universiti Brunei Darussalam. She has published in leading journals in migration and development and has also authored more than a couple of books. She is an assistant editor for *South Asia Research* (Sage) and serves as a peer reviewer for a number of refereed journals.

11 The Impact of SDG-11 on Building Sustainable Cities: PPP Project Scenarios for Urban Infrastructure Development in India

Nimisha Jha

Abstract

The twenty-first century has been nicknamed the "urban century", with urban responses to global challenges shaped by "smartness". Scholars have spent the last several years trying to explain "why, how, for whom, and with what consequences" the smart city paradigm emerges in different urban situations (Evans et al. 2019). The traditional urban development debate focuses on either urban development or sustainable cities and typically ignores people-centred development options. The former addresses economic growth challenges, whereas the latter emphasises environmental considerations at the expense of poor development concerns. India's urban population of more than 450 million people accounts for less than one-third of the country's total population. It is anticipated that by 2045, its cities will house 800 million Indians, more than the whole population of modern-day Europe.

Cities are now confronted with the escalating negative consequences of society's unsustainable path. Climate change, biodiversity loss, and increasing geographical segregation are all indicators of this. The repercussions of these issues frequently outpace individual urban house developers' management abilities. As a result, collaborations such as public-private partnerships (PPP) must be developed to counteract the current neoliberal trend. This chapter examines the flaws and limitations of PPP and its potential to address the problem of unsustainable urban development. However, fundamental differences like difficulties that connect the global sustainability model to the harsh reality of the local environment must first be addressed. The chapter outlines such a vision for sustainable cities in India and actions to achieve it. This gap threatens both the city's social foundation and its ecological ceiling. The chapter investigates the inherent hurdles that restrict efforts and potential institutional strategies to overcome or avoid them. This chapter's unified content will assist current researchers and motivate future investigators to integrate SDG-11 in their planned study in India.

Keywords

Sustainable Development Goals · Inclusive and sustainable cities · India

11.1 Introduction

A public-private partnership (PPP), according to the National Public Private Partnership Policy 2011, is an agreement between the government/

N. Jha (✉)
NITI Aayog, Government of India, New Delhi, India
e-mail: nimishajha.cg@gmail.com

V. Dutta, P. Ghosh (eds.), *Sustainability: Science, Policy, and Practice in India*, Sustainable Development Goals Series, https://doi.org/10.1007/978-3-031-50132-6_11

statutory body/government owned firm on one side and a private sector entity on the other. PPP is an agreement for the provision of public assets and/or public services for a certain period of time through investments and/or administration by a private sector organisation (GLOSSARY-Department of Economic Affairs – InfrastructureIndia.gov.in 2019). PPPs have a well-defined risk distribution between the private and public sectors. The private entity is paid based on performance requirements that are measured by the public entity or its representative. To address the country's growing infrastructure deficit, significant efforts have been made to increase investment in physical infrastructure, particularly since 2004, when several measures were implemented to encourage private investment in infrastructure both directly and through public-private partnership (PPPs). The PPP framework emerged in the 1980s, when governments in countries such as the United Kingdom and Chile sought to extend the benefits of privatisation to industries that were thought to be extremely difficult to privatise, such as transportation, schools, and hospitals, by utilising appropriate PPP models.

PPPs were first introduced in India between the mid-1990s and 2004; the priority industries were electricity and transportation. The lack of a well-established PPP framework, on the other hand, contributed to the country's poor performance (Laksmanan 2008). From 2004 to 2012, the country experienced an increase in the number of PPP projects awarded as a result of the government laying the basis for a mature PPP framework by developing laws and standardising documentation for PPP projects. Hans (2017) states that a total of 1539 PPP contracts have been granted in India thus far. About half of them are now operational, with the remainder either abandoned or in different stages of implementation. According to a sector breakdown, 58% of these projects are in the transportation sector, 24% in the energy sector, 9% in the social and commercial infrastructure sector, and 8% in water and sanitation. However, due to a lack of a strong PPP enabling ecosystem, specifically the "twin balance sheet problem," a lack of alternative sources of credit, an ineffective framework for dispute resolution, and a lack of bankable PPP projects, a large number of projects awarded between 2004 and 2012 did not take off. Following that, private sector investments fell precipitously, severely hampering India's PPP goals. Furthermore, due to India's federal structure and the absence of a uniform PPP legislation or policy, as well as a national level PPP body, the country's PPP framework remains fragmented. Only around 32% of projects are carried out by the government or its agencies; the remainder is carried out by state governments. This emphasises the importance of state-level governance and policy to the overall performance of PPP in India. PPP success is dependent on the optimum risk allocation among stakeholders, an environment of confidence among stakeholders, and strong institutional ability to carry out project grooming and implementation. A robust PPP supporting ecosystem, which includes liquid and diverse financial institutions, good regulatory and arbitration frameworks, mature developers, competent consultants, and so on, is also critical to the successful implementation of a PPP project.

11.2 SDG 11 as a Foundation for Development

SDG 11 is a spatially oriented development framework that strives to solve a variety of interconnected challenges with an emphasis on urban sustainability. It has numerous parallels with other SDG targets. The SDGs are built on a systems approach, with a significant emphasis on links with other developmental goals. It attempts to achieve the desired level of result by 2030 by establishing a set of attainable goals and indicators that will guide policy activities at the local level.

Achieving a balance of social, economic, and environmental goals is the cornerstone of "sustainable development". Rather than maximising benefit through a single system, trade-offs between numerous systems and objectives are promoted. According to Babier and Burges

(2017), the capital approach, which guides the vast majority of real development programmes, and the concept of sustainable development are fundamentally different. For example, most urban development policies strive to improve the functional efficacy of public infrastructure and attract new investments. These policies occasionally achieve their economic aims of reducing poverty and providing jobs, but they also have the unintended consequence of increasing social divide and pollution. It should be noted that the actual relevance of sustainability as a conceptual paradigm for policy creation is limited. To negotiate the point of trade-off between conflicting purposes and priorities, several layers of the governance hierarchy and power relationships among the involved parties are employed. As a result, the weight attributed to concerns of fairness, the environment, and the economy differs according to location. The political-economic climate unique to a specific region shapes the urban design through zoning laws and planning standards.

Because of the adaption of the Sustainable Development Goals, which include the standalone urban aim of making cities safe, inclusive, resilient, and sustainable, urbanisation is now firmly at the centre of international development policy (SDG11). This viewpoint sees urbanisation as more than just a demographic issue but as a transformational process with the potential to catalyse several aspects of global development. However, because both goals are intertwined, accomplishing one typically necessitates addressing issues that are more commonly linked to the other. Thus, Goal 11 is the most relevant, with the goal of eliminating slum-like conditions in order to establish inclusive, safe, resilient, and sustainable cities and human settlements. It will also provide accessible and affordable transportation systems, reduce urban sprawl, increase participation in urban governance, enhance cultural and heritage preservation, address urban resilience and climate change challenges, improve urban environment management (pollution and waste management), and provide a safe environment for all.

11.3 Basis of Public-Private Partnership

The National Policy on PPPs' definition of a public-private agreement includes some of the prerequisites. Despite the fact that each PPP contract is unique based on the circumstances, they nonetheless have some essential characteristics. A few more regulations are also desired to guarantee that the private sector is held to the same standards as service delivery. Some of the criteria include risk assumption definitions, mechanism of payment, and whether the payment would be made through cash flows from the public sector based on performance or by charging customers who use the service. Identification of minimum technical requirements without limiting the possibility of innovation, incentive-, and penalty-based structures in the agreement to ensure that the private sector is benchmarked against service delivery, and a time frame for either project transfer or private sector entity implementation.

11.4 The Scope of PPP and the Current Enabling Environment in India

Despite the existence of sporadic examples of private participation in the British and early post-independence eras, the true PPP movement began in the 1990s, when electricity generation was opened to the private sector and licenses were granted to eight mobile circular telephone service operators. Following this, in 1995, a substantial revision to the National Highways Act 1956 allowed for private participation. However, the Infrastructure Development Money Company was founded in 1997, which demonstrated how seriously the government was taking use of private sector finance, administration, and experience in the nation's infrastructure development (Priya and Jesintha 2011). According to Vinayak Chatterjee, while China's GDP is four times larger than India's, India's PPP market is ten times greater. In fact, it has the potential to be the world's largest PPP market.

The government's interest in developing such collaborations helps to the industry's success. Since the 2008 global financial crisis, the large amount of investment required for infrastructure projects, combined with soaring inflation in India, has repeatedly rendered it hard for any business to function independently. In addition to the efficiency and managerial skills it provides and the budgetary restrictions it faces, the public sector occasionally seeks investment advice from the private sector (Gurtner 2010). As a result, opportunities for such businesses abound. Building and development of urban as well as social infrastructure are two areas with enormous PPP prospects. The formation of urban infrastructure includes the construction of urban highways, urban transportation, water supply, sewage, solid waste management, traffic support infrastructure, slum regeneration and redevelopment, and so on. The local municipal administration in cities and metropolitan regions is continually in need of resources to carry out these obligations.

11.5 Challenges and Risks in PPPs

As previously stated, public-private partnerships in India are just 12–15 years old (or new), with the majority of activity occurring in the last 7–10 years. According to publications and analyses of the PPP business in India, the country's economy regards such collaborations favourably. The enabling environment that the federal and state governments have developed to attract private sector investments in such projects has created an especially favourable atmosphere for these collaborations. PPPs have not been a total success on different occasions and in various settings. According to the literature and particular instances of research on the subject, the bulk of the difficulties that have developed in these have occurred over time; nevertheless, neither party could have expected or recognised these concerns at the time the contract was written. PPPs are more vulnerable to bottlenecks due to their lengthy lifetime. It is because the possibility of an issue emerging grows as time passes. One of the most commonly mentioned PPP-related difficulties is a lack of openness. Despite substantial attempts to increase openness throughout the bidding process and contract award, people tend to have a general concern about PPPs' lack of transparency, which is not without reason.

The process of creating a PPP partnership is lengthy and laden with legal jargon. There have been countless cases where the private party has profited politically unjustly from their public sector counterpart, throwing question on the entire process. The issue of transparency extends to user-based programmes, where the entire contract smells like a clandestine alliance or partnership between the government and private companies (Laksmanan 2008). Another difficulty has been the failure to respect the moral code of conduct. It should always be remembered that the government prioritises the development of social and economic infrastructure with the goal of enhancing social welfare (Laksmanan 2008), while it is just another business opportunity for the private sector. Even if a social goal is stated, it can never be the other way around and must always come after the primary goal of generating profits has been satisfied. To make a long tale short, PPPs face problems and bottlenecks at nearly every stage, including conception, planning, execution, and implementation. The sectors involved are characterised by their diametrically opposed goals of profit maximisation and societal good, making finding solutions to these problems a time-consuming effort. Furthermore, the entrenched interests and inflexible attitudes of bureaucrats, administrators, and local politicians result in additional artificial impediments that impede the initiatives. Along with the benefits of PPP projects, there have been drawbacks in the form of bottlenecks and obstacles. These must be addressed with more concentration and policy formation by bringing more transparency, developing national sense, and rising above stakes.

11.6 Barriers to PPP Projects in the State and Urban Level

For a country like India, the PPP model is the most effective alternative. For infrastructure growth, the Smart City idea mainly relies on the PPP method. It does not, however, consider the experience of ongoing PPP enterprises. Initiatives with shorter time horizons, technologically uncomplicated output, and risks shared mostly by the public sector, such as bus shelters and solid waste management, likely to be mainly successful, according to Seddon and Mahalingam. Projects with erratic funding sources, a vested interest in equal access, and higher transaction costs, on the other hand, are unlikely to succeed. In Hubli-Dharwad, Belgaum, and Gulabarga, only 10% of the population is serviced by 24 × 7 water supply systems, while the Nagpur programme saw a 46% cost overrun due to delays in execution. Due to an incorrect calculation of demand and a land acquisition conflict, PPP on urban motorways such as the Karur Toll Bridge and Coimbatore Bypass ran into complications, while risk sharing between the public and private sectors proved troublesome in the case of the Chennai Ring Road. The following three impediments impede PPPs in Indian cities: institutional, organisational, and project level.

11.7 Institutional Barriers

Many PPP ideas for urban development have been thwarted due to a lack of supporting PPP legislation. Current regulation demands that only the public sector provides a specified set of infrastructure services in certain instances. In many situations, the absence of enabling laws on PPPs indicates that state government officials and bureaucrats who urge private sector participation do so at their own discretion based on their personal beliefs about the benefits of working with the private sector.

11.8 Barriers at the Organisational Level

PPPs are a new phenomenon in India, and both the public and private sectors are unfamiliar with them. As a result, public officials are undertrained in areas such as financial and legal structuring, which are critical to PPP transactions, and are unfamiliar with the new types of contractual arrangements that are common in PPPs, in which risk and responsibility are shared by the private and public sectors. As a result, governments submit a disproportionately small number of PPP project proposals. However, few projects under this strategy have yet to be realised through PPP, with state and municipal governments choosing to cover the financing gap using budgeted funds. Some states lack clarity on the practical value of PPPs in infrastructure development at the political level. Several failed attempts, as well as a lack of study or documentation on the initiatives undertaken, have raised questions about whether PPPs are as efficient or optimal as they are supposed to be. Furthermore, in several states with involvement from left-wing groups, such as Tamil Nadu, Kerala, and West Bengal, administrations or coalitions are typically antagonistic to the entry of the private sector to supply fundamental services. A lack of trust between the business and governmental sectors is a final organisational issue that has hampered state-level PPP attempts. As a result, the bulk of PPP ventures are private-public partnerships rather than true PPPs.

Barriers at the Project Level

The most major project-specific challenge is the lack of economic feasibility of several proposed PPPs. A bypass road project suggested on a build-operate-transfer (BOT) basis around the city of Coimbatore recently failed to attract private sector participation due to an inadequate rate of return on toll revenues. Surmounting societal pressures in the form of protesting activists battling for problems such as adequate resettlement, environmental deterioration, and other factors hamper PPP project completion. In other cases, the project's targeted users may protest to tariff increases resulting from privatisation. In the case of the Coimbatore bypass route, the Tamil Nadu government elected to toll a nearby bridge and factor the toll money into the project's financial calculations. Bridge users, on the other hand, were furious that a toll was being charged for a facility that they had previously used for free, and they refused to pay. Typically, many local bodies are involved in project implementation, and private enterprises usually need licences, approvals, and the like from a variety of agencies. This greatly increases a project's transaction costs and frequently necessitates the providing and acceptance of bribes in order for the project to progress. All of these difficulties have hampered the creation of PPPs at the state and municipal levels. Furthermore, the following constraints impede the successful execution of PPP initiatives. They are as follows:

11.9 Obstacles Resulting from the Project's Design and Structure

Due to a lack of rigorous feasibility studies and project analysis, urban PPP projects are typically poorly organised, resulting in renegotiations and failure. Demand was significantly lower than expected, project income had not materialised, and the project sponsor was unable to break even in the case of the Tirupur water supply project, where service had only recently begun. In the case of Chennai's solid waste management, the city initially picked a number of zones for privatisation. The amount of rubbish created by these zones, however, was inadequate for the private sector to function properly. As a result, no private bids were received in the first round of bidding. The project had to be reorganised at the expense of the exchequer. Even when significant study was conducted, governments failed to move quickly to design and auction projects, as in the case of an aviation project in the state of Gujarat. As a result, demand and other project-related information soon became out of date, making it difficult to bid on the project. In a series of planned land development PPPs spanning numerous cities, government agencies rejected to adjust the zoning patterns on the site, allowing all of these risks to be handled by the private developer. Developers avoided the project until the risks were spread.

11.10 Impact of Sustainable Development Goal (SDG 11) on PPP

When announcing the Sustainable Development Goals (SDG), former United Nations Secretary-General Ban Ki-Moon declared, "Cities are where the war for sustainable development will be won or lost" (Fabre 2017, p. 4). As we live in a more urban society, urbanisation has become a defining feature of the twenty-first century. Because just 30% of people lived in cities in 1950, the globe was mostly rural. The scenario is expected to alter by 2050, when urbanisation would have reached 70%.

The complex resource depletion challenges confronting decision-makers in the 1970s gave rise to sustainable development. After acquiring an understanding of all the main components of sustainable development, all nations recognised the need for a cohesive approach to bring about a shift in lifestyle. The historic vow to "spare no effort to relieve our fellow men, women, and children from the dismal and humiliating conditions of extreme poverty" has been developed into an inspirational framework of eight aims, which has then been translated into a range of meaningful practical actions. It had been concluded by the end of 2015 that many hands and brains would be required to work in order to achieve the Sustainable Development Goal of making cities sustainable by 2030. Partnerships including federal, state, and local government officials, business, financial institutions, civil society organisations, social movements, academic institutions, the media, and international organisations will be required to achieve this. Because cities house 40% of the world's population, sustainable urban development accounts for a large share of the sustainability goals. Cities are part of a sustainable strategy, since they consume a considerable amount of resources and display clearly unsustainable behaviour. Cities are multi-layered, complicated systems. The scenario in major cities with massive populations, which is driving cities to self-destruction, is especially troubling (Emilija 2016).

Cities are hubs of economic activity, employment, and innovation. Already, metropolitan areas account for 85% of global GDP and 55% of global population. However, the way urbanisation processes are evolving is also highly troublesome. 75% of global greenhouse gas emissions from cities and urban waste pollutes our air, water, and soil resources. As a result, it is difficult to solve global sustainability challenges without also tackling urban sustainability concerns (Laksmanan 2008). In acknowledgement of the importance of cities in the modern world, the United Nations General Assembly voted in 2015 to include "sustainable cities and communities" as a distinct objective (SDG 11) to the Agenda for Sustainable Development (2030). SDG 11's overarching objective is to make cities and human settlements more inclusive, safe, resilient, and sustainable. Recent UN-organised worldwide policy talks, such as COP 24, the Paris Agreement (UN Framework Convention on Climate Change), and the New Urban Agenda of UN Habitat, highlight the importance of a coordinated focus on the city. Community-scale development not only achieves long-term developmental goals but also provides direct concrete advantages to people's quality of life. Newly industrialising Asian and African nations are driving today's global urbanisation. In developing countries, the state typically fails to handle urban challenges appropriately, and gaps in municipal infrastructure persist. Because of population pressures caused by migration and natural expansion, demand for more infrastructure is quickly surpassing availability. While new housing development trailed well behind the rate of urban population growth, the number of people living in slums climbed from 807 to 883 million throughout this period. Globally, the proportion of the urban population living in slums fell by 20% between 2000 and 2014 (from 28.4% to 22.8%) (UN Habitat 2016).

11.11 Localising the SDG Goals in India

Although SDG 11 establishes broad objectives, it is vital to contextualise and integrate global goals into local development processes in order to meaningfully improve people's lives. Urban municipal governments play an important role in this respect. In order to ensure time-bound progress toward reaching sustainability goals, the SDG 11 aims and indicators must be aligned with the infrastructure delivery process via local development plans and budgetary priorities. However, incorporating global aspirations into local planning and policy discussions poses considerable governance and economic issues. Over the last two decades, India's economy has been among the world's largest, with the fastest expansion. India also has the world's second-largest urban system, with 377 million urban dwellers and a 31.14% urbanisation rate, as well as three megacities with populations of at least 10 million (Census 2011). India's urbanisation rate has begun to rise, but at a slower rate than that of other major Asian countries such as China or Indonesia. For the first time in history, urban net population growth outperformed rural net population growth during the census decade from 2001 to 2011. By 2031, the urban population is expected to reach 600 million (40%) and 850 million (50%) by 2051 (ibid).

The Atal Mission for Rejuvenation and Urban Transformation (AMRUT), Smart Cities Mission, Pradhan Mantri Awas Yojana (PMAY – Housing for All), and Swachh Bharat Mission are a few city-focused centrally funded missions launched in recent years by India's national government that are in line with the SDG 11 objectives for sustainable urban development (Clean India Mission). However, it is critical to recall that under the Indian constitution's multilevel governance system, state governments and urban local bodies are principally accountable for urban development. The national government's duties are limited to policy formulation and financial distribution. Direct implementation is the responsibility of the institutions and organisations involved in urban local governance, such as elected municipal governments, city development agencies, and parastatal organisations under state governments.

SDG 11 can be viewed as a thorough model to direct those in charge of urban policy. The aims and indicators (as shown in Table 11.1) are helpful for quantifying welfare gains and assisting in the development of evidence based policy. Additionally, it complements the systems approach to sustainable development and has important implications for a number of other SDG objectives. For instance, SDG 11's goals for inclusive planning and affordable housing are strongly related to SDG 1's goals for ending poverty.

Table 11.1 SDG targets and indicators

Target	Proposed indicators
1.1 By 2030, ensure access for all to adequate, safe, and affordable housing and basic services and upgrade slums	1.1.1 Proportion of urban population living in slums, informal settlements, or inadequate housing
1.2 By 2030, provide access to safe, affordable, accessible, and sustainable transport systems for all, improving road safety, notably by expanding public transport, with special attention to the needs of those in vulnerable situations, women, children, persons with disabilities, and older persons	1.2.1 Proportion of population that has convenient access to public transport, by sex, age, and persons with disabilities
1.3 By 2030, enhance inclusive and sustainable urbanisation and capacity for participatory, integrated and sustainable human settlement planning, and management in all countries	1.3.1 Ratio of land consumption rate to population growth rate
	1.3.2 Proportion of cities with a direct participation structure of civil society in urban planning and management that operate regularly and democratically
1.4 Strengthen efforts to protect and safeguard the world's cultural and natural heritage	1.4.1 Total expenditure (public and private) per capita spent on the preservation, protection, and conservation of all cultural and natural heritage, by type of heritage (cultural, natural, mixed, and World Heritage Centre designation), level of government (national, regional, and local/ municipal), type of expenditure (operating expenditure/ investment) and type of private funding (donations in kind, private nonprofit sector, and sponsorship)
1.5 By 2030, significantly reduce the number of deaths and the number of people affected and substantially decrease the direct economic losses relative to global gross domestic product caused by disasters, including water-related disasters, with a focus on protecting the poor and people in vulnerable situations	1.5.1 Number of deaths, missing persons, and persons directly affected by disaster per 100,000 people
	1.5.2 Direct economic loss in relation to global GDP, damage to critical infrastructure, and number of disruption of basic services, attributed to disasters
1.6 By 2030, reduce the adverse per capita environmental impact of cities, including by paying special attention to air quality and municipal and other waste management	1.6.1 Proportion of urban solid waste regularly collected and with adequate final discharge out of total urban solid waste generated, by cities
	1.6.2 Annual mean levels of fine particulate matter (e.g. PM2.5 and PM10) in cities (population weighted)
1.7 By 2030, provide universal access to safe, inclusive, and accessible, green, and public spaces, in particular for women and children, older persons, and persons with disabilities	1.7.1 Average share of the built-up area of cities that is open space for public use for all, by sex, age, and persons with disabilities
	1.7.2 Proportion of persons victim of physical or sexual harassment, by sex, age, disability status, and place of occurrence, in the previous 12 months
1.a Support positive economic, social, and environmental links between urban, peri-urban, and rural areas by strengthening national and regional development planning	1.a.1 Proportion of population living in cities that implement urban and regional development plans integrating population projections and resource needs, by size of city

(continued)

Table 11.1 (continued)

Target	Proposed indicators
1.b By 2020, substantially increase the number of cities and human settlements adopting and implementing integrated policies and plans towards inclusion, resource efficiency, mitigation, and adaptation to climate change, resilience to disasters and develop and implement, in line with the Sendai Framework for Disaster Risk Reduction 2015–2030, holistic disaster risk management at all levels	1.b.1 Number of countries that adopt and implement national disaster risk reduction strategies in line with the Sendai Framework for Disaster Risk Reduction 2015–2030
	1.b.2 Proportion of local governments that adopt and implement local disaster risk reduction strategies in line with national disaster risk reduction strategies
1.c Support least developed countries, including through financial and technical assistance, in building sustainable and resilient buildings utilising local materials	1.c.1 Proportion of financial support to the least developed countries that is allocated to the construction and retrofitting of sustainable, resilient, and resource-efficient buildings utilising local materials

Source: UN Habitat

Urban areas have significant difficulties in providing basic infrastructural services and affordable housing. Although the percentage of slum households decreased from 18.3 to 17.4 during the censuses of 2001 and 2011, there was a net increase of 13,920,191 overall (Census 2011). Similar to this, Table 11.2 demonstrates that socio-economic conditions and access to necessities had improved, but gaps still existed. The Atal Mission for Rejuvenation and Urban Transformation (AMRUT), Smart Cities Mission, Pradhan Mantri Awas Yojana (PMAY – Housing for All), and Swachh Bharat Mission are a few city-focused centrally funded missions that India's national government has launched in recent years that are in line with the SDG 11 objectives regarding sustainable urban development (Clean India Mission). National missions are specifically focused on two of the SDG 11 targets, with an emphasis on affordable housing (PMAY), sanitation, and access to clean water supply are specifically covered by national missions (Swachh Bharat Mission), as indicated in Table 11.3.

Table 11.2 Socio-economic indicators of Indian cities (in percentage terms)

Indicators	2001	2011
Literacy rate	73.08	84.1
Urban sex ratio	841	929
Infant mortality rate	42	29
Population below poverty line	25.5	13.7
Household with safe drinking water	90	91.4
Household with electricity	91.6	97.9
Household with septic tank and flush	70.7	81.6

Source: Handbook of urban statistics (2019)

Table 11.3 Indicators selected for SDG India index

SDG global targets		Indicators selected for SDG India index		National target value for 2030
3.1	By 2030, ensure access for all to adequate, safe, and affordable housing	1	Houses completed under Pradhan Mantri Awas Yojana for houses as a percentage of net demand assessment for houses	100
		2	Percentage of urban households living in slums	0
3.2	By 2030, reduce the adverse per capita environmental impact of cities, including by paying special attention to air quality and municipal and other waste management	3	Percentage of municipal wards with 100% door to door waste collection	100
		4	Percentage of waste processed	100

Source: SDG India Baseline Index (2018)

However, it is crucial to remember that under the multilevel governance system of the Indian constitution, state governments and urban local bodies are primarily responsible for urban development. The national government's functions are restricted to establishing policies and allocating funding. The institutions and organisations involved in urban local governance, such as elected municipal governments, city development bodies, and parastatal organisations under state governments, are responsible for direct implementation. Following the 74th Amendment Act of 1992, elected local governments have constitutional standing with established functions and responsibilities in relation to the state government. The local administrations are required to carry out 18 tasks that support urban sustainability under the constitutional devolution.

11.12 Further Research and Action

SDG 11 of the 2030 Sustainable Development Agenda recognises the worldwide relevance of cities. Combating climate change, reducing poverty, getting access to clean water, regulating energy usage, fostering social inclusion, and guaranteeing spatial justice are all opportunities to pursue urban sustainability. The New Urban Agenda, which was established in 2016, and SDG 11 have a strong link. The NUA requires member countries to adopt their own National Urban Policies to serve as a framework for attaining urban sustainability goals. The National Urban Policy is expected to manage and monitor urban development priorities in order to carry out the Sustainable Development Agenda 2030. SDG 11's success will be determined by how well it is incorporated into local urban planning frameworks and financial constraints, as well as contextualised to local conditions. Given the interdependence of SDG 11, an integrated urban development plan that crosses sectoral boundaries is required. Employees in government must be capable of fostering synergy within institutional boundaries and building new coalitions while supporting current institutions. The elected ULB, municipal development authority, and other parastatal organisations administering water supply or transportation networks assume main responsibility for SDG 11 implementation. In India, state governments wield considerable influence over urban development, and the vast majority of organisations responsible for delivering urban infrastructure work directly for them, frequently providing services across many administrative domains. Despite the constitutional necessity, elected ULBs are frequently ineffectual. As a result, the major obligation falls to governmental bureaucracy. The federal government has lately begun to place a greater emphasis on cities and direct support through mission-based programmes. Although the initiatives of the mission-based programmes meet a number of SDG objectives, they are carried out by specific project management teams and special purpose vehicles.

11.13 Conclusion

Cities substantially alter people's mobility, exposing them to new challenges while also affording possibilities that were previously unavailable. Cities, on the other hand, may be brutal to the weak and needy. The city is supported by the poor and the weak, who accept incredibly cheap wages while living in deplorable conditions. Perhaps more important is the opportunity for women to enter the labour field, progress their education, achieve financial independence, and participate fully in social and political life as equal citizens. Cities may play an important part in the tale of sustainable development by reimagining past injustices, existing challenges, and future potential. The most major social changes have been established and completely realised in urban zones over the last 250 years or so. Smaller towns and villages, on the other hand, lack the size, population, institutional density, and connectedness that bigger cities do in constructing and organising conceptions of growth and progress, as well as a diversity of future-oriented imaginations. Surprisingly, many of the movements and ideas that create the moral, legal, and practical basis for the SDGs emerged in difficult conditions and in cities all across the world. The conceptual and practical underpinnings of the SDGs, as well as the linkages between SDG 11 and the New Urban Agenda and other SDGs, must be made known to mayors, municipal council members, and senior administrators. There is an urgent need to improve knowledge and develop the requisite abilities among administration's middle and higher management. For integrated planning to incorporate such a diverse collection of stakeholders at many governance levels, a strong institution with technical expertise in planning (economic, social, and physical), management (project, financial, operational, and organisational), and governance would be required (stakeholder engagement, inter-agency coordination). Because the SDG 11 incorporates many sets of objectives and indicators, a database must be built to track progress and compare services offered in order to assist decision-making based on inputs from the bottom up. Urban observatories might be developed to disseminate data and information.

References

Babier EB, Burges JC (2017) The sustainable development goals and the systems approach to sustainability, discussion paper. Economics 11(28):1–23

Census (2011) Population census 2001–11, Registrar General and Census Commissioner. Government of India, New Delhi

Emilija S (2016) Improving sustainable concepts in developing countries: relevant factors in sustainable urban development of urban planning methodology and implementation of concepts for sustainable planning. Procedia Environmental Sciences, Science Direct by Elsevier, pp 140–155

Evans J, Karvonen A, Luque-Ayala A, Martin C, McCormick K, Raven R, Palgan YV (2019) Smart and sustainable cities? Pipedreams, practicalities and possibilities. Local Environ 24(7):557–564

Fabre EA (2017) Local implementation of the SDGs and the New Urban Agenda – towards a Swedish National Urban Policy. Global Utmaning, Stockholm

GLOSSARY – Department of Economic Affairs – InfrastructureIndia.gov.in (2019, December 6) GLOSSARY – Department of Economic Affairs – InfrastructureIndia.gov.in. Retrieved November 20, 2022, from https://www.pppinindia.gov.in/infrastructureindia/glossary

Gurtner B (2010) The financial and economic crisis and developing countries. Int Dev Policyl Revue internationale de politique de développement 1:189–213

Hans A (2017) Rebooting public private partnership in India. NITI Aayog. https://doi.org/10.1007/978-3-030-01488-9_4

Laksmanan L (2008) Public-private partnership in Indian infrastructure development: issues and options. Reserve Bank of India

Priya MS, Jesintha P (2011) Public private partnership in India. J Manag Sci 1(1):82–94

Habitat UN (2016) Sustainable development goal – 11: a guide to assist national and local governments to monitor and report SDG goal 11+ indicators. UN Habitat, Nairobi

Nimisha Jha is a civil servant working with the Government of India for almost 14 years now. She is a Chevening Scholar and an alumnus of London School of Economics and Delhi School of Economics. She has worked at the federal level in the Ministry of Finance, Ministry of Housing, and also at the state level in the government of Chhattisgarh and Jammu Kashmir. She is currently posted as Director at NITI Aayog. She has interests in areas of policy, governance, and culture of India; has published several articles and research papers in journals; and has to her credit a book recently published on the Kosa Silk of Chhattisgarh. She has great love for travelling and cooking and is a passionate marathon runner.

12 Water-Sensitive Urban Design as a Driver for Accelerating Sustainable Urban Development in India

Rajiv Ranjan Mishra, Jyoti Verma, and Manju Rajeev Kanchan

Abstract

The global water scenario has been witnessing a widening gap between demand and supply with rapid increase in population, urbanization, and adverse impact on health of water ecosystem. The urban water system is particularly critical. This has led to increased attention towards urban water management and development of several conceptual approaches and practices. Water-Sensitive Urban Design and Planning (WSUDP) is an important one of them. The chapter is an attempt to briefly explore these concepts and get broad understanding of how India is now envisioning to introduce these water-sensitive concepts at different level through various national initiatives and missions under overarching global sustainable development goals (SDGs) and climate agenda. The link between the application of WSUDP and SDGs in India is explored to underscore the critical role played by WSUDP in building urban resilience towards climate change and climate adaptation. Recognizing the key role of rivers in overall water management, economy, and ecology, this chapter explores Indian efforts in overall river rejuvenation at basin level and special initiatives towards managing urban rivers integrating the WSUDP, development of holistic Urban River Management Plan (URMP), river or water-sensitive master plans and River City alliance by National Mission for Clean Ganga (NMCG) and National Institute of Urban Affairs (NIUA). The chapter further illustrates the points of convergence of URMP and WSUDP along with relevant global and Indian case studies.

Keywords

WSUD · Climate change · URMP · River-sensitive planning · SDGs · Sustainability

12.1 Introduction

It is not a new or unknown to the world the role that water plays in sustaining life while acting as the key to maintaining a sustainable relationship between us and our natural environment. Yet growing population, rapid increase in water demand along with pressures of urbanization and a changing lifestyle have started to impose serious challenges to a water secure future. This is

R. R. Mishra (✉)
National Institute of Urban Affairs and Former DG, National Mission for Clean Ganga, New Delhi, India
e-mail: rajiv_ranjanmishra@yahoo.com

J. Verma · M. R. Kanchan
Water and Environment Vertical, National Institute of Urban Affairs, New Delhi, India
e-mail: jtverma60@gmail.com; manjurajeevkanchan96@gmail.com

V. Dutta, P. Ghosh (eds.), *Sustainability: Science, Policy, and Practice in India*, Sustainable Development Goals Series, https://doi.org/10.1007/978-3-031-50132-6_12

even more special and urgent for India with almost 18% of global population and only 4% of water resources.

India has witnessed a rapid increase in the urban population during the last few decades. This has led to the much demanding dual needs of managing urban water supply as well as sewage and sanitation of around 400 million urban population (2020).[1] All towns and cities currently face the problem of increasing gap between water supply and demand, which elevates the pressure imposed on water resources and its supply requirements. By 2030, India will have 68 cities with populations of over 1 million. Additionally, India will also add two more cities (Hyderabad and Ahmedabad) to the existing list of 5 megacities (10 million population), thereby accounting for 39% of the population living in urban India by 2036.[2] Growing urban centres, with the concurrent process of urbanization, have brought several issues to the fore, from governance and management of these areas to the provision of basic civic services. In the last two decades, built-up area has grown faster than population in nearly all of India's largest cities.

The need for eco-friendly alternatives is growing as the demand-supply gap widens and environmental conditions worsen. To lessen anthropogenic impact on the planet, we must rise to the challenge of regulating and conserving our natural resources. Judicious use of water resources in a sustainable manner is essential for effective water management. Maintaining the hydrological cycle through the management of supply and demand for water, stormwater, wastewater, and groundwater is possible when water sensitivity is incorporated into the design phase of new and existing developments.

Across the globe, cities that are particularly vulnerable to flooding have inspired a number of related terms, most of which have to do with urban drainage. Today's cities require stakeholder coordination, institutional support, and community engagement to tackle the challenges of climate change and rising populations brought on by industrialization and urbanization. As a result, new conceptual frameworks have emerged to support the technical, institutional, and social integration of water resource management. Several paradigms have developed and spread; all with the same goal in mind: to address the complicated issues of urban water supply without compromising the natural environments. These paradigms emerged in response to a number of challenges that cities have in common, such as climate change, rising demand and urbanization, ageing infrastructure, depleted and polluted natural resources, and an increase in the frequency of disruptive and extreme events. These made water infrastructure more susceptible to risk. There have been a number of novel approaches developed to address these root causes since decades. Several ideologies have converged over the past decades, including Integrated Water Resource Management (IRWM), Integrated Urban Water Management (IUWM), Water-Sensitive Urban Design (WSUD) with features like Sustainable Urban Drainage Systems (SUDS), Green Infrastructure (GI), Best Management Practices (BMP), and Water-Sensitive Cities (WSUC), and Water Wise Cities (WWC)[3] out of which India has started to experience waves from some of these for some time now.

12.2 Water-Sensitive Urban Design and Planning

Water-sensitive urban design (WSUD) is an approach to the planning and design of urban environments that support healthy ecosystems, lifestyles, and livelihoods through smart management of all our waters. The water-sensitive

[1]Projected Population. Statistical Year Book India 2018. MOSPI GOI 2018. Retrieved from: http://mospi.nic.in/statistical-year-book-India/2018/171 (Accessed on: 21st June 2021)

[2]Report of the Technical Group on Population Projections, Census of India.

[3]G Mishra, G Acharya, M Iyerand S Doshi 2020, Deconstructing water sensitivity: experiences from global cities, CEPT University, Kasturbhai Lalbhai Campus, University Road, Navrangpura, Ahmedabad 380009, Gujarat, India.

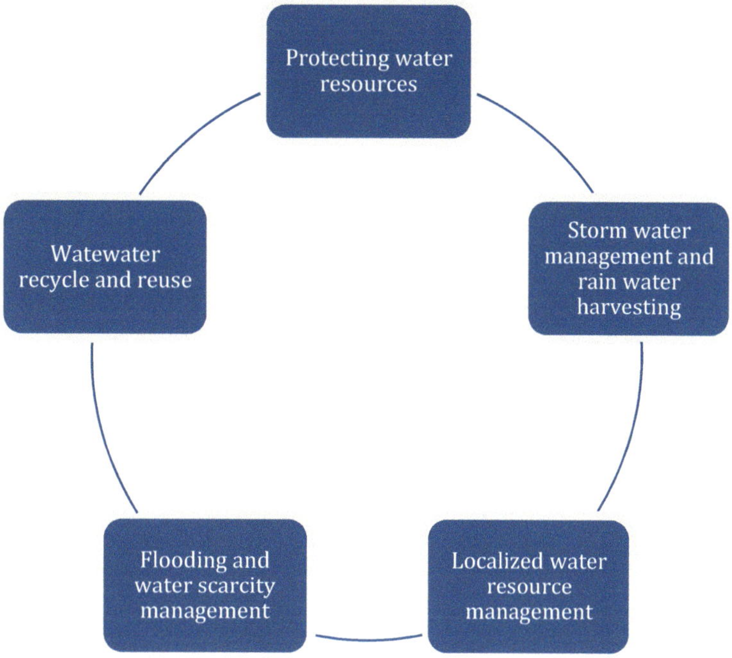

Fig. 12.1 Key principles of WSUDP. (Source: Author)

urban design also leads to a land planning and engineering design approach which is compatible with the urban water cycle and integrates stormwater, groundwater, wastewater management, and water supply into urban design to cut down adverse ecological impacts while also improving aesthetic and recreational appeal. The necessity for water-sensitive urban design applications in developing countries like India is quite high due to the unplanned urban sprawl.[4]

Grassed or landscaped swales, infiltration trenches and bio-retention systems, gross pollutant traps, wetlands and sediment ponds, rainwater tanks – stormwater harvesting and reuse, grey water harvesting and reuse, rain gardens, rooftop greening and urban forests, porous pavements, and aquifer recharge and reuse are just some of the many applications that fall under the umbrella of "watershed system design and development" (WSDD). The introduction of "Ecological Urbanism" (Stearns and Montag 1975) was the starting point of a process that later developed the concept of water-sensitive urban design and planning (WSUDP). Natural hydrogeological systems, such as lakes, ponds, wetlands, etc., can act as a sponge during heavy rainfall events, reducing the impact of urban flooding. This is the central idea behind the water-sensitive design (Fig. 12.1). When considered as part of a larger urban strategy, WSUDP, which is the integrated design of the urban water cycle, can make a positive impact on sustainability and liveability by addressing issues such as water supply, wastewater, storm-water, and groundwater management, urban design, and environmental protection (Niekerk et al. 2018).

12.3 Integrated Water Resource Management (IWRM)

Integrated Water Resources Management (IWRM) is a process that promotes the coordinated development and management of water, land, and related resources in order to maximize economic and social welfare in an equitable

[4]Suresh Kumar Rohilla, Mahreen Matto, Shivali Jainer and Chhavi Sharda 2017, Water-Sensitive Urban Design and Planning: A Practitioner's Guide, Centre for Science and Environment, New Delhi.

Fig. 12.2 IWRM is based on three principles: social equity, economic efficiency and environmental sustainability. (Source: Author)

manner without compromising the sustainability of vital ecosystems[5] (Fig. 12.2).

Integrating Water Resource Management (IWRM) strategies into water and development planning require drawing on a wide range of expertise and perspectives. Consequently, IWRM is a thorough, collaborative planning and implementation instrument for managing and developing water resources in a way that strikes a balance between social and economic needs and guarantees the protection of ecosystems for future generations. Coordination is the key here due to the diverse applications of water usage in agriculture, ecosystem health sustenance, supporting human life and economic activity, etc. An IWRM approach is a transparent, adaptable procedure that brings together decision-makers from all sectors that have an impact on water resources, as well as all relevant stakeholders, in order to formulate policy and make well-considered, well-informed decisions in response to the unique water challenges that each sector must address.

12.4 Sponge City

In 2013, a formal proposal for the "Sponge City Program" (SCP) was developed in China as a novel approach to environmentally responsible city planning. The core thought process is though not completely new; it has its roots in traditional "Water Town" practises in ancient China. Farmers in Ningbo and Suzhou in China, for instance, have historically used agricultural land to create flood buffer zones around cities. The Sponge City concept can be explained by comparing it to a sponge, which soaks up precipitation during storms before releasing the water it has purified through infiltration and percolation to the urban run-off. Through the implementation of green or blue-green/grey policies and integrated infrastructure systems, the SCP hopes to restore the urban hydro-system by collecting, managing, and reusing urban water.

A similar concept which could be considered as a subset of SCP is the sponge basin concept developed by an India-US-based firm called Sponge Collaborative that works on the basis of the functions delivered by natural landscapes (Fig. 12.3). The principles of this concept were tried out using the case of Chennai city whose hydrology has been altered drastically by its urban fabric.

12.5 Room for the River, Netherlands

Protecting riverside communities from frequent flooding and enhancing water management infrastructure in delta regions are at the heart of the Dutch government's flagship project "Room for the River". Due to its low elevation, the Netherlands has a long history of river flooding and its management. The majority of the landmass is below sea level. The country sits at the confluence of the Rhine, the Meuse, and the Scheldt rivers. The "Room for the River" project's central idea is to expand the river's channel to allow it to accommodate unprecedentedly high floodwaters (Fig. 12.4). Several sites across the Netherlands participate in the project along with river-specific strategies.

[5]Definition provided by UN Water.

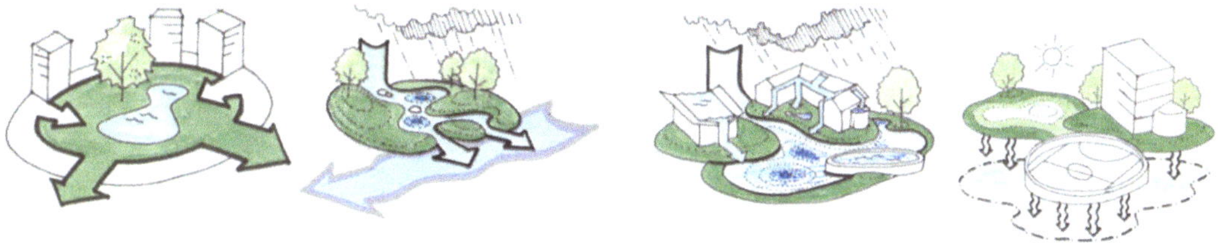

Fig. 12.3 The four principles of Sponge Basin concept. (Source: The Sponge Handbook by Sponge Collaborative)

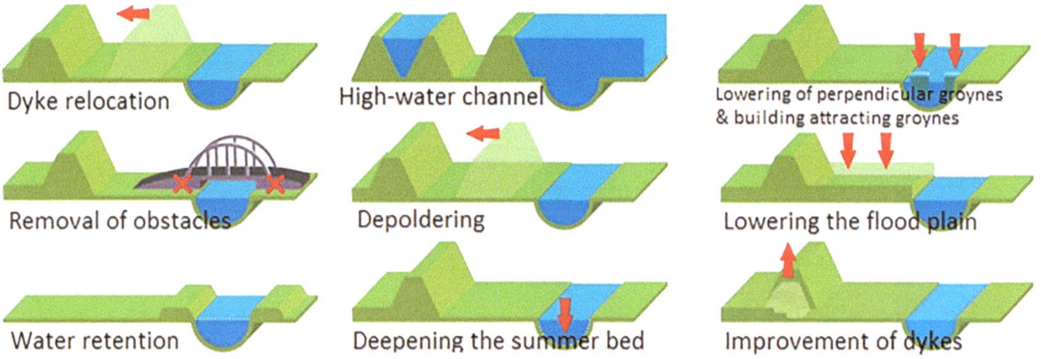

Fig. 12.4 The nine measures of Room for River, Netherlands, project. (Source: Ministry of Infrastructure and Water Management, Netherlands)

12.6 Green Infrastructure

The term "green infrastructure" refers to a system of natural and semi-natural areas, as well as other environmental features, that are planned and managed strategically to provide a variety of ecosystem services, such as water purification, air quality, recreation space, and climate mitigation and adaptation. The health and well-being of locals can be enhanced, thanks to this system of parks, trails, and other open areas, both on land and in the water. It also supports a green economy, creates job opportunities, and enhances biodiversity.[6] Green infrastructure planning (Fig. 12.5) is a successfully tested tool to provide environmental, economic, and social benefits through natural solutions. It has the potential to lessen the demand for "grey" infrastructure, which is often more expensive to construct and maintain, and which can have negative effects on the natural systems. In fact, a carefully throrough green infrastructure can reduce the money shelled out by sewer authorities by reducing the volume of water treated by STPs. The European Commission has developed a Green Infrastructure Strategy with the aim to ensure that the protection, restoration, creation, and enhancement of green infrastructure become an integral part of spatial planning and territorial development whenever it offers a better alternative, or is complementary, to standard grey choices.

[6]Definition as per European Commission.

12.7 WSUD: Multidimensional Integration

Integrating urban water management with urban planning and design has multiple dimensions such as the following:

Integrated management of the three urban water streams of potable water, wastewater, and stormwater

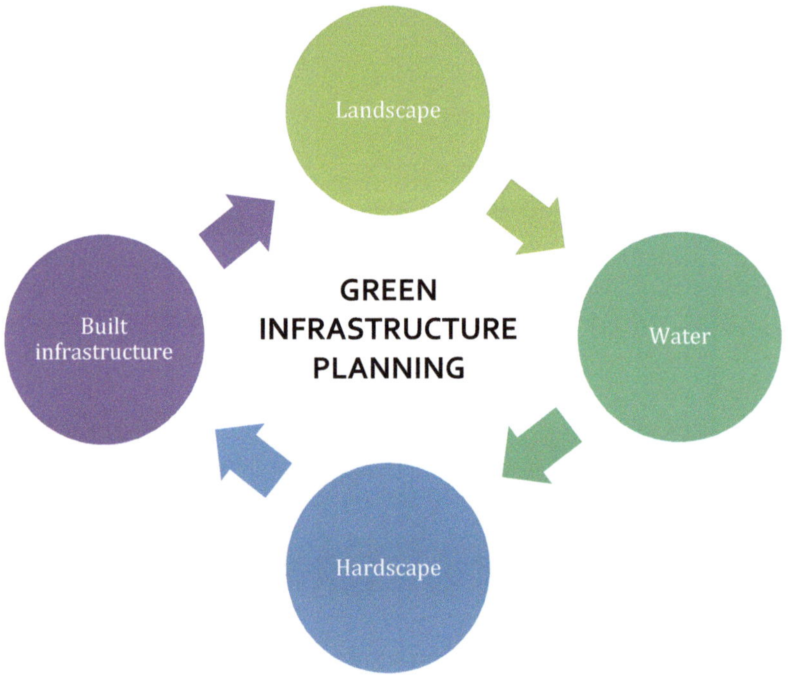

Fig. 12.5 Key components of Green Infrastructure Planning. (Source: Author)

Integration of the spatial scale of urban water management (local, precinct, regional)

Integration sustainable urban water management into the Built Form incorporating building and landscape architecture and public art

Integration of structural and non-structural initiatives (policies for sustainable operation of structured measures, physical and social sciences, compliance assessment, etc.)[7]

12.8 Role of WSUDP in Climate Change and Sustainable Development Goals (SDGs) in India

Rapid and often unplanned urbanization across India has triggered a mosaic of climatic issues across the country, out of which some get buried under the tag of natural disasters. Anthropogenically induced climate change is now pushing urban heat island effect (UHI), pluvial (rain), and fluvial (river) floods into realms beyond human survival. Urbanization and climate change are perhaps the primary triggers for the severity and frequency associated with urban floods in India. The potential climate change impacts on the urban water supply are not limited to reduced hydrological flow posing probable water shortages, increased demand for water, increased stress on future water supply, water contamination via sewage systems, a decrease in surface water and stormwater run-off, increased water tariffs, water originated health problems, etc. The floods of Bengaluru (2022), Mumbai (2020), Kerala (2018), and Chennai (2015) are a clear proof of this climate change-driven precipitation variability. A closer introspection, however, will reveal that terming these as "natural disaster" is perhaps incorrect as the factors causing them are also largely human made due to lack of water sensitivity and neglect of natural ecosystems while taking up construction and other activities leading to the climatic disasters of

[7]Wong, Tony HF, (2007), Water Sensitive Urban Design-the Journey Thus Far, August 2007. Des 11. Page 3, Environment Design Guide.

today such as floods, droughts, etc. Major metropolitan Indian cities such as Bangalore, Hyderabad, Mumbai, etc. face recurrent floods at increased frequencies. Preliminary findings by WRI India reveal that India's ten most populous cities (viz., Ahmedabad, Bengaluru, Chennai, Delhi, Hyderabad, Jaipur, Kolkata, Mumbai, Pune, Surat) that have seen multiple floods in the past 5 years show an increase of 35% in urban development within 20km of the city centre between 2000 and 2015 over low-lying, vegetated, and high recharge potential lands. Likewise built-up cover has increased by 47% (9% in Mumbai to 127% in Surat) and 134% in 20–50 km from the city centre (30% in Kolkata/Mumbai to 412% in Pune) (Goswami and Anand 2020).

The harsh reality is that most of our grey urban infrastructures are ill equipped to withstand the atrocities of both natural and anthropogenic climate change. The core focus of urban flood management in India is directed majorly towards enhancing grey infrastructure, rescue, and relief instead of nature-based sustainable solutions for increased resilience. These are all essential actions, but frequent and persistent floods are a clear-cut indicator of the fact that this is insufficient to battle the recurrent floods. Mitigating climate change impacts is critical to the growth of our cities as they steer towards liveable, resilient, and sustainable urban environments. This has therefore paved the way for a water prudent urban development comprising a hybrid approach of integrated green and grey infrastructure solutions. WSUDP is therefore an ecosystem-based adaptation (Eba) to the vulnerabilities of climate change that has developed along the lines of ecosystem services and resilience thinking.

The multidisciplinary yet integrated approach undertaken by WSUDP to promote resilience at grassroot levels via urban stormwater management is quite essential in increasing the defence of the aesthetic value of blue-green infrastructural nexus as well while also reiterating the importance of hydrological cycle to run its natural course. Destruction and encroachment of the natural flood defences of urban cores are bound to deteriorate the natural resilience power of the cities. Similarly increased ground impermeability directs urban areas towards UHI and droughts.

There are cities in India that are naturally drought prone due to their natural geography (like Udaipur, Jaipur, etc.), while there are cities that are moving towards drought and water insecurity in the near future (like Chennai, etc.). This sets the need for alternative approaches to manage the increased water demand versus the declining supply while also working out strategies for reducing urban climatic vulnerabilities. WSUD offers scope to reconfigure our cities via regenerative water services as sustainable drainage systems (SuDS) as an integral aspect of this design. This further helps our urban environments and specifically the vulnerable sections of our cities such as slums to be resilient against water-based climatic vulnerabilities. In fact, the Ministry of Housing and Urban Affairs (MoHUA)'s think tank – National Institute of Urban Affairs (NIUA) in association with Australian experts – is presently associated with a project titled Participatory and Inclusive Water Sensitive Urban Design for Sustainability and Resilience (PARIWAR) that aims to configure an enabling environment for implementing the principles of WSUDP in the challenging areas of Delhi such as dense slum settlements (Bakkarwala and Mubarakpur Dabas). The results of PARIWAR shall also satisfy SDG 13.1, 13.2, and 13.3 which are associated with climate action primarily. Further Climate Centre for Cities (C-Cube) – a division of NIUA in association with World Bank – has recently published a curated catalogue of best practices for building flood resilience as part of capacity building and knowledge dissemination.

As the common thread of link to nearly all SDGs, water-sensitive urban design and planning holds the power to establish a water-secure, sensitive, and sustainable city which is primarily SDG 11. A foresight that water shall be the primary medium through which the preliminary climatic impacts will be felt by the world makes the WSUDP concepts all the more imperative in the route to achieving climate-sensitive sustainable cities (Kumar et al. 2023).

The principles of WSUDP are not just integral for urban water management, but they may also be expanded to other urban systems that shape the growth of cities such as water, food, energy nexus, transport, used water reuse, and management and the built infrastructure which ultimately contribute towards holistic sustainable development of cities. Climate adaptive architecture for infrastructure development along with a platform for hybrid business models –government and private – is also set. The appropriate WSUD strategies can lead to the economic benefits including savings on capital costs and costs for water quality improvement, as well as increases in market value from the beautification projects measures that are an outcome of selective WSUD strategies. Social benefits include opportunities for the linkages of community nodes green-blue-grey infrastructure nexus. The research agenda for the possibility and scope of what WSUDP can offer cities is diverse and ambitious.

The country has made progress on the SDGs since its inception in 2015, specifically in SDG 1 (no poverty), SDG 6 (clean water and sanitation), SDG 8 (decent work and economic growth), SDG 11 (sustainable cities and communities), and SDG 13 (climate action) between 2019 and 2020 (NITI Aayog's SDG Index Report 2021). In fact, NIUA has undertaken numerous WSUDP initiatives which have linkages to SDGs as shown in Table 12.1 (Niuaadmin 2021).

Table 12.1 Initiatives by NIUA with respect to WSUDP

S. no.	Initiatives	Linkages to Govt. of India missions (directly/indirectly)	Linkages to SDGs
1.	Participatory and Inclusive Water-Sensitive Urban Design for Sustainability and Resilience (PARIWAR)	Jal Jeevan Mission; Atal Mission for Rejuvenation and Urban Transformation (AMRUT); Smart Cities Mission; National Urban Livelihood Mission	SDG 1.5, 6.1, 6.3, 6.B
2.	National Mission for Clean Ganga (NMCG) – Addressing the urban drivers of river health in Ganga Basin	NMCG, AMRUT 2.0, Swachh Bharat Mission (SBM 2.0), National Greening Mission, National Water Mission, Atal Bhujal Yojna, Compensatory Afforestation Fund Management and Planning Authority (CAMPA)	SDG 6.3, 6.5, 6.6
3.	Assessing the Efficacy and Efficiency of the Drink for Tap	AMRUT 2.0	SDG 6.1
4.	Developing an Urban Water Body Diagnostic Tool	Jal Jeevan Mission; AMRUT; Smart Cities Mission; National Urban Livelihood Mission	SDG 6.1, 6.3, 6.B
5.	Sustainable Cities Integrated Approach Pilot	Swacch Bharat Mission; AMRUT	SDG 6.3, 6.5,6.a, 6.b
6.	National Mission for Clean Ganga	NMCG	SDG 11
7.	Urban Resilience and Adaptation for India and Mongolia (URGENT)	No direct	SDG 13.1, 13.3

12.9 Water-Sensitive Urban Design and Planning in India

In recent years, there has been a growing awareness of the need to integrate WSUDP into the master plans for cities. The principles of WSUDP have already been incorporated into the planning processes of some cities around the world, and many more are giving serious consideration to doing the same thing. Incorporating WSUDP into city master plans has a number of advantages, some of which are the reduction of flooding and stormwater run-off, the improvement of water quality, the expansion of green space and habitat for wildlife, and the improvement of liveability and the availability of amenities for residents. There is a pressing need to incorporate WSUD into every facet of the planning process, from the planning of land use to the planning of transportation. Additionally, it is essential to think about how WSUD can be implemented at various scales, from individual buildings to entire neighbourhoods.

In India, sponge cities are being developed in response to the growing problem of urban flooding. Chennai, a city on the southeast coast of India, is one of the first cities in the country working on the sponge city concept. Sponge cities are designed to be more water-sensitive and flood resilient by managing stormwater through a variety of techniques, including green infrastructure, infiltration, and detention. Heavy rains and poor drainage have caused devastating floods in Chennai in recent years. The city was flooded in 2015 by over 2 feet of rain in less than 24 h, resulting in widespread destruction and loss of life. The city is developing a sponge city master plan to mitigate the effects of natural disasters. The plan incorporates a number of strategies to enhance drainage, decrease run-off, and store more water. Green infrastructure is a priority in the master plan for the Sponge City of Chennai. Plants and other natural elements are used in green infrastructure, a method of stormwater management. Rain gardens, green roofs, and tree planting are all part of the plan for Chennai's green infrastructure. These actions will lessen the amount of water that flows out of the city, thereby lowering the risk of flooding. The master plan for Chennai's "sponge city" incorporates green infrastructure and other strategies to increase infiltration and detention. When rain falls on land, instead of flowing off into gutters and sewers, it "infiltrates" or soaks into the ground. Detention is the practice of temporarily holding water in a basin or pond before releasing it gradually into the surrounding area (Goswami and Basak 2020).

The government of India is working to advance WSUD in urban areas through various plans and initiatives, such as the National Action Plan on Climate Change and the Atal Mission for Rejuvenation and Urban Transformation (AMRUT). Additionally, MoHUA is also working on National Guidelines for Sustainable Urban Development, which will function as a tool for the planning and design of urban infrastructure projects that are sustainable.

In the last few years, many Indian cities have started working on interventions aimed at flood-proofing and stormwater management at various scales. Bhopal's green-blue smart city plan, Delhi's Masterplan 2041, and Chennai's water as leverage initiative are all addressed at an urban policy and development guidelines scale. Other cities are experimenting with landscape design-led solutions to tackle the flood fury such as Chennai and Bengaluru with recharge wells.

Local governments in India are working on a city water action plan to address the country's dire water situation. In order to maximize water efficiency and minimize waste, this strategy will incorporate a number of different tactics. Rainwater harvesting is one of the plan's central tenets. The rooftop run-off will be collected and used to replenish aquifers as part of this plan. Water conservation measures, such as repairing leaking pipes and educating the public about the importance of conserving water, will also be implemented as part of these plans. The plan's focus on water management is meant to help cities weather the effects of climate change and a growing population. The city can ensure its citizens' continued access to potable water for decades to come by enhancing its water efficiency and management systems.

12.10 Making City Master Plans Water Sensitive

Existing planning of cities does not consider management of water bodies, rivers, and vegetation as an important component. There is an urgent need to integrate this thinking and processes in to the city planning, which can be achieved through the City's Master Plan having legal sanctity.

Under the Namami Gange Mission, emphasis was given to integrate urban planning which is also sensitive to water. After the comprehensive analysis of the master plans for major cities in India under this initiative with assistance of NIUA, it was found that most of these do not have adequate water bodies or river-specific provisions.

A strategic guideline has been developed for the town planners/city planners, on how to integrate the water-/river-sensitive thinking into the city's master plans. It seeks to leverage on the legal sanctity of the Master Plan to ensure a harmonious relationship between cities and water. The current Master Plan for Delhi has integrated these aspects, but the effectiveness is subject to its implementation on ground. We need to learn from our past experience not only for what worked but also for what did not work and what changes it require to be an effective on ground. This approach would be helpful in adopting water-sensitive urban design, an emerging urban development paradigm for overall sustainable urban development and improved climate resilience. Sensitization of the officials and city planners on this aspect should be the foundation step towards planning for water-sensitive cities.

Cities that are located along rivers are more susceptible to flooding and other climate-related disasters if their natural courses and ecosystem get disrupted by haphazard development. A river should be viewed not as an isolated feature but as part of a larger system. Along the same lines as WSUDP, NIUA along with National Mission for Clean Ganga (NMCG), has come up with a plan to manage urban river stretches in India. This plan is known as the "Urban River Management Plan" (URMP) (NIUA and NMCG 2020). It was launched as a response to the problem of the deteriorating health of rivers in India, beginning with the Ganga River. The overarching goal of the URMP is to propagate river-sensitive development in Indian cities, in tune with the idea of water-sensitive planning and design. Every Indian River city is required to include a ten-point agenda (Fig. 12.6) as part of the process of developing their own city-specific URMPs.

The components of this agenda are grounded in the principles of sustainable development and advocate interventions under the environment – economic and social categories while following a generic framework which offers the needed flexibility to be used by all river cities equally, regardless of the difference in their characteristics, needs, and aspiration. One can even notice how the 10-point agenda components harbour a link with the principles of WSUDP concept as well (Table 12.2).

Supporting the URMP, strategic guidelines were developed by NIUA and NMCG for making river-sensitive master plans (NIUA and NMCG 2021) with the sole intention of guiding and assisting city planners across the country at large on how they may integrate river-sensitive thinking into a Master Plan. Town and Country Planning Organization (TCPO) and Ministry of Housing and Urban Affairs as also come up with a guideline for river-centric urban planning in India.

Fig. 12.6 10-point agenda of the Urban River Management Plan framework. (Source: NIUA)

Table 12.2 Link between URMP 10-point agenda and WSUDP

URMP objectives	WSUDP principles
1 – Regulating floodplain activities	Aquifer recharge and help in base flow to river in the lean season, important in climate action too
2 – Pollution-free river 4 – Enhance riparian buffer	Storm-water management at public places, including open areas in cities through elements of landscape design (e.g. vegetated swales and buffer strips, bio-retention systems)
3 – Rejuvenate waterbodies and wetlands	Protecting local waterbodies (lakes, ponds, and wetlands) for supplementary water sources
5 – Increased use of treated wastewater	Recycling and reusing wastewater naturally (low cost/low energy) and not treating it as a liability
6 – Maximum good quality return flow	The attention on bringing back flow with good quality brings home focus on maintaining the minimum flow to sustain ecology and also to make urban ecosystem sensitive towards this
7 – Eco-friendly riverfront projects 8 – Leveraging economic potential of the river	Adding value to the social and ecological aspects of areas by planning and designing the built environment in accordance with community needs and water issues
9, 10 – Citizen awareness in river-sensitive behaviour and engage citizen in river management activities	The river is conceptualized in broader sense and support and proactive involvement of community is essential to promote the activities for water-sensitive design

12.11 Riverine Islands in WSUD Initiatives

A riverine island is a natural or artificial land mass surrounded by water, usually found in rivers or streams and can provide valuable habitat for wildlife. These can be small or large in size and in rural and urban areas. The benefits of conserving these islands in proximity to urban areas as part of WSUD can be in terms of (a) flood mitigation by acting as barrier, (b) improving water quality as natural filters trapping sediments and pollutants reducing some expensive treatment solutions, (c) enhanced habitat for biodiversity and inter-habitat connectivity, and (d) aesthetics and recreation:

Government of India has recently begun efforts aimed at the management and conservation of riverine islands. The Indian National Trust for Art and Heritage (INTACH) and NMCG are working together to develop a policy for the riverine islands on the Ganga. As part of riverfront development, Hooghly Chinsurah Municipality has sought technical advice from NIUA and INTACH on developing strategies for a nearby river island.

12.12 A Global Outlook

Being a place-based approach of sustainable integrated water management, there are no standardized solutions of WSUDP that can be readily transposed from one city to another. Yet there are numerous WSUDP-based innovative global multi-scalar interventions and practices worth adapting for increasing the resilience of cities to combat natural phenomena. At a grassroot level, using a multidisciplinary approach to develop a transport infrastructure that is an amalgam of engineering, landscape design and hydrology assist in increasing the overall permeability of the ground surface for recharge and discharge of stormwater. The Green Streets Program in the USA is an example of this. As a next layer, urban green acupuncture is another successful practice worth adapting. Cities like Paris, Singapore, Rotterdam, and New Orleans have transformed their existing open/vacant greenfield and brownfield zones with recharge and discharge potential into water-prudent and water-sensitive landscape intervention points. The Bishan-Ang Mo Kio Park, Singapore, and Benthenheim Water Square in Rotterdam are classic examples of the same.

Hybrid blue-green-grey solutions over dense urban fabric such as Chulalongkorn University Centenary Park and Thammasat University urban rooftop farm, when integrated with acupuncture interventions, can help ease the burden on citywide urban drainage and sewage infrastructure by forming unique clusters of micro-watersheds which act as decentralized off-grid systems for effective urban flood management.

A city-wide approach that integrates the above-discussed strategies is equally beneficial in achieving the tag of water-sensitive and secure cities as seen in case of both Vancouver and Copenhagen. Vancouver's hybrid rain city strategy focuses on adaptive and flexible stormwater infrastructure that adapts to the variable precipitation scenarios. Similarly, Copenhagen's cloudburst plan imbibes urban planning, blue-green-grey solutions, traffic and hydraulic analysis backed by strong investment strategies as part of urban flood reduction, and improved urban liveability. Provision of a clear passage of natural hydrological nexus to function at a regional level is yet another integral aspect of WSUDP. The Netherlands Room for the River program works along similar lines through means of regional interventions for managing the higher water levels across the four rivers and subsequently mitigate urban flooding along rivers.

While interventions and practices are the real deal, policies and frameworks are also equally important to ensure the success and longevity of these interventions. Singapore and China have green policies and programmes that support widespread urban flood-proofing via enabling financing and collaborative implementation of innovative hybrid solutions. The latter's ambitious sponge city initiative offers financial aid to its cities so that at least 80% of urban areas absorb and reuse at least 70% of stormwater by 2020 as observed in the case of pilot demonstration city Wuhan. Despite such policies and frameworks, key speed breakers do prevail when it comes to factors such as operationalization of water-sensitive city policies.

12.13 Conclusion and Way Forward

Importance of water is well known, but still water management has been a challenge. The idea of water-sensitive cities has started to gain traction, and progress has been made in the past decade globally as well as in India. The water literacy is improving, and people, especially younger generation, are getting conscious about water- and sanitation-related issues. The understanding of our ancestors was much deeper, and traditions believed in coexistence with nature. The thought of becoming water sensitive is nothing but reimagining these traditions in light of latest scientific and technological global best practices. Water is perhaps the only entity capable of linking all the aspects of sustainable development and connects the water-food-energy-biodiversity-forest-economy nexus. This is why the idea of WSUD becomes critically important for India's march to sustainably developed nation. It has also evolved from its earlier days of stormwater management to becoming a framework for integrating the three water streams holistically with the urban design. In this decade, we are likely to see water-sensitive practices in the mainstream of urban development and all stakeholders becoming active and proactive practitioner. Through this, cities will develop transformative capacities to sustain their broader liveability, resilience, and prosperity over the long term. City stakeholders can already draw on significant actionable knowledge to rapidly march towards their water-sensitive vision.

Upscaling and mainstreaming of water-sensitive cities practices, however, will not be easy – mainly impeded by institutional factors. Urban planning still remains rule-bound to traditional practices that are based on highly fragmented and single-dimension objectives, supported by outdated and narrow economic methods for assessing the benefit-cost of infrastructure investment. Overcoming these impediments and creating the enabling environment for transformations to a higher scale need to be a collective mission. There would be a need to apply composite measures adapted to the local context. Governments would need to strive for enhancing system capabilities to cope with the growth and changing paradigm. Infill developments are prominent in urban densification and are often spatially piecewise. These features make them ideal opportunities to combine decentralized solutions with existing trunk infrastructure as integrated hybrid systems. By creating these hybrid solutions, we can ensure that the existing infrastructure is not stranded as we embrace emerging technologies and integrated system thinking. This approach ensures optimal economic returns from past and future infrastructure investments. Present practice of installing grey infrastructure for urban flood resilience is no longer adequate to meet the erratic and increased run-off with changing climate. Hybrid blue-green-grey solutions possessing nature-based components, are also cost-efficient and scalable, and offer a multitude of ecosystem services.

WSUD needs to be considered at all stages of urban development, from planning and design to construction and maintenance and its success depends on the cooperation among multiple and diverse stakeholders such as government agencies, developers, engineers, planners, and the community. There would be a need for a combination of policies, strategies, regulatory and planning tools supplemented by area-specific projects and local interventions for conservation and restoration.

A robust policy framework can help ensure that WSUD principles are embedded into decision-making at all levels, from planning and design through to construction and maintenance. At a national level, policies and frameworks associated with inculcating a blue-green sensitivity through WSUD, SUDs etc. may be mandated through different urban and water missions. Trans-state boundary approaches such as watershed and basin management programmes need to be mainstreamed at a state level while strategies such as sponge city plan, IUWM, URMP plan for the city may be taken up as part of grassroot level initiatives. The future city planning would need to consider likely changes in the blue-green landscape infrastructure over time and its implications on the cities. Building the capacities of decision makers as well as community towards WSUD is critical. The challenge ahead is to continue the collaborative approach to urban design and to provide technically robust solutions that excite and engage communities and deliver improved environmental outcomes for today and future generations.[8] A strong policy foundation, proper capacity building and engaging communities would ensure that WSUD initiatives bring lasting change and don't remain ad hoc and piecemeal.

References

Goswami S, Anand A (2020, September 3) Living with water: integrating blue, green and grey infrastructure to manage urban floods. WRI INDIA. Retrieved November 11, 2022, from https://wri-india.org/blog/living-water-integrating-blue-green-and-grey-infrastructure-manage-urban-floods

Goswami S, Basak S (2020, August 5) Is urban India going underwater again? WRI INDIA. Retrieved November 10, 2022, from https://wri-india.org/blog/urban-india-going-underwater-again

Kumar A, Button C, Gupta S, Amezaga J (2023) Water Sensitive Planning for the Cities in the Global South. Water 15(2). Retrieved December 24, 2023, from https://www.mdpi.com/2073-4441/15/2/235

Niekerk WVAN, Duncker L, Maditse K, Davis C, Pieterse A (2018) Water sensitive urban planning as adaptation strategy. Retrieved November 11, 2022, https://isocarp.org

NIUA and NMCG (2020) A strategic framework for managing urban river stretches in the Ganga River Basin, Urban River Management Plan (URMP). Accessed on October 2022 from website https://nmcg.nic.in/writereaddata/fileupload/48_Urban%20River%20Management%20Plan%20framework.pdf

NIUA and NMCG (2021) Strategic guidelines for making river-sensitive master plans. Accessed on October 2022 from website https://nmcg.nic.in/writereaddata/fileupload/59_Mainstreaming%20Urban%20River%20report%20-%20compressed.pdf

Niuaadmin (2021, October 1) NiUA – SDG dashboard. niuaadmin. Retrieved November 16, 2022, from https://niua.in/niua-sdg-dashboard#sdg13

Stearns F, Montag T (1975) The urban ecosystem: a holistic approach. New York: Halsted Press. Retrieved September 12, 2023

Wong, Tony HF (2006) An overview of water sensitive urban design practices in Australia, Water practice & technology, vol I, no I. IWA Publishing. https://doi.org/10.2166/WPT.2006018

Wong, Tony HF (2007, August) Water sensitive urban design-the journey thus far. Environ Des 11:3

[8] Wong, Tony HF, (2006), An overview of Water Sensitive Urban Design Practices in Australia, Water Practice & Technology Vol I No. I, IWA Publishing 2006 doi 10.2166/WPT.2006018.

Rajiv Ranjan Mishra is an IAS officer of the 1987 batch and IIT Kanpur alumni, and he virtually changed the discourse on Ganga rejuvenation and its long-term sustenance during his two separate stints at the National Mission for Clean Ganga (NMCG), first as Mission Director and then as the Director-General. He expanded the scope of the mission to include the entire ecosystem of the river, its basin and the clean-up of Ganga, and its tributaries. He has made pioneering effort for developing a framework for planning for river cities by mainstreaming it in urban master plans, water sensitive urban design and developing first ever River City Alliance in India. He coauthored a book, Ganga: Reimagining, Rejuvenating, Reconnecting giving an account by a change maker, of the enormity of challenge, institutional processes, and reforms which led to momentum and impact on river health and how it could be sustained. He is continuing his work as chief advisor of National Institute of Urban Affairs (NIUA) and professor of Practice, IIT, Roorkee.

Jyoti Verma is a civil engineer and an environmental planner who holds her bachelor's degree in civil engineering from Gujarat Technical University and a master's degree in urban and regional planning, specialization in environmental planning from CEPT University, Ahmedabad. She has previously worked as an urban river planner at the National Mission for Clean Ganga (NMCG), where she was involved in the planning perspective of the Ganga River rejuvenation. Currently, she is working as a senior research specialist at the National Institute of Urban Affairs (NIUA), New Delhi. Her career inception in a small but rapidly growing organization offered her opportunities to handle various responsibilities, such as background study, data analysis, survey, tools development, strategy planning, ground-level implementation, and monitoring of urban sanitation. She did valuable research work for the book, *Ganga: Reimagining, Rejuvenating, Reconnecting.*

Manju Rajeev Kanchan graduated in Architecture from College of Architecture, Thiruvananthapuram. She completed her Masters in Landscape Architecture from the School of Planning and Architecture, Vijayawada, winning the Gold Medal for the best student in the department. Her postgraduate thesis, "The sombre case of an odd triad – dolphin, man and zoonosis," was adjudged the best thesis that year in SPA. She is also a recipient of numerous prestigious accolades, such as Sheikh Hamdan and Sharjah Award for overall excellence and first place in Namami Gange-sponsored thesis competition on "Reimagining Urban Rivers" by NIUA and NMCG. She has also held leadership roles such as school assistant head girl, SPA literary club coordinator, etc. She now works as a research associate under the Water and Environment Vertical at NIUA. Outside work, she is an avid bibliophile, orator, a growing polyglot with a penchant for research, and an aspiring global citizen with diverse interests and hobbies.

SDG 13 and Climate Change in India

13

Tania Chakravarty and Priyanka Ghosh

Abstract

Climate change—one of the significant environmental issues in today's world—has caused tremendous impacts on earth's physical and human systems. The problems associated with climate change such as change in precipitation and temperature, frequent tropical storms, floods, droughts, forest fires, rising sea level, and melting of glaciers are visible throughout developed and developing countries. In order to combat such global problem, the United Nations has recognized the role of more developed and less developed countries in climate change mitigation and adaptation under Sustainable Development Goal (SDG) 13. As a United Nations member, India has also adopted measures toward combating climate change within the country as well as in the South Asia and has shown significant commitment to mitigation efforts. India had pledged to reduce its carbon intensity, increase the share of renewable energy in its energy mix, and enhance energy efficiency. The country has set ambitious targets for expanding its renewable energy capacity, particularly in solar and wind energy. India is actively involved in international climate negotiations and has been part of global initiatives like the Paris Agreement.Moreover, India has recently moved away from its defensive traditional approach toward climate change issue and has adopted an active role in developing long-term strategies for mitigation and adaptation through bilateral, multilateral, and minilateral agreements with other nations. This chapter examines the climate change issue in India with a larger global context. Additionally, it demonstrates India's contemporary role in global climate diplomacy.

Keywords

Climate change · Global climate politics · Climate policies · India · SDG 13

13.1 Introduction

Climate change is one of the significant environmental challenges which is looming large worldwide, and it has caused an enormous impact on natural and human systems. In many parts of the world, the problem of climate change is associated with changing precipitation pattern, change in water resource availability, and shrinking of gla-

T. Chakravarty
Vivekananda College (Affiliated to University of Calcutta), Kolkata, India
e-mail: taniachakravarty@gmail.com

P. Ghosh (✉)
VIT-AP School of Social Sciences and Humanities (VISH), VIT-AP University, Amaravati, Andhra Pradesh, India
e-mail: priyanka.ghosh@vitap.ac.in

V. Dutta, P. Ghosh (eds.), *Sustainability: Science, Policy, and Practice in India*, Sustainable Development Goals Series, https://doi.org/10.1007/978-3-031-50132-6_13

ciers. Climate change is also impacting terrestrial, marine, and freshwater species as these species are shifting their geographical range, migration pattern, and seasonal activities in order to cope up with the issue (IPCC 2014). In 2016, the United Nations introduced its 17 Sustainable Development Goals (SDGs) as a global agenda to end poverty, protect the planet, and ensure prosperity for all inhabitants by 2030. Among these 17 goals, SDG 13 addresses the issue of climate change. The subgoals under SDG 13 include strengthening adaptive capacity toward climate change-induced hazards and natural disasters in all nations; integrating climate change measures into national policies; raising awareness among citizens and institutions on climate change mitigation, adaptation, and impact reduction; and promoting effective climate change-related planning and management in less developed countries centered around women, youth, local, and marginalized communities (United Nations 2015). The United Nations has acknowledged the importance of socioeconomic development of each country and the role played by respective national governments in it. However, such development measures should be carried out in a sustainable manner with considerable national and international financial support (United Nations 2015).

13.2 Climate Change and Global Scenario

Prior to 1850, global surface temperatures were relatively stable. Post 1850 rapid industrialization led to an increase in greenhouse (GHG) emissions which in turn led to rise in global surface temperatures. According to the World Meteorological Organization, years between 2015 and 2022 were recorded as the hottest years (WMO, 2022). At present, the global surface temperature is 1.14° C above preindustrial levels. Scientists and policy-makers are making efforts to limit global warming at 1.5° C to avoid catastrophic impacts of climate change. However, over the past years, the dangers of climate change are becoming more and more visible (ibid.). Over the past 50 years, climate-related natural disasters have increased from less than 100 (1970) to over 300 (2021) per year (International Disaster Database 2022). Frequency of extreme weather events like heat waves, hurricanes, droughts, heavy rainfall, flooding, and wildfires is increasing around the world. The devastating 2022 flood in Pakistan is one such example of extreme weather events. Around 1700 people died, more than 2.2 million houses were damaged or destroyed, and over 33 million people were affected by the flood. The flood started around mid-June 2022 due to heavy monsoon and continued till November displacing over 15% of the country's population (UNICEF 2023). Climate-related natural disasters do not affect everyone equally. Children suffer more than adults, and underprivileged communities are affected worse than the political and economic elite (Ibid.). Although natural disaster-related fatalities have reduced over the years due to early warning system, the economic losses associated with climate change have increased sevenfold over the past 50 years (World Meteorological Organization 2021; World Bank 2023).

On an average, the global sea level has increased 8–9 inches since 1880 (Lindsey 2022). This rapid surge of economic damage is indicating the increasing effects of climate change on our environment and society. Since the 1870s, high greenhouse gas emissions, rapid melting of glaciers and ice sheets, and thermal expansion of seawater have led to sea level rise. In 2021, global sea level rose to 3.8 in (above 1993 levels) setting a record (Ibid.). According to NOAA (2022), the global mean sea level would rise at least 0.3 m or 1 ft above 2000 levels by 2100; even the greenhouse gas emissions take a lower pathway. Urban areas along the world's coastline are increasingly threatened by rising sea level. Rising sea levels pose considerable threat to coastal ecosystems, including destruction of natural habitat of marine creatures, coastal erosion, loss of human settlements and livelihoods, and contamination of freshwater aquifers. Higher sea level also means destructive tropical storms can travel further inland than earlier (Ibid.).

Climate change can cause extreme damage to infrastructure and push people into long-term poverty. It has been estimated that by 2030, 132 million people will face poverty due to climate change-induced food insecurity and poor labor productivity. At the global scale, food security was affected by degradation of at least 100 million hectares of productive land each year in 2015–2019 (UN SDG Report 2023). Certain areas such as Sub-Saharan Africa, Western Asia, Latin America, and Southern Asia experienced faster land degradation than the global average (Ibid.). Lack of accessible and affordable resources will force over 200 million people to migrate within their respective countries (World Bank 2021). The sixth Assessment Report of IPCC (2022) projected the impacts and risks of climate change at a global scale, observing that in the future, multiple climate hazards will occur simultaneously increasing overall risk to the ecosystems and human societies. The risks have been projected for three time periods: near term (2021–2040), midterm (2041–2060), and long term (2081–2100) based on different global warming levels for multiple decades. Additionally, the report examines the interconnected nature of such risks by focusing on the "non-climatic global trends" like overconsumption of natural resources, rapid urbanization, socioeconomic inequalities, biodiversity loss, and environmental degradation (IPCC 2022, p. 5).

13.3 Climate Change in India

India being a country of varying physiographic diversity is highly vulnerable to the impacts of climate change (Athal 2022). Average temperature in India has increased by 0.7° C between 1901 and 2018. The sea surface temperature in Indian Ocean has also increased by 1° C between 1951 and 2015 (The Tribune 2020; India Today 2022). In 2022, the annual global temperature was about 1.15° C above the level of preindustrialization. According to the World Meteorological Organization (WMO), 2022 was the eighth consecutive year (2015–2022) which experienced at least 1° C temperature rise above the preindustrial levels (WMO 2023). In India, 2022 was also the fifth hottest year since the commencement of recording climatic data in 1901. During the months of March and April 2022, India experienced record-breaking temperature rise than the normal affecting western Himalayas, the plains of Punjab, Haryana, Delhi, Rajasthan, and Uttar Pradesh. Several states such as Odisha, Madhya Pradesh, Gujarat, Chhattisgarh, Telangana, and Jharkhand experienced heat waves with temperatures ranging from 40° C to 44° C toward the end of March. Such heat waves prevailed until the end of April, and almost 70% of India was affected. The coastal areas and eastern parts of India also experienced heat waves toward the end of April. 2022 heat waves also led to forest fires in India. On April 28, 2022, about 300 large forest fires occurred in India due to heat waves, and at least one third of those fires occurred in Uttarakhand. India's vulnerability to climate change can also be clearly understood from the increase in extreme weather events such as cyclones, heavy rainfall, floods, landslide, lightning, thunderstorm, and droughts. In 2022, 15 cyclonic disturbances (3 cyclonic storms and 12 depressions) formed over north Indian Ocean which was against the normal.

In 2022, Southwest monsoon season was above normal. Among 36 meteorological subdivisions, 12 subdivisions recorded excess rainfall, 18 recorded normal rainfall, and the remaining 6 recorded deficient rainfall. Excessive rainfall (over 2000 mm) was recorded in parts of Arunachal Pradesh, Assam and Meghalaya, Sub-Himalayan West Bengal and Sikkim, the entire west coast, East Madhya Pradesh, and Andaman and Nicobar Islands. The west coast, parts of Assam and Meghalaya, also received more than 3000 mm of rainfall (IMD 2022 Annual Report). In contrast to this high rainfall, southern parts of India received its lowest rainfall for the month of June in 122 years. The region recorded only 88.6 mm rainfall for the same month which is 45% less than the recorded rainfall between 1971 and 2020 (Sangomla 2023). Telangana (53% deficit), Kerala (52% deficit), Karnataka (44% deficit), and Andhra Pradesh (26% deficit) experienced huge deficit in rainfall. Due to huge pre-monsoon (34% deficit) and monsoon deficit, the rainforests of Kerala did not receive much rainfall (Ibid.). Northern part of India also experienced the effects of climate change when floods hit Delhi on July 19, 2023. Due to excessive rainfall, the water level in Yamuna River reached 208.66 meter breaching its embankments and flooding the city. The flood caused massive damage to businesses and property, forcing over 27,000 people to evacuate their homes (Ibid.). Although Irrigation and Flood Control Department identified east Delhi as a flood-prone region, massive floodplain encroachments and rapid development have occurred in this area. Since 2009, 2,480 hectares of land in the floodplains of Yamuna have been encroached (The Hindu 2023).

Recently Forest Survey of India collaborated with BITS Pilani (Goa campus) to map the climate hotspots in Indian forest areas which are likely to face considerable impact of climate change. These climate hotspots were categorized as high, very high, extremely high, and critical based on the temperature and rainfall change parameters in three different time periods, 2030, 2050, and 2085. By studying these projections, it has been found that Ladakh, Jammu and Kashmir, Himachal Pradesh, and Uttarakhand are likely to experience highest temperature increase, whereas West Bengal, Tamil Nadu, Goa, and Andhra Pradesh are likely to experience least temperature rise. Northeastern states and upper Malabar coast of India will witness the highest increase in rainfall. In contrast, northwestern parts of India like Ladakh, Jammu and Kashmir, and Himachal Pradesh will witness least increase or even decline in rainfall (Govt. of India, Ministry of Environment, Forest, and Climate Change, India State of Forest Report 2021). By the year 2030, about 315,667 km^2 of forests will be identified as climate hotspots which will cover 45% Indian forests. Himalayan ecosystems and forests (western and northeastern regions) will be exposed to extreme impacts of climate change. These forests and ecosystems will include Himalayan dry temperate forests, Himalayan moist temperate forests, Sub-Alpine forests, Moist Alpine scrub, and Dry Alpine scrub (Ibid.).

Increased anthropogenic activities and climate change have led to severe land degradation in India. As per the Desertification and Land Degradation Atlas of India (ISRO 2018), nearly 97.85 million hectares (mha) of land have been degraded of which 3.32 mha has been added in the last 15 years. Most of the degraded land in India is unirrigated rainfed agricultural land. In case of irrigated land, there is a vegetation cover on top. However, in case of rainfed farmland, the top soil is exposed to water and wind erosion in between two crop cycles (Sengupta 2021).

13.4 Climate Change and Global Policies

In recent past, climate diplomacy has become an important part of addressing climate change at the international level. The Montreal Protocol of 1987 is one of the earliest multilateral environmental agreements which paved the way for future climate diplomacy. The protocol was intended to reduce ozone depleting substances (e.g., CFCs) for both developed and developing countries in a phase-wise manner. The Montreal Protocol successfully eliminated 99% of all ODS from the atmosphere (UN Environment Program n.d.; Maizland 2022). The UN Framework of Convention on Climate Change (UNFCCC) was commissioned at the United Nations Conference on Environment and Development (UNCED) in 1992, and it entered into force as well as ratified by 197 countries in 1994 (Kuh 2018). UNFCCC defined climate change as "a change of climate which is attributed directly or indirectly to human activity that alters the composition of the global atmosphere and which is in addition to natural climate variability observed over comparable time periods (UNFCCC 1992a)." The objective of the convention is to stabilize greenhouse gas concentrations into the atmosphere "at a level that would prevent dangerous anthropogenic interference with the climate system. Such a level should be achieved within a time-frame sufficient to allow ecosystems to adapt naturally to climate change, to ensure that food production is not threatened and to enable economic development to proceed in a sustainable manner (Article 2, UNFCCC 1992b)." The first principle of

UNFCCC states that all countries who are Parties to the UNFCCC "should protect the climate system for the benefit of present and future generations of humankind, on the basis of equity and in accordance with their common but differentiated responsibilities and respective capabilities (Article 3, UNFCCC 1992c)." Though the convention was not binding, it was expected that the developed countries would take the lead in combating climate change. The UNFCCC became the first convention to provide the protocols and mandates that formed the basis for the future climate agreements (Kuh 2018). Under the convention, a Conference of the Parties (COP) was established to promote and facilitate information on measures adopted by the parties to address climate change and its effects taking into account the different circumstances and capabilities of the parties. Following the enforcement of UNFCCC, at the third COP held in 1997 in Kyoto, Japan, a protocol commonly known as Kyoto Protocol was adopted. It recognizes the six greenhouse gases (GHGs) that contribute the most toward global warming: carbon dioxide (CO_2), methane (CH_4), nitrous oxide (N2O), hydrofluorocarbons (HFCs), perfluorocarbons (PFCs), and sulfur hexafluoride (SF6). The Kyoto Protocol divided the COPs into two categories: Annex-I countries, namely, the developed countries and non-Annex-I countries, primarily the developing countries (Gupta 2016). The Kyoto Protocol facilitates the Annex-I countries to limit and reduce greenhouse gas emissions as per their agreed individual targets. The protocol sets legally binding emission reduction targets for 37 industrialized countries and economies in transition as well as the European Union (https://unfccc.int/kyoto_protocol). The Protocol placed the onus on Annex-I countries as (1) the developed nations can successfully carry the financial burden of reducing the GHG emissions and (2) historically Annex-I countries have been responsible for greater amount of GHG emissions as compared to developing countries. The countries were urged to meet their individual targets through three mechanisms, viz., joint implementation (JI), international emissions trading (IET), and clean development mechanism (CDM). These mechanisms were based on the principles of emissions trading—countries could purchase emission rights from other countries (JI and IET) or gain carbon credits by investing in ecofriendly projects in developing nations (CDM). According to the contemporary mainstream economists, these mechanisms were supposed to generate benefits for both Annex-I and non-Annex-I countries. Developed countries could offset their carbon emissions by purchasing carbon credits and developing countries would receive the future benefits of sustainable infrastructure. However, CDM failed to achieve its goal (Laing et al. 2013).

The Paris Climate Agreement, an international treaty on climate change, was adopted at COP21 on 12 December, 2015, and came into force on 4 November, 2016. The agreement was signed by 196 Parties. It was for the first time all nations made a common pledge to combat the global climate crisis. The primary aim of Paris Agreement is to limit global temperature increase under 1.5 °C. The agreement required all parties to formulate respective "long-term low GHG emissions strategies" or nationally determined contributions (NDCs) and make their best efforts to achieve it (Clemencon 2016, p. 9). Reviews were to be conducted after every 5 years when countries will be given the opportunity to update or renew their NDCs. In this context, the Paris Agreement proposed appropriate mobilization of financial resources, new technological frameworks, and enhanced capacity building to support developing and vulnerable nations to reach their own national objectives. However, the Paris Agreement was not legally binding emission commitments. Additionally, it abandoned the policy of "common but differentiated responsibilities and capabilities" as constituted by Kyoto Protocol (Article 3), making both developed and developing countries equally responsible for sharing the burden of reducing global carbon emissions. During the Copenhagen conference in 2009, it was decided that developed countries would provide a financial support of 100 billion US dollars to developing countries for climate change mitigation by 2020. When an assessment of Copenhagen financial pledge was made at

Paris conference, it was found that there was a lack of transparency regarding the flow of fund from developed to developing countries (Clemencon 2016).

Conference of Parties (COPs) is a decision-making body of the UNFCCC which reviews the emission inventories of the Parties and the progress made by them toward climate change mitigation (UNDP). COP26 was held in Glasgow in 2021 in which about 200 countries participated. It was organized by the United Kingdom in partnership with Italy (Arora and Mishra 2021). Four goals were set by the United Kingdom ahead of hosting COP26 (The National Archives, UK 2021):

1. Secure global net zero by mid-century, and keep 1.5° C within reach by phasing out coal, curtail deforestation, accelerate use of electric vehicles, and investment in renewable energy. In order to achieve this goal, the global emissions need to be reduced by 45% by 2030 (Arora and Mishra 2021).
2. Adapt to protect communities and natural habitats by protecting and restoring ecosystems and building warning systems and resilient infrastructure to avoid loss of life and property.
3. Mobilize at least 100 billion US dollars in climate finance per year with the help of the developed countries and international financial institution.
4. Accelerate climate actions through collaboration between governments, businesses, and civil societies.

Following the COP26, COP 27 was held in Sharm el-Sheikh, Egypt, between 6 and 18 November, 2022. One of the historic milestones reached in COP27 was to mobilize "loss and damage fund." The decision of creating this fund was to help vulnerable countries deal with the impacts of climate change. However, there is no consensus among member countries how the fund will be set up and from where it should come. During COP27, some countries wanted to reverse the goal of 1.5° C and terminate the ratchet ambition. Although they were not successful, a resolution to peak emissions by 2025 was dismissed (Harvey 2022). Instead of phasing out all fossil fuels as proposed by India in COP27, member nations decided to maintain the Glasgow resolution of phasing down of coal emission. Additionally, COP27 made provisions to facilitate further usage of "low emission energy" like natural gas as opposed to boosting the production of renewable energy (Ibid.). In COP 27, developed and developing countries also pointed out an urgent need of structural changes in providing funding by the World Bank and other similar institutions for climate change adaptations (e.g., building embankments, mangrove forests restoration, and afforestation) as less developed countries often struggle to receive fund for strengthening resilience against climate change. Nevertheless, out of 100 billion USD as promised by developed countries since 2020, only about 20 billion USD goes to adaptation (Ibid.).

13.5 India in Current Global Climate Politics

India faces unique challenges, balancing the need for economic development and poverty eradication with the imperative to reduce carbon emissions. The government has emphasised sustainable development and has launched various initiatives to promote clean energy, afforestation, and sustainable agriculture practices (Dutta et al. 2016). Due to contemporary severe ecological crisis, the cooperation of the Global South is increasingly becoming important in the context of global climate politics. As part of the Global South, India has always played a key role in global climate negotiations. In the global climate diplomacy, India is known for establishing common but differentiated responsibilities (CBDR) principle which recognizes the right of developing countries to increase their GHG emission, pushing developed countries to take responsibility of emissions and eradicating North-South economic inequity (Joshi 2013). Since 2007, India has moved away from the traditional defensive approach to more of a cooperative internationalist approach where Indian government is prepared to work with bilateral and multilateral bodies to secure climate finance and technology (Joshi 2013). Since the 1960s, India has been a member of G77, a group of 77 countries of the Global South. India became an active member of BASIC (Brazil, South Africa, India, and China) and BRICS (Brazil, Russia, India, and China and South Africa) since 2009. Being a member of BASIC, India played a key role in Copenhagen Accord adopted in 2009. In COP 27, India again played an assertive role when it supported the establishment of Loss and Damages Fund. However, Indian representatives stated that the country will not contribute to it but lay its claim (Oguntuase 2023). In the same conference, India prevented an attempt of use of terminologies such as "major emitters" and "top emitters" as proposed by the developed countries in the context of global climate justice (Ibid.). It has also called for developed nations to provide financial and technological support to developing countries for climate change adaptation and mitigation.

In November 2015, India and France jointly launched the International Solar Alliance (ISA) to promote solar energy in developing countries. With its headquarter in New Delhi, the alliance consists of 114 countries. The ISA provides a platform for coordination and cooperation of solar-resource-rich countries through which they can increase the use of solar energy (ISA 2023). In the Asia-Pacific region, Japan, India, and Australia share 96.4% of total solar PV capacity. This region has a population of 4.4 billion spread across a diverse economy ranging from small island nations to some of world's largest energy consumers. In the past 5 years, India has made considerable growth in total solar PV installed capacity from 17,923 MW (2017) to 62,804 MW (2022). This growth was made mainly by focusing on local manufacturing (Ibid.). Despite this growth, India still depends on fossil fuel to fulfill its energy need. Recent data published by Central Electricity Authority (CEA) shows that as of 31st July, 2023, 56.3% of India's energy requirement has been fulfilled by fossil fuel (i.e., coal, lignite, gas, and diesel), whereas 43.7% has been fulfilled by non-fossil fuel (wind, solar, hydro, etc.). The energy industry has been traditionally male dominated and ignored women's career advancement. Recent data shows percentage of women working in solar PV industry is 40%, the highest share in any renewable subsector. The Ministry of New and Renewable Energy (MNRE), Govt. of India, is also focusing on bringing gender equity in renewable energy sector by providing regular training and capacity building among women entrepreneurs (Govt. of India, Ministry of New and Renewable Energy 2022).

The Coalition for Disaster Resilient Infrastructure (CDRI) was initiated by Government of India in 2019. The formation of CDRI was driven by India's periodical experience with different types of natural disasters such as floods, tropical storms, earthquakes, landslides, wildfires, etc. As of 2023, this global coalition has 39 members consisting of 31 national governments and 8 organizations (CDRI). The aim of this agency is to perform a risk analysis of critical infrastructure at national level from natural disasters. As climate change is causing extreme weather events like tropical storms, floods, and draughts, it often has a crippling effect on the national infrastructure. In 2020, India ranks 89 out of 181 in World Risk Index (WRI) with a vulnerability of 52.94% (https://www.cdri.world/member-country-profile-india). An estimate shows that nearly 3–10% of GDP is lost due to losses to infrastructure from natural disasters. CDRI is trying to create a global database studying natural disaster-related risks to national infrastructure and set up a global framework to address these issues (Tripathi 2023). To this effect, India is focusing on the following sectors: power sector infrastructure resilience, transport sector infrastructure resilience, telecommunication sector infrastructure resilience, health sector infrastructure resilience, urban sector infrastructure resilience, finance for resilient infrastructure, technical standards for disaster resilient infrastructure, infrastructure and governance, and capacity enhancement initiatives among youth, communities, and organizations (CDRI Annual Report 2021–22). The CDRI will continue to work toward the needs of most disaster-prone regions with special emphasis on women, children, elderly, and differently abled people. It will closely work with disaster-stricken economies and promote instant actions to minimize risks at all levels (https://www.yuvamanthan.org/cdri).

13.6 SDG 13 and India

Since the launch of 17 sustainable development goals by the UN, national and state governments of India have become proactive in addressing the climate change issue in the broader development context. In 2008, India launched the National Action Plan on Climate Change (NAPCC) integrating climate change with larger development goals (Dubash et al. 2018) through sustainable pathways. On one hand, it acknowledged the need for raising the living standards of marginal communities through economic development; on the other, the focus was to protect them from the increasing impacts of climate crisis. NAPCC introduced eight national missions to understand the impacts of climate change, its mitigation, and adaptation strategies. These eight national missions are (1) National Solar Mission, (2) National Mission for Enhanced Energy Efficiency, (3) National Mission for Sustainable Habitat, (4) National Water Mission, (5) National Mission for Sustaining the Himalayan Eco-system, (6) National Mission for a Green India, (7) National Mission for Sustainable Agriculture, and (8) National Mission on Strategic Knowledge for Climate Change. The focus of these missions was to protect underprivileged communities vulnerable to climate change while achieving national growth through the development of ecologically sustainable and cost-effective technology and innovations (Ministry of Information and Broadcasting, Government of India 2021). The central government instructed 34 states and union territories to prepare a state action plan on climate change (SAPCC) in line with NAPCC, taking into account the states' specific issues related to climate change. The SAPCCs outlined sector-specific and cross-sectoral priority action, including adaptation and climate resilient infrastructure (MoEFCC, Government of India 2023). National adaptation fund for climate change (NAFCC), a central government scheme, was set up in 2015–2016 to support adaptation activities under SAPCCs. Till November 2022, INR 847.5 crore has been spent on 30 projects from different states of India focusing on agriculture, water, forestry, livestock, and restoring ecosystems (Govt. of India, Ministry of Finance, Economic Survey 2022–2023). However, it has been observed that states with well-established institutional structures like Tamil Nadu and Madhya Pradesh have been able to deliver their SAPCC projects more effectively (Ibid.). Various achievements have been made across several domains under the eight national missions of the NAPCC. Under the National Solar Mission, India installed solar power capacity of 61.62 GW by October 2022. Under National Mission for Enhanced Energy Efficiency, PAT Cycle-VII notified in October 2021 for energy saving target of 6.63 million tons of oil equivalent (MTOE). As per National Mission on Sustainable Habitat, 721 km of metro rail network has been made operational by August 2022. 62.79 lakhs individual household toilets and 6.21 lakhs community/public toilet have been constructed until April 2022. National mission for a Green India has allotted INR 626.96 crores for afforestation targets over an area of 2.1 lakh hectors (Ibid.). In 2022, National Water Mission introduced Jalshakti Abhiyan: Catch the Rain Campaign to encourage the states and stakeholder to create rainwater harvesting structures (RWHS) suitable for the local climatic conditions and subsoil strata before monsoon (Govt. of India, National Water Mission 2022). The campaign encouraged the building of check dams, water harvesting pits, rooftop RWHS, etc. and repair of unused wells and step-wells to recharge the aquifers and increase the groundwater level. States were requested to open district level "rain centers" where technical guidance will be provided to the common people for their active participation in "Catch the Rain" campaign (Govt. of India, National Water Mission 2022). Under the National Mission on Strategic Knowledge for Climate Change, 12 centers of excellence for climate change were created by June 2021. These centers of excellence (CoEs) include four IITs—

Delhi, Bombay, Kharagpur, and Madras, which are working toward creating improved and accurate climate projection models (Govt. of India, Ministry of Finance, Economic Survey 2022–23). National Mission for Sustaining Himalayan Ecosystems (NMSHE) has initiated eight major research and development programs to conserve biodiversity, forest cover, glaciers, and other ecological sources in the Himalayan region. To implement these programs, NMSHE has collaborated with various institutions and constituted six task forces (TF)—TF1, Natural and Geological Wealth; TF2, Water, Ice, Snow including Glaciers; TF3, Forest Resources and Plant Diversity; TF4, Micro Flora and Fauna and Wildlife and Animal Population; TF5, Traditional Knowledge System; and TF6, Himalayan Agriculture (Government of India, Ministry of Science and Technology, NMSHE 2017). Finally, National Mission for Sustainable Agriculture (NMSA) aims to make agriculture more productive and sustainable by promoting region-specific integrated farming systems. Farm water management was implemented to promote technological interventions to enhance water use efficiency. Soil health management was implemented to regulate usage of chemical fertilizers and promote use of organic manures and bio-fertilizers. For the year 2022–2023, NMSA has set their target to cover 0.15 lakh hectares of land under organic farming and 10 lakh hectares under micro-irrigation (Govt. of India, Ministry of Finance, Economic Survey 2022–23).

13.7 Discussion

India is a signatory to the Paris Agreement and has submitted its Intended Nationally Determined Contributions (INDC) outlining its climate action targets. The country has committed to reducing its emissions intensity and increasing the share of non-fossil fuel capacity in its energy mix. In the recent past, the country has shifted from its defensive position in global climate diplomacy and taken up a lead role in asserting long-term strategies for climate change. To this effect, India has entered into bilateral, minilateral, and other plurilateral climate agreements with other partner nations. India has established bilateral climate cooperation with the European Union, the United States, Germany, France, Japan, and the United Kingdom (Nachiappan 2023). For instance, India and France have established International Solar Alliance (ISA) to promote solar energy access among its member nations. India is also part of minilateral agreements like Quadrilateral Security Dialogue (Quad). On May 24, 2022, Quad members met in Tokyo and launched The Quad Climate Change Adaptation and Mitigation Package (Q-CHAMP) to promote climate mitigation and adaptation for its Indo-Pacific partners. Mitigation policies include decarbonizing the shipping and port sectors, enhancing and accelerating the clean energy transition through knowledge sharing, enabling participation in high integrity carbon markets, and support capacity building in the region. Adaptation policies include advancing climate information services through international organizations like CDRI, promote climate-smart agriculture, and enhance ecosystem-based adaptation through various initiatives like the International Coral Reef Initiatives (Quad Cooperation in Climate Change 2022).

India has taken up a "neighborhood first" policy to address the issue of climate change in South Asia. India is a member of Bay of Bengal Initiative for Multi-Sectoral Technical and Economic Cooperation (BIMSTEC). Being a sector-driven grouping, cooperation within BIMSTEC includes trade, technology, energy, transport, tourism, agriculture, public health, security, environment, climate change, etc. In 2021, these sectors and subsectors were reorganized among its member nations, and India was put in charge of security, counterterrorism, and transnational crime, energy, and disaster management (https://bimstec.org/bimstec-history/). India is also tapping into its blue economy potential by building climate resilience infrastructure within the Bay of Bengal region.

At the time of writing this article, the G20 Summit 2023 is coming to an end. Being presided over by India and scheduled ahead of COP28, the summit became the platform to showcase India's ambitious global climate goals. As a positive outcome of G20, India received consensus on thematic priorities on reducing land degradation, accelerating the pace of ecosystem restoration, and increasing biodiversity. Additionally, priority was given on creating sustainable and climate-resilient blue economy. Climate change was the core agenda of G20 summit, and scaling up the need of climate finance from billions to trillions of dollars globally was recognized. The summit also stressed the importance of voluntary reduction of land degradation by 50% by 2040 as pledged under the G20 Global Land Initiative (GLI) and recognized the areas under forest fire and degraded mining areas as priority landscapes. In conclusion, the commitment of G20 countries in preserving the environment and combating climate change was reflected through the implementation of high-level principles on lifestyles (HLPs) for sustainable development and understanding the support to implementation through international cooperation, finance, and development of technology (Nandi 2023). Finally, G20 countries encouraged international organizations to incorporate the HLPs into their programs to make a significant impact on preservation of natural environment at the global scale (Ibid.).

References

Arora NK, Mishra I (2021) COP26: more challenges than achievements. Environ Sustain 4:585–588

Athal K (2022) Never-ever-seen climate changes in India – the concerns and solutions. The Times of India, February 4. https://timesofindia.indiatimes.com/blogs/krishna-athal/never-ever-seen-climate-changes-in-india-the-concerns-and-solutions/. Accessed 31 July 2023

CDRI (2020–21) Coalition for dster resilient infrastructure annual report 2020–21. https://annualreport.cdri.world/. Accessed 11 Sept 2023

Clemencon R (2016) The two sides of the Paris climate agreement: dismal failure or historic breakthrough? J Environ Dev 25(1):3–24. https://doi.org/10.1177/1070496516631362

Dubash NK, Khosla R, Kelker U, Lele S (2018) India and climate change: evolving ideas and increasing policy engagement. Annu Rev Environ Resour 43:395–424

Dutta V, Dasgupta P, Hultman N, Gadag G (2016) Evaluating expert opinion on India's climate policy: opportunities and barriers to low-carbon inclusive growth. Climate and Development 8(4):336–350

Government of India, Ministry of Environment, Forest and Climate Change (2021) India State of Forest Report. https://fsi.nic.in/forest-report-2021-details. Accessed 27 Nov 2023

Government of India, National Water Mission (2022) https://nwm.gov.in/objective-national-water-mission. Accessed 26 Nov 2023

Government of India, Ministry of Environment, Forest and Climate Change (2023) SAPCC. https://pib.gov.in/PressReleaseIframePage.aspx?PRID=1895857. Accessed 8 Sept 2023

Government of India, Ministry of Finance (2023) Economic survey 2022–23. Department of Economic Affairs Economic Division, Government of India, Ministry of Finance, New Delhi

Government of India, Ministry of Information and Broadcasting (2021) National action plan on climate change (NAPCC). https://static.pib.gov.in/WriteReadData/specificdocs/documents/2021/dec/doc202112101.pdf. Accessed 8 Sept 2023

Government of India, Ministry of New and Renewable Energy (2022) Women in renewable energy (RE): call for action, February 17. https://mnre.gov.in/img/documents/uploads/file_s-1663738338619.pdf. Accessed on 10 Sept 2023

Gupta A (2016) Climate change and Kyoto protocol: an overview. In: Ramiah V, Gregoriou GN (eds) Handbook of environmental and sustainable finance. Elsevier, pp 3–23

Harvey F (2022) What are the key outcomes of COP27 climate summit. The Guardian, November 22. https://www.theguardian.com/environment/2022/nov/20/cop27-climate-summit-egypt-key-outcomes. Accessed 6 Sept 2023

ISRO (2018) Desertification and land degradation atlas of India. ISRO, Ahmedabad

Joshi S (2013) Understanding India's representation of north–south climate politics. Glob Environ Polit 13(2):128–147. https://doi.org/10.1162/GLEP_a_00170

Kuh KF (2018) The law of climate change mitigation: an overview. Encycl Anthropocene 2:505–510

Laing T, Sato M, Grubb M, Comberti C (2013) Assessing the effectiveness of the EU emissions trading scheme (working paper no. 126). London School of Economics, Centre for Climate Change Economics and Policy. http://www.lse.ac.uk/GranthamInstitute/wp-content/uploads/2014/02/WP106-effectiveness-eu-emissionstradingsystem.pdf

Li J, Lee K, Wadhwa D, Lambrechts M (2023) From climate science to global action. In: Pirlea AF, Serajuddin U, Thudt A, Wadhwa D, Welch M (eds) Atlas of sustainable development goals 2023. World Bank, Washington, DC. License: Creative Commons Attribution CC BY 3.0 IGO. https://doi.org/10.60616/vg01-7787

Lindsey R (2022) Climate change: global sea level. NOAA, April 19. https://www.climate.gov/news-features/understanding-climate/climate-change-global-sea-level. Accessed 21 Aug 2023

IMD (2022) Annual Report 2022. https://mausam.imd.gov.in/imd_latest/contents/ar2022.pdf. Accessed 28 Nov 2023

India Today (2022) Indian Ocean hit by six heatwaves in 2021, temp rising by 0.15 degrees Celcius per decade. https://www.indiatoday.in/science/story/climate-change-ocean-temperature-rising-heatwave-bay-of-bengal-indian-ocean-1929111-2022-03-24. Accessed 30 Nov 2023

Internatioanl Disaster Database (2022) 2022 Disasters in Numbers. https://www.google.com/url?sa=t&rct=j&q=&esrc=s&source=web&cd=&ved=2ahUKEwj1q_LlsrGDAxVFcmwGHT2DB8QQFnoECA0QAQ&url=https%3A%2F%2Fcred.be%2Fsites%2Fdefault%2Ffiles%2F2022_EMDAT_report.pdf&usg=AOvVaw0F38Wsj1GDK6RMZqfLw_ds&opi=89978449. Accessed 29 Nov 2023

IPCC (2014) Climate Change 2014 Synthesis Report: Summary for Policymakers. https://www.ipcc.ch/site/assets/uploads/2018/02/AR5_SYR_FINAL_SPM.pdf. Accessed 28 Nov 2023

IPCC (2022) Summary for Policy Makers. https://www.ipcc.ch/report/ar6/wg2/downloads/report/IPCC_AR6_WGII_SummaryForPolicymakers.pdf. Accessed 28 Nov 2023

ISA (2023) 6th Assembly of International Solar Alliance. https://pib.gov.in/PressReleaseIframePage.aspx?PRID=1968881. Accessed 27 Nov 2023

Maizland L (2022) Global climate agreements: successes and failures. Council on Foreign Relations. November 4. https://www.cfr.org/backgrounder/paris-global-climate-change-agreements. Accessed 31 July 2023

Nachiappan K (2023) India's sprawling climate diplomacy. May 8. https://www.isas.nus.edu.sg/papers/indias-sprawling-climate-diplomacy/

Nandi J (2023) Persistence, balance helped India achieve climate consensus at G20: Bhupender Yadav. September 11. https://www.hindustantimes.com/india-news/persistence-balance-helped-india-achieve-climate-consensus-at-g20-bhupender-yadav-101694398385488.html. Accessed 11 Sept 2023

NOAA (2022) Climate Change: Global Sea Level. https://www.climate.gov/news-features/understanding-climate/climatechange-global-sea-level. Accessed 28 Nov 2023

Oguntuase OJ (2023) India and Africa leverage climate diplomacy, ORF Issue Brief No. 628, March 2023. Observer Research Foundation. https://www.orfonline.org/research/india-and-africa-leverage-climate-diplomacy/. Accessed 10 Sept 2023

Quad Cooperation in Climate Change and Launch of the Quad Climate Change Adaptation and Mitigation Package (2022) https://www.mofa.go.jp/files/100348057.pdf

Sangomla A (2023) South India records its lowest June rainfall in 122 years. Down to Earth, July 5. https://www.downtoearth.org.in/news/climate-change/south-india-records-its-lowest-june-rainfall-in-122-years-90421. Accessed 21 Aug 2023

Sengupta R (2021) Land degradation in India hurts farmers and forest dwellers the most. Down to Earth, August 27. https://www.downtoearth.org.in/news/environment/land-degradation-in-india-hurts-farmers-and-forest-dwellers-the-most-78701. Accessed 21 Aug 21, 2023

The Hindu (2023) Yamuna water level breaches danger mark amid rains in Delhi. The Hindu, July 19. https://www.thehindu.com/news/cities/Delhi/yamuna-water-level-breaches-danger-mark-amid-rains-in-delhi/article67096789.ece. Accessed 31 July 2023

The National Archives, UK (2021) What did the UK Presidency aim to achieve at COP26?. https://webarchive.nationalarchives.gov.uk/ukgwa/20230311034236/https://ukcop26.org/cop26-goals/. Accessed 27 Nov 2023

The Tribune (2020) Average temperature over India projected to rise by 4.4 degrees Celsius: Govt report on impact of climate change in country. https://www.Tribuneindia.Com/news/nation/average-temperature-over-india-projected-to-rise-by-4-4-degrees-celsius-govt-report-on-impact-of-climate-change-in-country-99583. Accessed 30 Nov 2023

Tripathi S (2023) How vulnerable is India's critical infrastructure to disasters? CDRI to release key report at G20. India Today, May 23. https://www.indiatoday.in/environment/story/how-vulnerable-is-indias-critical-infrastructure-to-disasters-cdri-to-release-key-report-at-g20-2383041-2023-05-23. Accessed 10 Sept 2023

UN Environment Program (n.d.) About Montreal Protocol. https://www.unep.org/ozonaction/who-we-are/about-montreal-protocol. Accessed 01 Dec 2023

United Nations (2015) Transforming Our World: The 2030 Agenda for Sustainable Development. https://sdgs.un.org/sites/default/files/publications/21252030%20Agenda%20for%20Sustainable%20Development%20web.pdf. Accessed 27 Nov 2023

United Nations (2023) The Sustainable Development Goals Report 2023: Special Edition. https://unstats.un.org/sdgs/report/2023/The-Sustainable-Development-Goals-Report-2023.pdf Accessed 30 Nov 023

UNFCCC (1992a) Article 1: Definitions. https://unfccc.int/resource/ccsites/zimbab/conven/text/art01.htm Accessed 29 Nov 2023

UNFCCC (1992b) Article 2: Objective. https://unfccc.int/resource/ccsites/zimbab/conven/text/art02.htm. Accessed 29 Nov 2023

UNFCCC (1992c) Article 3: Principles. https://unfccc.int/resource/ccsites/zimbab/conven/text/art03.htm. Accessed 29 Nov 2023

UNICEF (2023) Devastating floods in Pakistan. https://www.unicef.org/emergencies/devastating-floods-pakistan-2022. Accessed 19 Aug 2023

WMO (2022) Past eight year confined to be the eight warmest on record. https://wmo.int/media/news/past-eight-years-confirmed-be-eightwarmest-record. Accessed 29 Nov 2023

WMO (2023) Past eight year confined to be the eight warmest on record. https://wmo.int/media/news/past-eight-years-confirmed-be-eightwarmest-record. Accessed 29 Nov 2023

World Bank (2021) Millions on the move in their own countries: the human face of climate change, September 13. https://www.worldbank.org/en/news/feature/2021/09/13/millions-on-the-move-in-their-own-countries-the-human-face-of-climate-change. Accessed 27 July 2023

Tania Chakravarty completed her PhD in History from the School of Oriental and African Studies (SOAS), University of London. She did her BA, MA, and MPhil from Department of History, Jadavpur University, Kolkata. She is an Assistant Professor at Vivekananda College (Affiliated to University of Calcutta), Kolkata, India. Her areas of expertise include social and environmental history of modern India, urban history of late nineteenth and early twentieth century India, and history of constitutional reforms in early twentieth century India.

Priyanka Ghosh is a Sr. Assistant Professor of Geography at VIT-AP University, Amaravati, Andhra Pradesh, India. Her areas of interest are political ecology, biodiversity conservation, protected area management, human-animal conflicts, traditional ecological knowledge, tourism, and sustainable development. Her research on the Indian Sundarbans has been published in leading international journals such as *GeoJournal*, *Geographical Review*, and *Environmental Management*. She has also published several chapters on environmental conservation, water resources, and tourism governance in peer-reviewed international books.

Unsustainable, Imperial Dreams: What India's Blue Economy Portends for "Life Below Water"

14

Adam Jadhav

Abstract

Sustainability of Indian seas has been submerged almost entirely within a national Blue Economy discourse and politicized ambitions of limitless wealth beneath the waves. In this imaginary, shipping, fisheries, marine minerals, naval might, coastal commons, and even tourism serve to bolster national development in various ways. Such expansionary and even imperial aims, however, conjure doubts about India's engagement with international development policy and commitments to economic, social, and ecological sustainability. In this political economic context, this chapter interrogates India's performance at achieving the United Nation's Sustainable Development Goals (SDGs) and specifically Goal 14: "Life Below Water." First, I historicize sustainable development in order to unsettle a concept frequently deployed as though self-explanatory, uncontested, or ahistorical. I also briefly review the SDGs and their evolution into discursive objects and exemplars of development's fetishization of quantification. The analysis then turns to the specifics of India's performance of the SDG process which I argue produces an abstracted, flexible notion of sustainability made available for other projects, namely, the Hindutva-infused Blue Economy agenda. Contra this development imaginary, I present evidence of actually existing unsustainability at the scale of an estuary on India's southwestern coast. The chapter concludes with a provocation: India's oceanic agenda—underwritten by its performative embrace of "sustainable" development—appears in the process of articulating both ethnonationalism and neoliberalism into a neoimperial dream.

Keywords

Development · Sustainability · Blue Economy · India · Ethnonationalism

14.1 Introduction

Development discourse today in India and internationally routinely hails a two-faced, almost duplicitous "nature" to the ocean. According to many policy-makers and (increasingly nationalist) economic planners, the world's seas constitute an almost unfathomable trove of resources waiting for scientists, workers, capitalists, and, above all, nation states to claim them. And yet, we are told, marine geographies from the shallow estuary

A. Jadhav (✉)
University of California, Berkeley, CA, USA

Panchabhuta Conservation Foundation, Kumta, India
e-mail: ajadhav@berkeley.edu

V. Dutta, P. Ghosh (eds.), *Sustainability: Science, Policy, and Practice in India*, Sustainable Development Goals Series, https://doi.org/10.1007/978-3-031-50132-6_14

to the deepest trench are simultaneously vulnerable to reckless consumption and despoilation. Sometimes speakers offer both perspectives at once, as did then Science and Technology Minister Harsh Vardhan during a 2015 speech at the National Institute of Oceanography (NIO) in Goa.

> The ocean and its resources are increasingly seen as indispensable in addressing multiple challenges the planet is facing in decades to come. By 2050 enough food, jobs, energy, raw materials and economic growth will be required to sustain a world population of nine billion people… But it (ocean) is already under stress from over exploitation, pollution, declining biodiversity and climate change. Hence realizing the full potential of the ocean will demand responsible, sustainable approaches to its economic development… Therefore, I urge you all to see urgency in our situation. We must move fast to grab India's legitimate share in the emerging ocean-based industries, which also include seabed mining for metals and minerals, marine aquaculture, marine biotechnology, ocean-related tourism and leisure activities and most importantly, ocean monitoring, control and surveillance (PTI).

NIO is one of an alphabet soup of acronym-laden government agencies in India—e.g., CIFT, CMFRI, INCOIS, ISRO, MOEFCC, MOES, MPEDA, NCSCSM, NFDB, NIOT, NITI-Aayog, among others—committed to the proposition that India urgently requires aggressive oceanic development, at once nationalist and capitalist. The very expanse of the ocean underwrites dreams of wealth waiting to be tapped—or grabbed, per Vardhan—from a kind of *aqua nullius*, an underutilized, three-dimensional watery geography of fantastic proportions.[1] In this so-called Blue Economy, sea lanes, fish, minerals, tides, currents, breakwaters, the beach and even carbon-sequestering and cyclone-buffering mangroves are reimagined as new inputs to a program of national development (cf. Barbesgaard 2017; Carver 2019; Childs and Hicks 2019; Mallin and Barbesgaard 2020; Silver et al. 2015). Accordingly, India's future "success" depends on claiming and making best use of its exclusive economic zone (EEZ) and coastal space as rightful territory and property of the nation. The Blue Economy then serves to imbue national and capitalist development ambitions with an expanded territoriality, despite the ocean's resistance to both property regimes and the jurisdiction of the nation-state system (cf. Germond-Duret 2022).

This dominant imaginary of limitless oceanic resources is not so much contradictory as relational (perhaps even dialectical) with a twinned narrative of an ocean under threat. Consider that presumptions of the ocean's inexhaustibility—an echo of colonial-era science and policy still persistent today—have prompted all manner of overharvesting, unchecked dumping, and widespread chemical and biological runoff. Scholars and activists have repeatedly brought attention to the plight of coastal communities, and especially fishers, facing environmental degradation, appropriation of ecological commons, and the loss of livelihoods (Ali et al. 2021; Kanan et al. 2020; Research Collective 2018). Ignoring such criticism, Narendra Modi, prime minister and suzerain of India's ruling Hindu nationalist Bharatiya Janata Party (BJP), hails his Blue Economy "vision for a sustainable and prosperous future" (2015) and pads policy and campaign speeches with blandishments such as "our lifestyle, transport systems and trading behavior should not spoil the ecology of the oceans" and "there is a need for holistic and sustainable development of coastal regions and communities" (2016). Other Indian planners and politicians parrot a flexible mantra of "sustainability," switching deftly among abstract proclamations and fantastic promises that development on the coasts or at sea *should be, already is or will shortly be* sustainable while remaining profitable (cf. Thakur 2022).

Yet the meaning of oceanic sustainability—in science, policy or practice—is too often unspecified or glossed over. This problematic—the submergence of critical sustainability within an oceanic agenda both nationalist and capitalist—serves as my point of departure. In this chapter, I read the political and economic agenda of ocean development through and against India's practical engagement with the United Nations'

[1] I repeatedly hail the fantastic, dream-like quality of India's oceanic ambitions—contradictory, wishful, unreal, inchoate, and/or fleeting—inspired by a range of critical scholarship engaging late capitalism's dreamscape (e.g., Anand et al. 2018; Benjamin 2002; Buck Morss 1995; Cross 2014; Larkin 2013).

Sustainable Development Goals (SDGs), and specifically SDG 14, which concerns "Life below Water." Part aspiration and part pledge, the "goals" are to be measured by dozens of statistical indicators. Like the Millennium Development Goals (MDGs) before them, the SDGs treat "development" as a discrete condition that can be measured and tinkered with. In this framing, individual countries fit national data to internationally agreed indicators and then track and report their progress. Architects and proponents of the SDGs tout this voluntary system of monitoring and feedback as transformative of the (implicitly) unsustainable trajectory of global capitalism. Yet the meaning of sustainability remains ambiguous and flexible, subject to expedient definition for political (and political economic) ends, while development in actual practice remains shot through with fantasy and wishful thinking. Just listen to then Vice President of India, M. Venkaiah Naidu, in 2019: "The Blue Economy will aid in achieving the 14th goal of United Nations sustainable development goals." Here, the discourse borders on tautology, where simply hailing the ocean economy as "blue" assumes sustainability by default.[2]

Two disclaimers are appropriate. First, India's oceanic development ambitions are hardly novel. Whatever its intensity, contemporary Blue Economy discourse (cf. Silver et al. 2015) echoes longue durée histories of capitalist development and territorialization at sea (Campling and Colás 2021). Likewise, framing nature and ecology as crucial assets for nation-building, economic growth, and expanded/improved (re)production repeats policy/praxis of both British colonial and postcolonial national governments.[3] Second, not all claims of "sustainability" are necessarily specious. A range of actors and interests collaborate on and contest the agenda of development (and the related keyword, improvement), sustainable or otherwise, sometimes with radical intent. Li (2007) cautions "the rush to identify hidden motives of profit or domination narrows analysis unnecessarily." Rather than pronounce oceanic development plans and policies as only tragedy or farce, this chapter seeks to unpack how sustainability—and the SDGs as discursive objects reinforced by performance and practice—accomplish work for other agendas and ideologies. In the case of India, I will ultimately argue that abstracted "sustainability" helps suture together a muscular nationalism and a neoliberal capitalism under the sign of Blue Economy.

The remainder of the chapter proceeds as follows. In Sect. 14.2, I historicize sustainable development in order to unsettle a concept frequently deployed as though self-explanatory, uncontested, or ahistorical. Section 14.3 briefly presents the SDGs and their evolution into discursive objects and exemplars of development's fetishization of quantification. Section 14.4 examines India's performance of the SDG process—focusing on Goal 14—which effectively produces an abstract notion of oceanic sustainability made available for other projects, namely, the Hindutva-infused Blue Economy agenda. Contra this development imaginary, Sect. 14.5 highlights actually existing *un*sustainability at the scale of an estuary. Finally, Sect. 14.6 reiterates the claims of the paper and concludes that India's oceanic agenda articulates both nationalism and neoliberalism into a neoimperial dream.

14.2 Sustainable Development as Keyword

Though historically contested, the origin stories, meanings, and methods of development have "rarely broken free from organicist notions of growth and from a close affinity with teleological views of history, science and progress" (Watts 1993). A robust intellectual history—beyond the

[2]Note how the Blue Economy is an oceanic mirror to the so-called Green Economy (cf. Bratman 2015; Cavanagh and Benjaminsen 2017).

[3]A wide range of scholarship concerns the intersections of British colonialism, scientific management, and the development/improvement of "nature" (Cederlöf and Sivaramakrishnan 2014; Grove 1997, 2003; Grove et al. 2000; Damodaran et al. 2015; Drayton 2000; Nandy 1999; Rangarajan and Sivaramakrishnan 2011), though comparatively little attends explicitly to marine spaces and places amid empire.

scope of this chapter[4]—would cover the rise of modernist, Orientalist binaries about "the West and the Rest" (Hall 1996) coincident with the expansion of Western colonialism/capitalism and paradigms of economism, scientific management (of people and resources), and notions of linear progress. Development carries this baggage and "inheritances" of imperial and colonial projects that sought to build a hierarchical world: "What obtained prior to development, either in the form of pure nature or non-Western subsistence, did not have, we were told, the rationality, slickness, and efficiency of modern science. People, societies, nature itself were backward because of its absence" (Alvares 2010). Varying accounts, more and less critical, trace development's rise as a project of state intervention to the twilight of formal empire and the British Colonial Development and Welfare Act, both placation from above and a demand from below (Cooper 1997; Furnivall 1948; Hart 2001, 2010). Late colonial interventions in social and economic space continued postwar, and Sachs (2010a) suggests "the age of development" proper begins with US president Truman's 1949 inauguration speech calling for: "a bold new program for making the benefits of our scientific advances and industrial progress available for the improvement and growth of underdeveloped areas." In this formulation, development becomes a discrete condition conflated often with high economic growth and income. Geographies can possess or lack development. States can measure and then bolster levels of development (e.g., caloric intake, household income, or "modern" technologies). This dominant framing implies a process of catching up, overcoming primitivity, and progressing linearly through stages (cf. Rostow 1960). Here development is neither relational nor the outcome of historical, uneven global capitalist relations, but instead a technical, even apolitical exercise.

Guided by central planning, elite interests, and "expert" rule, postcolonial leaders in the Global South often embraced so-called modernization, depeasantization, and demographic transition and the like as crucial to becoming a "developed" nation-state. Poverty, famine, and inequality—themselves often colonial legacies (cf. Davis 2002; Watts 1983)—amplified the urgency for material gains. In India, the development regime started under British colonialism would turn to central economic planning, cooperative producer societies, public corporations, state science agencies, government food procurement/distribution, new extractive industries, and more (Ludden 2005), all guided by calculations of India's economic assets and deficiencies.[5] Nominally disconnected from the international economy, India practically began relying on foreign aid, technology, and expertise, in projects ranging from Green Revolution agriculture to massive, multi-scalar dams to the spread of commercial fishing trawlers (Baviskar 1995; Kurien 1985; Patel 2013). Consequent destruction of traditional livelihoods, mass-scale land appropriation, rising inequality, and ecological devastation would engender peasant resistance and counter claims for alternative forms of development (Peet and Watts 1996; Rangan 1996; Sachs 2010b).

By the 1980s in India, "neo-liberal free-market orthodoxy conquered the economic mainstream, where harsh critics of state planning, provisioning and regulation become most influential" (Ludden 2005, p. 4047). The multiplying burden of international debt and a foreign currency shortfall forced the country into economic "liberalization" in 1991, as a condition of International Monetary Fund lending. Nehruvian state-managed capitalism gave way to neoliberal policy: new land grabs, falling tariffs, reduced import-export restrictions, and the gutting of state social safety nets (see also Harvey 2003). Despite the discourse of free markets and enterprise, this period of development in "New India" was underwritten by active policy, interventionist

[4]Watts' series of articles in *Progress in Human Geography* (1993, 1994, 1996) are a good starting point. See also Chari and Corbridge (2008) and Allen (2021).

[5]For example, the National Council of Applied Economic Research produced state-by-state techno-economic surveys in the 1950s and 1960s largely repeating colonial narratives of India as deficient in capital, industry, infrastructure, and entrepreneurial spirit.

planning, and "regimes of dispossession" (Levien 2018; Sud 2021) often worsening inequality, environmental degradation, and loss of (especially rural) livelihoods (Le Mons Walker 2008).

Sustainability, meanwhile, appeared as a term of science in the early eighteenth-century context of long-term forest management (Scoones 2007). Fears over environmental degradation—often driven by dubious Orientalist prejudice (cf. Davis 2007)—crept into imperial aims of environmental management alongside goals of "improvement" (Drayton 2000). Peasants and other subjugated often rural producers were frequently blamed for what colonial rulers thought of as under-utilization; in India, for example, Francis Day (1873), the first British inspector general of fisheries in India, insistently promoted management of "waste" and "destruction" of resources by fishers (1873).[6]

The idea of sustainable development enshrined in the SDGs draws specifically from a "Malthusian moment" (Robertson 2012) in the latter half of the twentieth century.[7] Recognition of the ecological cost of industrial capitalism merged with critiques of colonial and postcolonial development, generating new forms of environmental politics from above and below in Global North and South (Adams 2001; Conca and Dabelko 2010; Guha and Martinez-Alier 1997; Peet and Watts 1996). From the 1960s, international development agendas turned to fertility and birth control (cf. Murphy 2017), while concerns over "limits to growth" (e.g., Meadows et al. 1972) and "overpopulation" (e.g., Ehrlich 1968) produced specious Global North fears of postcolonial masses devouring global ecology (despite their actually meagre consumptive footprints). At the 1972 UN Conference on the Human Environment, India's prime minister Indira Gandhi—one of the few Global South leaders to attend—famously pushed back, decrying uneven, "reckless exploitation of man and earth in the name of efficiency" driven disproportionately by industrial, capitalist nations (Robbins 1972).

The dyad term of "sustainable development" entered policy debates alongside a flowering of critical literatures: mathematical ecology, political ecology, critical development studies, environmental sociology, and agrarian political economy to name a few. In 1987, the UN's World Commission on Environment and Development published *Our Common Future* with a now-popular definition: "Sustainable development is development that meets the needs of the present without compromising the ability of future generations to meet their own needs." Concurrently, new environmental movements increasingly resisted mega projects: hydroelectric dams, forest-clearing highways, or nuclear weapons alongside quotidian expropriations of land and livelihood. This intellectual and political foment would crescendo until the 1992 UN Conference on Environment and Development (popularly known as the Earth Summit) in Rio de Janeiro, one of the largest civil society gatherings ever. Despite grassroots involvement, neither sustainability nor development would shed ideologies of ecological modernization and liberalism, techno-utopian innovation, market efficiency, free trade, environmental cost accounting, and private and individual responsibility all combined in the sustainable development paradigm (Bratman 2019), deployed from the smallest community NGOs to the commanding heights of the global economy.[8]

[6]Notably, European "science" of the period considered the possibility of fisheries collapse an open question, with no less than Thomas Huxley arguing that "probably all the great sea fisheries are inexhaustible" (Huxley 1883).

[7]Thomas Malthus' famous *Essay on the Principle of Population* (1798) foreshadowed contemporary sustainability politics. Malthus proclaimed a "natural law" that geometric population expansion would outstrip land under food production, but the tract also defended existing class hierarchy and argued that poverty reduction would only encourage the masses in their improvidence (Harvey 1974). Malthus advocated "benign neglect" of the poor, forcing them to serve emergent industrial capitalism or starve. Implicit in Malthusian political economy, then, are concerns about relationships among demography, modes of development, state policy, environmental scarcity, and overpopulation—all elements of "liberal government of life" (Dean 2015) and enduring keywords today.

[8]The World Bank began to use the term "sustainable" by 1989 (see Landell-Mills et al. 1989), though often in the business sense of sustained operation rather than as environmental and ecological concern.

By the mid-2000s, "there is nothing, it seems, that cannot be described as 'sustainable': apparently everything can be either hyphenated or paired with it. We have sustainable cities, economies, resource management, business, livelihoods – and, of course, sustainable development" (Scoones 2007, p. 589).

Continued policy failures globally would also spur promulgation of the UN Millennium Development Goals (MDGs), a set of big-picture ambitions for tackling poverty, hunger, education gender inequality, public health, and, yes, sustainability (Hulme 2009; MacArthur 2014). Proponents hailed the "power of numbers" to mobilize attention urgently; the 8 "goals" with 21 targets tracked by 60 indicators were ostensibly to be achieved by 2015. The MDGs' emphasis on metrics, however, reduced complex problems and multivalent social relations to that which is quantifiable (Fukuda-Parr and Yamin 2013), and they can be critically understood as "an ambitious and broadly hegemonic attempt to rearticulate the development project and produce entrepreneurial neoliberal subjects" (Death and Gabay 2015).

MDG environmental targets/indicators were largely vague; in some cases, assessment only captured spectacular failure to transition economies away from fossil fuels and carbon emissions (Liverman 2018). Their ultimate failings to achieve "development" would provide impetus for creating the SDGs at a new international agreement at 2012 UN Conference on Sustainable Development (also known as Rio + 20), 25 years after the Brundtland Commission report (Carant 2017). The 8 MDGs were reborn as 17 SDGs with 169 targets and 231 unique indicators as of 2022 (United Nations Statistics Division). Thus expanded, the SDGs officially proclaimed that development required healthy ecologies and environmental life support systems (Fukuda-Parr and McNeil 2019; Griggs et al. 2013) while further enmeshing development in the production of measures and metrics.

14.3 The SDGs, Discourse, and Quantification

Like the MDGs, the SDGs idealize development as a discrete condition—that of being developed—which is achieved by meeting goals (e.g., "No Poverty") composed of targets (e.g., "by 2030, eradicate extreme poverty for all people everywhere") tracked by indicators (e.g., "proportion of the population living below the international poverty line"). In theory, data produced transparently will identify "deficiencies" and needed interventions. Thus the SDGs reinscribe a colonial imaginary of development: some*thing* a population desperately needs in order to catch up, civilize, or modernize. Continuity with the past, however, casts doubt on whether SDGs truly contain radical potential to challenge existing development inequalities (Brown 2016; Death and Gabay 2015; Mawdsley and Taggart 2022). Despite methodological, ideological, political, and practical critique (Blythe et al. 2018; Esquivel 2016; Liverman 2018; Sexsmith and McMichael 2015; Sultana 2018 among others), the SDGs approach hegemony in influence. They underpin multilateral agreements; justify policy across scales; condition development aid, lending, and investment; but also create opportunities for political struggle.

The SDGs sit alongside dozens of indexes and rankings—what scholars have come to call "global performance indicators" (Kelley and Simmons 2020)—that purport to measure development. An ever-expanding cast of NGOs, think-tanks, state agencies, and individual scholars churn out measures and metrics to the point of competition and redundancy. UN programs routinely calibrate activities to the Human Development Index or Oxford's Multidimensional Poverty Index. To attract capital, governments promote their rankings in the World Bank's Ease of Doing Business survey. Other popular indicators compare democratic institutions, gender inequality, corruption, contemporary slavery, violent conflict, pollution, and more. This proliferation is testament to the power—perceived or actual—of indicators to produce a frame of reference and understanding that Foucault called a "regime of truth" (Hall 1996; Foucault 1980, p. 131). "By naming an issue, coining a vocabulary for describing it, and creating categories for its assessment, promulgators hope to affect discourse and, ultimately, policy" (Kelley and Simmons 2020). Put another way, the SDGs and other performance indicators constitute discursive practices that provide (and privilege) language and evidence to comprehend and narrativize development (cf. Watts 2003).

Though typically constructed by scientists and experts outside popular democratic venues, these indexes and indicators should not be understood as *apolitical* or purely technocratic. Fischer (2019) identifies three forms of politics undergirding the world of development as measures, metrics, and numbers: first, politics "of representation," in struggles over narratives of the history of capitalist development; second, "of prioritization," in construing (often reductively) development as a limited set of discretely measured conditions; and third, "of conception and production," in generating statistics that result from conscious, contestable decisions. In other words, politics suffuse the SDGs process from start to finish, prompting questions: Who and what are counted? By whom, how, when, and where? Most of all, why? Here we glimpse the performative character of projects like the SDGs that construct development as that which is assessed, while the details of the assessment configure what is (or can be) developed. As Watts (1993) critically notes, "a part of the armoury of development is… the measures (and the practices of measurement) of development itself."

Development by quantification is hardly a radical break from the past. The collection of statistics on population, economy, environment, and society—emergent concerns even in Malthus' day (Dean 2015)—have long rendered people, activity, and geography legible for management. Foucault (1991, pp. 92–93) suggests contemporary government is rooted in "meticulous attention" to a state's population "in their relations, their imbrication with … things which are wealth, resources, means of subsistence, the territory with its specific qualities, climate irrigation, fertility, etc. …customs habits, ways of acting and thinking, etc.… accidents and misfortunes such as famine, epidemics, death, etc." What Foucault called biopolitics would mature amid empire as ruling governments routinely measured, ranked, and intervened in the lives of colonized subjects and geographies (Drayton 2000; Tilley 2011; Parenti 2015).[9]

The fetishization of quantification is particularly trenchant in environmental domains, synergizing with neoclassical economics which frames ecological, hydrological, and other "natural" processes and systems as "stocks," "flows," or "natural capital" (Hébert and Brock 2017; Höhler and Ziegler 2010). Measuring productive natures has also yielded a heterodox literature around "ecosystem services" and their valuation (Costanza et al. 1997; Costanza et al. 2014; Jadhav et al. 2017; Kareiva et al. 2011). In short, measurement, counting, and mapping—all of which underpin the SDGs—are political technologies at the heart of governing people, places, and "natures" while (re)producing state and non-state power.

Quantification serves as development methodology (Li 2007, p. 7) through "problematization," or "identifying deficiencies that need to be rectified," and "rendering technical," a range of representational practices for (following Rose 1999) "defining boundaries, rendering that within them visible, assembling information about that which is included and devising techniques to mobilize the forces and entities thus revealed." Problems rendered technical find a ready industry of experts, planners, activists, and entrepreneurs; British colonial and postcolonial Indian history offers numerous examples of expert-driven, top-down attempts at with high modern, transforma-

[9] Foucault labels contemporary population governance as biopolitics, distinct from sovereign rule which emphasizes growing/protecting the specific wealth and authority of the ruler.

tive projects (Bhattacharya 2019; D'Souza 2006; Scott 1998). This production of development problems and solutions—though shot through with latent politics and ideologies—also forecloses democratic or inclusive participation (Li 2007; Sultana 2018). Indicator selection, calculation, and measurement are the terrain of scientists and bureaucrats who populate the government ministries and NGOs able to engage elite policy (Sénit 2020). Values and interests are "obscured behind the veil of a technocratic process driven by (supposed) objectivity" (Fukuda-Parr and McNeil 2019, p. 7) which serves to "cloak normative agendas in languages of neutral and technocratic assessment" (Broome and Quirk 2015, p. 814). Development discursively and performatively produces "a target for a particular sort of intervention: the technical, apolitical, 'development' intervention" (Ferguson 1994, p. 28).

14.4 India's "Life Below Water"

To see this discursive performance, I turn now to India's engagement with the SDGs, and specifically Goal 14, dubbed "Life Below Water" for its concern with the ocean. Technical consultations following Rio + 20 yielded an initial "global indicator framework" for 17 SDGs adopted by the UN General Assembly 2017. That year, the Modi government released its first Voluntary National Review (VNR) report on India's SDG implementation (NITI Aayog 2017). The VNR signaled seemingly enthusiastic engagement with development by measurement, though without effort to plot national data against international indicators.

While official SDG indicators were revised through UN processes, India produced and updated its own indexes to track state and national "progress." The political thinktank and pseudo-planning agency NITI Aayog[10] has primary responsibility for aggregating and communicating SDG statistics. NITI Aayog produced the 2017 VNR and published in 2018 a first-pass Indian SDG index (refined in 2019 and again 2021). NITI Aayog's 2018 index excluded metrics on SDG 14 but reproduced boosterish pages on government fisheries policy, port construction, and marine protected areas. Subsequent versions presented a handful of SDG 14 indicators, calculated only for India's coastal states. In parallel, the Ministry of Statistics and Programme Implementation (MoSPI) in 2019 generated its own indicator list mapped to SDG targets and now hosts an online "dashboard" with periodically updated data.[11] A 2020 VNR from NITI Aayog repackaged limited data in graphics alongside narrative text. Table 14.1 charts the crystallization of India's SDG 14 index alongside the UN indicators as of 2022.

I do not wish to exhaustively dissect goals/targets/indicators or suggest alternative specifications.[12] Instead, this chapter aims to problematize development by quantification and demonstrate how slippery conceptions of sustainable development—ostensibly measured by the SDGs—enable India's articulation of muscular nationalism to neoliberal capitalism in the name of Blue Economy. While more radical or critical metrics could make SDG measurement more self-reflexive or better capture unsustainability, perfection of quantification would not automatically (or even likely) challenge the SDGs' discursive and performative power. In this analysis, four key points arise.

[10]The National Institution for Transforming India, known as NITI Aayog, is a recent invention understood to promote/justify the BJP political and economic agenda. Modi created it as a slimmed-down corporate thinktank in 2015 abolishing the vaunted Planning Commission, a top-heavy but influential economic planning agency that dated to early Independence. In neoliberal India, however, central planning and the commission's ponderous Five-Year Plans had become moribund (while also remaining outside Modi's control). Roberts (2021) reviews recent works on NITI Aayog's birth from the ashes of the commission.

[11]The online portal—http://www.sdgindia2030.mospi.gov.in/dashboard/—is managed by a private Big Data company, iTech Mission, whose slogan "for a better world" emphasizes how data sit at the heart of development.

[12]Nor do I wish to dwell on basic mathematical problems known to even undergraduate students of statistics (e.g., the pitfalls of calculating percentage change on short-term basis).

Table 14.1 SDG 14 in India

Goal 14. Conserve and sustainably use the oceans, seas, and marine resources for sustainable development			
2019 National Indicator Framework MoSPI	2019 SDG Index 2.0 by NITI Aayog	2021 SDG Index 3.0 by NITI Aayog	2022 Official U.N. SDG indicator list
Target 14.1 By 2025, prevent and significantly reduce marine pollution of all kinds, in particular from land-based activities, including marine debris and nutrient pollution			
(a) Coastal Water Quality Index (b) Number of sewage treatment plants installed along the coast and construction of toilets under Swachh Bharat Mission[a] (c) Percentage change in the use of nitrogen fertilizers in the coastal States	(a) Percentage increase in the use of nitrogen fertilizers in the coastal states (b) Coastal Water Quality Index	(a) Mean shore zone coastal water quality: biochemical oxygen demand (b) Mean shore zone coastal water quality: total nitrogen	(a) Index of coastal eutrophication (b) Plastic debris density
Target 14.2 By 2020, sustainably manage and protect marine and coastal ecosystems to avoid significant adverse impacts, including by strengthening their resilience, and take action for their restoration in order to achieve healthy and productive oceans			
(a) Percentage change in area under mangroves (b) Implementation of Coastal Zone Regulation Notification of 2011[a] (c) Percentage change in Marine Protected Areas[a]	Percentage increase in area under mangroves	Percentage increase in area under mangroves	Number of countries using ecosystem-based approaches to managing marine areas
Target 14.3 Minimize and address the impacts of ocean acidification, including through enhanced scientific cooperation at all levels			
Coral health index of Exclusive Economic Zone[a]	Average marine acidity (pH) measured at representative sampling stations in the shore zone	Average marine acidity (pH) measured at representative sampling stations in the shore zone	Average marine acidity (pH) measured at agreed suite of representative sampling stations
Target 14.4 By 2020, effectively regulate harvesting and end overfishing, illegal, unreported and unregulated fishing and destructive fishing practices and implement science-based management plans, in order to restore fish stocks in the shortest time feasible, at least to levels that can produce maximum sustainable yield as determined by their biological characteristics			
Maximum Sustainable Yield (MSY) in fishing	No indicator	No indicator	Proportion of fish stocks within biologically sustainable levels
Target 14.5 By 2020, conserve at least 10% of coastal and marine areas, consistent with national and international law and based on the best available scientific information			
(a) Coverage of protected areas in relation to marine areas[a] (b) Percentage change in area under mangroves	No indicator	No indicator	Coverage of protected areas in relation to marine areas
Target 14.6 By 2020, prohibit certain forms of fisheries subsidies which contribute to overcapacity and overfishing, eliminate subsidies that contribute to illegal, unreported and unregulated fishing and refrain from introducing new such subsidies, recognizing that appropriate and effective special and differential treatment for developing and least developed countries should be an integral part of the World Trade Organization fisheries subsidies negotiation			
No indicator	No indicator	No indicator	Degree of implementation of international instruments aiming to combat illegal, unreported and unregulated fishing
Target 14.7 By 2030, increase the economic benefits to small island developing States and least developed countries from the sustainable use of marine resources, including through sustainable management of fisheries, aquaculture and tourism			

(continued)

Table 14.1 (continued)

No indicator	Percentage of available potential area developed under coastal aquaculture	Percentage of available potential area developed under aquaculture	Sustainable fisheries as a proportion of GDP in small island developing States, least developed countries and all countries
Target 14.a Increase scientific knowledge, develop research capacity and transfer marine technology, taking into account the Intergovernmental Oceanographic Commission Criteria and Guidelines on the Transfer of Marine Technology, in order to improve ocean health and to enhance the contribution of marine biodiversity to the development of developing countries, in particular small island developing States and least developed countries			
Allocation of budget resources for research as per the EEZ or coastal line[a]	No indicator	No indicator	Proportion of total research budget allocated to research in the field of marine technology
Target 14.b Provide access for small-scale artisanal fishers to marine resources and markets			
Assistance to the traditional/artisanal fishers for procurement of FRP boats and other associated fishing implements	No indicator	No indicator	Degree of application of a legal/regulatory/policy/institutional framework which recognizes and protects access rights for small-scale fisheries
Target 14.c Enhance the conservation and sustainable use of oceans and their resources by implementing international law as reflected in the United Nations Convention on the Law of the Sea, which provides the legal framework for the conservation and sustainable use of oceans and their resources, as recalled in paragraph 158 of "The future we want"			
Percentage compliance of international laws[a]	No indicator	No indicator	Number of countries making progress in ratifying, accepting and implementing through legal, policy and institutional frameworks, ocean-related instruments that implement international law, as reflected in the United Nations Convention on the Law of the Sea, for the conservation and sustainable use of the oceans and their resources

[a]Proposed indicator, data under compilation

First, many U.N. indicators have no national corollary. This *persistent incompleteness* is not surprising as national (and subnational) monitoring and evaluation have long been targeted for "capacity building."[13] Yet India is not "data-poor"[14] and has numerous government bureaucracies devoted to statistical collection and analysis, including NITI Aayog and MoSPI; Modi himself encourages his subordinates to govern via data (Arnimesh 2021). India's omissions of indicators raise a question: Does the government actually desire transparent evaluations? Selective reporting or withholding information controls development narratives and minimizes accountability, as "no data" supports abstract claims about sustainability better than explicit evidence of unsustainability. "Data opacity and a lack of accountability have been the two glaring leitmotifs of the statistical ecosystem" during Modi's tenure, producing just "enough smoke so that a discussion as fundamental and important as poverty levels remains ensnared in confusion" (Shroff 2022).

Second, variable/indicator choices reflect the SDGs' tendency toward *synoptic reduction* of multi-layered systems or complex phenomena to supposedly straightforward statistics (cf. Scott 1998). Take the example of SDG Target 14.4, which envisions future sustainability in fishing, defined as harvest regulation and scientific man-

[13]SDG17 itself calls for a national "statistical capacity indicator for Sustainable Development Goal monitoring."

[14]This commonplace descriptor repeats a political ideological of poverty (Fischer 2019) in the ethereal realm of data.

agement to end "overfishing, illegal, unreported and unregulated fishing and destructive fishing practices and implement science-based management plans." In India, this goal would require command and control of diverse, multi-scalar marine fisheries,[15] composed of dozens of overlapping stocks, hundreds of catch landing centers, thousands of fishing villages, tens of thousands of fishing craft, hundreds of thousands of active fishers, and millions of fisher household members (Jadhav 2018). The SDGs reduce "sustainability" of this complex assemblage to a single official UN indicator: the "proportion of fish stocks within biologically sustainable levels." This measurement would calculate annual harvests relative to a fisheries concept known as maximum sustainable yield (MSY), a quantitative estimate of fish that can be harvested without threatening the reproduction of the remaining population. Multiple mathematical methods have been tested for assessing theoretical MSY, but all calculations rely on estimation, extrapolation, and abstraction from other proxy/sampled data, each with their own conceptual problems and margins of error (cf. Vase et al. 2019). Errors compound when aggregated to the millions of square kilometers of that make up India's EEZ, making national MSY impossibly abstract and fictionally precise; put another way, MSY may be only myth that reproduces fantasies of scientific management (Ramesh and Namboothri 2018). Fish populations and their growth rates fluctuate wildly with myriad factors ranging from climate and ecology (e.g., nutrients, water quality, upwelling strength, or ocean temperatures which condition survival rates) to social and techno-economic considerations (net mesh sizes, spatial distribution of effort, or seasonal fishing intensity). Even crude MSY calculations for many species in India remain in progress, and government agencies lack the monitoring or enforcement capacity (scientifically, politically, or practically) to constrain harvests at any given level, sustainable or otherwise.[16] One internationally published estimate suggests that out of 223 identified stocks in Indian waters, only 34 percent are not currently threatened by present fishing effort while also still having a population capable in the abstract of sustaining the fishery (Sathianandan et al. 2021).

In place of MSY relative to harvest (again, a mainstream if contestable metric), MoSPI's online SDG dashboard reduces fishing sustainability to a static figure—5.3 million metric tons of "potential yield" from the EEZ.[17] Produced by a government committee in 2018, this number is significantly higher than previous estimates, which held Indian marine fisheries capable of sustaining between 3.7 million and 3.9 million metric tons of annual harvest (Desai et al. 1990; Srinath and Balan 2003). Even these estimates were higher than the average marine catch over the last decade and a half, stagnant at about 3.3 million metric tons (Lakra and Gopalakrishnan 2021). The drastic increase by the 2018 committee of "potential yield" could reflect higher resolution assessment of individual fish stocks, though the needed science—detailed for individual species/stocks by region—has not been published on such a grand scale. Meanwhile, the absence of publicly available documentation from the committee or its deliberations (e.g., in digital science repositories) raises the possibility of political influence to ensure current levels of fish harvest and expansionary development ambitions appear within supposed limits. Critical scholars and fishers themselves for years have suggested that India's marine fisheries are actually under severe stress (Bhathal and Pauly 2008, interviews 2012, 2014, 2015, 2016).

Third, India's chosen indicators reflect *geographic confusions* that follow from the SDGs' reliance on measurement within demarcated

---

[15]Except where explicitly mentioned, this chapter concerns fisheries carried out at sea or in coastal estuaries and backwaters. India also has considerable, vibrant inland fisheries in rivers, ponds, and ancient human-made tanks.

[16]In India, fishing regulations are often shifting, contradictory and contested. Arguably, the most successful regulation—a seasonal ban on large-scale fishing during the monsoon season (Gunakar et al. 2017)—is more a politically accepted annual brake on fishing intensity than rule designed from scientific certainty.

[17]Notably, NITI Aayog indexes sidestep Target 14.4 entirely.

spaces (idealized boundaries within which development operates). Reporting of aggregate statistics within a given space obscures nuance and variation within that geography, potentially hiding unsustainability at smaller scales. For example, India uses remotely sensed mangrove forest canopy to proxy coastal ecosystem health (Targets 14.2 and 14.5).[18] Yet celebrating aggregate canopy increase conceals rampant clear-cutting and in-filling of mangrove wetlands for industrial and real estate projects (Ghosh et al. 2015; Sud 2021; Vijay et al. 2005). Also problematic is the presentation of coastal statistics while ignoring marine impacts of terrestrial ecologies and economies far upstream. Consider that in the absence of more robust water quality measurement, MoSPI suggests that the number of toilets and sewage treatment plants built in coastal states could serve as indirect proxy to assess marine pollution. Problematic conflations aside—neither toilets nor treatment plants automatically prevent sewage release[19]—this choice of metric also presumes that oceanic water quality is only a coastal responsibility and discounts pollution across India that still ultimately washes out to sea. Conceptualizing oceanic and coastal sustainability as domains disconnected from inland activity problematically overlooks both hydrological and sociohistorical relations in India between the ocean's waves, current, winds, and climatology and a landed territory soaked in snowmelt, monsoon rainfall, and flowing rivers (cf. Amrith 2018).

These twinned geographic confusions or insensitivities—obscuring intra-scalar variation and ignoring inter-scalar relationships—demonstrate a non-relational approach to development and sustainability. Put another way, this performance of the SDGs fails to account for development as an uneven geographic process both horizontally (interrelation across space) and vertically (interplay of scales within space). This understanding of development also draws on and reinforces methodological nationalism. Sexsmith and McMichael (2015) note that the SDG process reifies a "state-centric framework for global development planning" inadequate to diagnosing unsustainable development when the root causes or motive forces transgress administrative and political boundaries (e.g., fish that swim, mobile capital, far-away consumer demand, pollution that always flows downstream or globalized climate change).

Fourth, SDG targets and indicators remain *ambiguous and flexible*, available for (re)interpretation by dominant politics. To start, note the recursive, near-tautology in SDG 14 as a whole: States are to "sustainably use" marine environments and resources for "sustainable development." Some individual targets also reference sustainability without definition, further transmuting the concept into open-ended jargon, buzzword, or even floating/empty signifier (cf. Brown 2016). While I would critique attempts to precisely delimit sustainability indicators across divergent geographies, as they stand, the SDGs allow flexible (re)definition expedient for ruling political economic interests—e.g., India's Blue Economy agenda. India's indicator selection for Target 14.7 provides the object lesson, conflating aquaculture expansion with "the sustainable use of marine resources" (Target 14.7). Across the country, farming marine products is hailed today as a new "blue revolution," underscoring a fantastic imaginary of the Blue Economy while presaging a transformation of agrarian life and economy that echoes the Green Revolution (Newman 1981; Patel 2013). Though coastal communities have long practiced aquaculture-like rearing of trapped fish in combination with rice paddy that tolerates brackish water, low-input local traditions have increasingly been subordinated to industrial, commodity fish farming. Social-ecological effects are manifold: commons dispossessed, waterways fouled with pollution, wild stocks diverted into feed for farmed fish, and conflicts between local communities and would-be capitalist or "mafia" elements over wetland destruction (Ali et al. 2021; Bhat and Bhatta

[18] Mangrove forests serve as global charismatic megaflora, and their protection arguably is understood as a near-universal conservation target globally (cf. Spalding and Leal 2021).

[19] These statistics also repeat fetishization of the toilet by development policy generally (cf. Menga and Goodman 2022) and the Modi government specifically.

2004; Bhatta 2019; Bhatta and Bhat 2002; Kanan et al. 2020; Roy 2019). Yet aquaculture serves as a key plank in the national Blue Economy project's dreams of limitless production, especially of prawns, cobia, pompano, and other export species (Ranjan et al. 2017). Against this political economic backdrop—Modi (2021) himself touts the future of aquaculture—the SDGs appear more a project of incentivizing and justifying capitalist (and in this case specifically nationalist) expansion than of tracking sustainability.

Taken together, these four features of India's SDG efforts—persistent incompleteness; synoptic reduction of complexity to statistics; multiple geographic confusions; and an ambiguous and flexible character—author and authorize a notion of sustainable development that is uncritical, abstracted, and blurred in meaning. Put another way, the performance of development by measurement has produced sustainability as a mutable signifier available for articulation to/within broader political and political economic projects. In India, this reconfigured discourse of sustainability has become hitched to and submerged within Hindutva's vision of oceanic capitalism, at once nationalist and neoliberal.

Against this fantasy-laden discursive production of sustainable seas and coasts, I turn now to the concrete experience of lightly developed estuary in southwestern India where the Blue Economy agenda is hard at work.

14.5 Actually Existing (Un) sustainability of the Blue Economy

The sheltered backwaters of the Aghanashini River present a contrasting image to the frenetic, glitzy, ultramodern megacities and urbanizing industry of New India's policy imaginaries and development praxis. There are no heavy industries, high-rise buildings, or planned subdivisions. Instead, several thousand households live scattered in small hamlets around this estuary nestled on northern Karnataka's rural coastline (see Fig. 14.1). Such is the contemporary biodiversity and ecological health of "lightly touched" estuary (Bhat et al. 2014; Jadhav et al. 2017) that scientists, activists, residents, and even a collective of fishers have formally and informally proposed its designation as a protected area (Chandran et al. 2012; Hegde 2014, interviews 2015, 2016).

The red clay tile roofs of traditional bungalows and brightly colored box construction of newer homes contrast with tropical greenery. Mangroves sprawl in swamps and line the embankments along fields and ponds. Coconut and areca palms tower over homesteads and gardens. Fish, shrimp, crabs, oysters, clams, and mussels are seasonably in abundance, and petty producers also extract salt, sand, shell, and leaflitter from the river. Small-scale and subsistence fishers ply shallowest reaches of the estuary, while small-but-ocean-going trawl and purse seine boats shelter at jetties along the deeper riparian channel (see Figs. 14.2, 14.3 and 14.4). Paddy rice grows in submerged fields that are also, sometimes, used for traditional fish trapping and ranching. Cashew, mango, and a dozen other tree species produce fruit and fodder on higher ground away from the brackish water table. Notably, many households straddle subsistence and commercial production, producing one crop to sell (e.g., peanuts) while reserving another for the dinner table (e.g., rice). Survey data suggest most households avoid dire poverty by combining diverse income sources with self-provisioning from livestock, gardens, or the commons.[20]

However, this (wet) landscape should not be imagined purely as a rural idyll locked in a vestigial past disconnected from India's postcolonial "modernity." Lengthy supply chains insert peasant producers and harvesters into regional and national political economies. Farm consolidation, plantations, and aquaculture signal increasing commodity production that drives class

[20] These depictions of life, livelihoods, and politics in and around the Aghanashini Estuary are drawn from more than 18 months of research and activism in academic, NGO, and government projects. This work included hundreds of interviews, two large-*n* household surveys conducted by a team and practical community development programs.

Fig. 14.1 The Aghanashini River estuary, situated in Karnataka on India's southwestern coast. (Adapted from Jadhav et al. 2017)

Fig. 14.2 The Aghanashini main channel (left) bordered by ponds for fish trapping and aquaculture (right). (Author photograph)

Fig. 14.3 Oyster harvesters at extreme low tide. (Author photograph)

Fig. 14.4 An Aghanashini bag net fisher. (Author photograph)

stratification and swells the ranks of landless labor. Small-scale fishers also increasingly beach their own boats to crew ocean-going vessels locally or seasonally in distant, city harbors. Caste, class, gender, and religion all further configure socioecological relations and "development." In short, I suggest the Aghanashini estuary constitutes an exemplary (as opposed to exceptional) coastal political ecology (Bennett 2019; Jadhav 2015), one that sits at the intersection of multiple Blue Economy (un)sustainabilities. In this section, I highlight four "developments" looming around the estuary: a state-capital partnership to construct a shipping port, speculative investments in aquaculture, takeovers of space by coastal tourism, and an aggressive mangrove afforestation campaign.

Port Development

For decades, development planners have overlooked socioecological complexity, focusing instead on the river's narrow mouth, guarded by rocky headlands, which shelters the estuary from ocean waves and winds. Industrial boosters have long envisioned this "safe harbor" as an ideal site for economy by the sea. In the 1970s, the state government seized more than 1400 acres of wetlands to build a caustic soda plant utilizing abundant clam, oyster, and mussel shell discards. Nehruvian state-led developmentalism of the day would also reinforce dike walls along the river and pave roadways—ostensibly to secure agricultural production and expand marketing. The factory failed so quickly that many estuary residents have no memory of its operation (interviews 2015, 2016); the only remainders of that dream are weathered pylons jutting from the water. Other designs for the acquired land—a coal-fired power plant, a Special Economic Zone, various harbor configurations, even a shipbreaking yard—have repeatedly stalled in the face of local backlash (Arun and Habbu 2010; Kohli and Sekhsaria 2000; Menon 2004; Menon and Kohli 2015). Some dispossessed residents have sued in court for additional compensation, and the failure to bring promised development hangs over local politics. In the late 2000s, the Karnataka State Industrial and Infrastructure Development Corporation (KSIIDC) began planning the latest iteration of an industrial shipping port, one with "total automation and 24 × 7 operations," to export coal and import steel (NEERI 2015). Feasibility reports, cost-benefit analyses, and speculative plans are shot-through with boosterism, wishful assumptions, and bad math. Dreaming of "public-private partnership," officials would hand the estuary over to a yet unnamed private corporation to build and operate the port for an unfixed term before ultimately ceding the development back to state control.

Completion would take decades and require the sacrifice of at least 500 acres of mangroves—protected habitat under Indian law—as well as fishing grounds, oyster beds, and mudflat ecologies. The estuary, currently only a few feet deep in many locations, would also be dredged down to at least 16 meters to accommodate industrial-scale traffic, shuttering commercial fishing jetties and banishing small-scale fishers from the river. Residents, ecologists, and environmental activists note that tides would spread pollution and debris far away from the port itself, despoiling many more acres of mangroves, cultivation ponds, decades-old salt pans, oyster beds fish nurseries, and beaches (interviews 2015, 2016). Additional land use change would likely follow knock-on industrialization in the region. At risk, according to even coarse estimates, are hundreds of millions of dollars annually in "ecosystem goods and services" (Jadhav et al. 2017).

Though the port proposal predates Modi's rule, boosters pitched the project for inclusion in the BJP's flagship Blue Economy program, Sagarmala ("ocean necklace" in Hindi). Under this scheme, the Ministry of Ports, Shipping, and Waterways envisions cargo facilities, highway connections, rail lines, relocated fishing harbors, and other "modernized" maritime infrastructures as a garland draped over the neckline of the Indian peninsula.[21] This agenda doubles down on

[21]The name conspicuously also reflects a Hindu-ization of national development policy; the *mala* features in Hindu rites and rituals (e.g., wedding garlands) as well as individual religious practice.

India's neoliberal export economy dreams,[22] and the government claims to have completed or started projects worth $77.5 billion (more than 600,000 crore rupees).[23] Not included in that figure is the $556 million (4300 crore rupees) price tag of the new port imagined for the Aghanashini estuary. And though Modi (2016) touts port development as a return to India's "glorious maritime heritage," the uneven distribution of environmental, social, and economic impacts of oceanic capitalism and infrastructure continues to produce opposition across the country (Research Collective 2018; Rodriguez and Sridhar 2010).

In the Aghanashini case, some residents from communities who lost land decades ago have backed the port project; at a project hearing in March 2015, a contingent of supporters heckled all opposition—a loose coalition of residents, fishers, activists, NGOs, and scientists who decry the port's outlandish claims of sustainability. A government review committee ultimately recommended environmental clearance only if planners could satisfy two-dozen riders related to environmental management, worker safety, impact monitoring, and pollution mitigation (MoEFCC 2017). The central government Ministry of Environment, Forests and Climate Change has since withheld a final decision on permits, and activists reported in 2018 that the project was unofficially, quietly cancelled.[24] However, proponents and critics alike say they expect the government to continue pitching industrial development as long as it continues to hold the acquired wetlands (interviews, 2016, 2018, 2020).[25]

Aquaculture Expansion

The port project is only one example of the multivalent Blue Economy. For years, development institutions in India have dreamt of aquaculture as an alternative to stagnating capture fisheries (Krishnan and Birthal 2002). Oceanic harvests face fundamental (and fluctuating) ecological limits and declining returns to fishing effort, but aquaculture represents, for state scientist and capitalist investor alike, the dream of expanded fish production through agro-industrial methods: private property regimes, financing tied to land, consolidation of small holder wetland tenures, new commodity chains of inputs/equipment, scientific management, vastly scaled-up production, and foreign exchange from fish export. A cadre of development and science agencies—the Coastal Aquaculture Authority (CAA), National Fisheries Development Board, Central Institute of Brackishwater Aquaculture, National Centre for Sustainable Aquaculture, and Central Marine Fisheries Research Institute, to name a few—run training, incentive, demonstration, or subsidy schemes to industrialize the once-nascent sector. The result has been an almost 900-percent increase from 1988 to 2013 (Jayanthi et al. 2018) in the area under coastal aquaculture (notably, a statistic conflated with sustainability in the SDG process).

Modi's government will continue the trend; the central government in 2020 announced a 5-year plan to boost annual fish production by seven million tons and double marine product exports. This neoliberal leap in output will require the spatial expansion of fishing into the farthest reaches and depths of the EEZ, and the continued transition to a capitalist, debt-financed industrial fishing fleet that exploits smaller-scale fisher labor and depletes the marine commons (Jadhav 2020). According to Modi, the government wants fishers to take loans "so they can buy

[22] More than a decade ago, the Congress Party-controlled central government announced a "maritime agenda" called for some $41 billion (by today's exchange rates) of port infrastructure by 2020 along with a 54-point ocean commerce development plan (Ministry of Shipping 2011).

[23] Karnataka alone is slated for "development" worth almost $2 billion (more than 15,000 crore rupees), including upgrades and expansions at existing ports in Karwar and Mangalore, as well as a coastal petrochemical facility. Figures computed using May 2022 exchange rates (Ministry of Ports, Shipping and Waterways 2022).

[24] The port still features prominently on the KSIIDC official website. http://www.ksiidc.com/tadadi_port.html. Accessed September 2022.

[25] The repeated political/economic reanimation of mega-projects in the acquired wetlands land presents a case of actually existing zombie neoliberalism (Peck 2010).

bigger boats" (ANI 2017). However, with annual marine fish harvest stagnant at roughly 3.3 million tons, this fantasy of fisheries growth will depend more crucially on the intensification/expansion of aquaculture. In this development dream, India's millions of acres of backwaters, wetlands, and rivers must be repurposed and exploited, despite devastating consequences for water quality, wetland commons, biodiversity, and social economy. At the same time, wild caught fish—consumed daily in many coastal households—are discursively and materially reimagined as feedstock for fish farming.

In the Aghanashini, aquaculture represents both imminent future and echo of the past. Decades ago, the mosaic of small fields surrounding the estuary supported a traditional land-use system known as *gajni*, combining salt-tolerant *kagga* rice, shrimp culture, and sometimes fish trapping (Bhatta 2019; Jadhav et al. 2017; Stone et al. 2008). Much of this land locally was consolidated and converted to full-time aquaculture during a shrimp boom in the 1990s. The bust that followed—in part because disease and pollution killed off stocks—left much gajni land too degraded, deep, or large for traditional methods. These fallow ponds are now again fluid terrain for experiment and speculation. In recent years, multiple aquaculture ventures have appeared, often shrouded in quasi-secrecy. One facility with pipes running directly into the ocean hides behind large walls, while security guards at another site along the river's edge prohibit visitors from approaching. One entrepreneur from Bangalore, however, was both candid and condescending about the opportunity to (re)organize coastal space and livelihoods: "These people, they can't do anything like this," he said, gesturing to floating fish cages in ponds adjacent to the river. "They need to be taught. If you don't teach them, they'll spend their money on alcohol. I can come here and bring some economic development, and of course everyone can get rich" (interviews 2015).

Coastal Tourism

The Blue Economy dream of reconfiguring coastal natures and livelihoods into revenue also motivates hype surrounding eco-tourism—short for ecological tourism. Dreaming of profitable conservation, promoters seek guests willing to pay premiums to experience natural spectacle—from the mangrove swamp to the sun-soaked beach. The Blue Economy's performative mantle of sustainability, as with industrial shipping and aquaculture, provides cover for many forms of development, from high-end resorts to low-budget backpacker beach shacks. Modi himself in a visit to Kerala in 2021 inaugurated a new large-scale cruise ship terminal while simultaneously hailing the sector as space for entrepreneurs and speculators: "I urge our young start-up friends to think about innovative tourism-related products" (ANI 2021a). However, packaging remnant "wilderness" for wealthy visitors often consumes or encloses local environments while fueling inequality and social conflicts (Fletcher 2014); put another way, nature tourism trends toward "exporting paradise" (Clancy 2001; Hampton 2013).

As with other Blue Economy ambitions, this vision of tourism is hardly novel. Decades ago, India's small state of Goa helped pioneer the marketing of the sensual sea: sumptuous meals, fresh-caught fish, stunning sunsets, sandy expanses, bathtub-warm waters, beach parties, flowing alcohol, sex, drugs, and rock and roll. The 1960s scene catering to "Western hippy travelers" has morphed through a mix of entrepreneurial capital, speculative real estate markets, and state intervention into an industry of guest houses, hotels, entertainment, and transport services (Routledge 2000; Wilson 1997). Tax and foreign exchange receipts have boomed but at a cost: overburdened infrastructure, pollution, drug/sex trades, and social conflicts (Lakshmi 2013; Menezes 2014). Tourism's takeover of the state's beaches and coastal towns has also coincided with the displacement of small-scale fishers and the growth of a mechanized, commercial fleet (Newman 1981; Jadhav 2018).[26] Goa's story also increasingly repeats along much of India's coastline, from houseboat tours in Kerala to scuba resorts in the Andaman Islands to innumer-

[26]This transition to mechanized fishing is a national trend that will only intensify under Modi's Blue Economy (ANI 2021b).

able ventures capitalizing on coastal forts, golden sands, or charismatic wildlife (Equations 2019; interviews 2015, 2016).

This transformation has also begun around the Aghanashini, where just two decades ago the tourism economy concentrated on Hindu pilgrims visiting a fourth-century temple complex. Today, a new generation of resorts, cottages, and beach shacks has proliferated along beachfronts north and south of the river mouth, offering a vibe akin to Goa's earlier days. Overnight rail and bus services connect the rural coast to major cities, especially Bangalore. Some fishers now take guests aboard boats for dolphin-spotting or cruises upriver through mangroves. New adventure companies, Airbnb hosts, and NGOs market varying visions of eco-tourism around the estuary and in the nearby Western Ghat mountains.[27] As of June 2022, KSIIDC, the state industrial development agency backing the seemingly stalled port project, had begun pitching development of an "eco-tourism hub" to attract private developers to the land appropriated decades ago by the government.

Downsides have been predictable: pollution, clashes over social taboos (e.g., beach attire, parties, public drinking, drugs), and enclosures of the commons. Consider an incident when many fears became manifest: In the 2016 dry season, tourists exploring the ruins of a stone fort overlooking the ocean sparked a fire that burned through acres of grassland commons used by a nearby village for cattle fodder. Like elsewhere in India (Sud 2021), a hyper speculative real estate markets also increases pressure to sell to developers, further commodifying land or incentivizing outright enclosure (interviews 2016). Meanwhile, residents say the eco-tourism's primary beneficiaries are elites who command labor, capital, and micropolitical power. One member of the local elected council, the *panchayat*, provides an example. Congenial, eager, and broadly interested in local "development," he has secured rights to harvesting cashews from a state plantation while acting as a gatekeeper and intermediary for NGO projects (interviews 2015, 2016). Most recently he has parlayed his clout and coastal land ownership into partnership in an Airbnb-operated small-scale resort, which has since expanded onto beach commons.[28]

Mangrove Plantation

As industrial projects repeatedly failed, portions of the estuary seized by the state reverted to a fishing commons. In the mid-2000s, these wetlands were targeted for a different mode of development: conservation. The state Forest Department, local environmental NGOs, and community activists began aggressively planting the "unused" government "land" with mangrove saplings, and a kind of mangrove monocropping has since spread into the same development dreamscape of ports, eco-tourism, and aquaculture. This "successful" afforestation of mangroves and the preservation of biodiversity (Kumar and Singh 2022; Mesta et al. 2014) exemplifies a contradiction of the Blue Economy agenda. As charismatic megaflora, mangroves are legible to planners and policy-makers as highly productive wetlands, generating multiple ecosystem goods and services. They receive strict de jure protection under state coastal regulations and forest laws, though de facto enforcement remains uneven (Jadhav et al. 2017; Kohli and Menon 2019). Increased mangrove cover—a remotely sensed indicator in India's SDG assessment—remains integral to claims that the Blue Economy agenda is or can be sustainable. In short, mangrove plantation around the Aghanashini is framed in the abstract as a conservation "win."

Yet Blue Economy developments today threaten mangroves and their social ecologies. The industrial port would explicitly despoil old and new mangrove cover in the estuary. Aquaculture, too, threatens mangrove swamps, even as it depends indirectly on their existence as

[27]The NGO Panchabhuta Conservation Foundation (PCF) has promoted eco-tourism experiments such as a farm-to-table restaurant and homestays. PCF provided institutional support and assistants for this research in 2015 and 2016.

[28]This micropolitical powerbroker has been a general supporter of and occasional collaborator with PCF.

breeding grounds for future wild feedstock. Unregulated or aggressively developed tourism also poses dangers to mangrove swamps—directly from the inevitable damage caused by boat excursions and indirectly through the increased pollution (to say nothing of the social inequalities likely to arise). Yet these specific actually existing *unsustainabilities* are routinely dismissed summarily in planning documents or otherwise reframed by policy-makers and elites as a necessary cost of the dream national development. Indeed, the success of mangrove afforestation in the estuary paradoxically smooths this process as planners then argue that new mangroves can be replanted elsewhere (NEERI 2015, interviews 2015, 2016).

These and other transformations, dispossessions, and sacrifices at sea and along the coast are nonetheless increasingly fantasized as sustainable development through recourse to India's Blue Economy agenda (Ali et al. 2021; Kanan et al. 2020; Research Collective 2018). The case of the Aghanashini demonstrates what can happen when an uncritical vision of sustainability—performed through the SDG process—provides justification for the construction of a shipping port, the expansion of aquaculture, the creep of ecotourism, or even the sacrifice of mangroves habitats. In sum, what India's Blue Economy dream portends is a drastic reworking of social and ecological space that would upend livelihoods, lifestyles and natures for the twinned goals of nationalist development and neoliberal profit.

14.6 Conclusion: Oceanic Neoimperialism?

This paper began with the observation that the Blue Economy agenda drowns out many critical questions about sustainability—ecological, social, and economic. To understand how this came to be, I first historicized the concept of sustainable development and its contemporary (fetishized) praxis of quantification and measurement. Development today operates within a discursive regime of so-called "global performance indicators," of which the SDGs are just one exemplar. India's cranking of the gears of SDG assessment remains incomplete, reductive, geographically confused, and ambiguous, which authors and authorizes an abstracted, mutable vision of sustainable development. Emptied of critical content, sustainability becomes a signifier available for ambitions of capitalism in and around the sea. Contra this performance of sustainability, I have also highlighted a case of the actually existing Blue Economy in southwestern India that foretells drastic transformations: the construction of shipping ports, expansion of fisheries and aquaculture, marketing/packaging of space for tourists, and the conservation/destruction/replantation of mangroves. In short, the subsumption of sustainability within Blue Economy dreams (re)produces watery ecologies and coastal places as new sources of (neoliberal) capital, profit, and employment for the nation.

I also argue that within the broader politics of Hindutva and the dominant BJP, the Blue Economy has still more work to do. Consider the 2020 comments of Amitabh Kant, then the chief executive of NITI Aayog, lamenting India's political landscape as a basic constraint on capitalism. During an online event organized by rightwing magazine *Swarajya* (and sponsored by mining giant Vedanta), Kant praised Modi's "political determination and administrative will," because "tough reforms are very difficult in the Indian context. We are too much of a democracy. For the first time the government has had the courage and the determination to carry out very hard-headed reforms across sectors." Kant's words could be dismissed as clumsy bureaucratic prattle, but his thinktank produces India's development policy and oversees SDG assessment. Though he retired in 2022, Kant was a policy wonk, friend of corporate India and Modi acolyte (Chatterji 2021). Any vision of development expounded by Kant—including one that is skeptical of democratic governance—cannot be easily distinguished from BJP rule.

Neither Kant's endorsement of strong-armed "reform" (Narrain 2021; Roy 2021) nor his implication that democracy is an obstruction to development produced much public controversy.

After all, Kant had only voiced an already open secret: India is in the grips of a decades-long slide away from its postcolonial democratic ambitions toward the authoritarian, ethnonationalist rule of Hindutva. This project of hegemony aggressively reimagines the space, norms, beneficiaries, and constituents of the nation as Hindu-first, supplanting a multicultural, secular constitutional liberalism (Bardhan 2022; Jaffrelot 2021; Vanaik 1997, 2001, 2018). Simply put, India is witnessing the rise of a neoliberal-neofascist combine, one that I suggest has decidedly *neoimperial* pretensions and proportions.

Under the sign of the Blue Economy, oceanic and coastal spaces (and the life, livelihoods, and relations therein) are reconfigured, repossessed, or repurposed for national "improvement"—with all the coloniality that term implies (Drayton 2000; Li 2007). Simultaneously, the improved nation is redefined as capitalist, sustainable, and modern while still the bearer of an ancient Hindu lineage and a "glorious maritime heritage" (Modi 2016). The capitalist, Hindu nation centers the mostly Hindi-speaking heartland and subordinates peripheral people and places (coastal communities and traditional fisheries, but also Dalits and Muslims nationally). This also reinscribes the historical marginalization of rural coasts and seas (cf. Subramanian 2009) producing them as available for sacrifice (and possible reincarnation). Ultimately, these reconfigurations—both material and discursive—serve as the foundation for the further accumulation of capital, the expansion of Hindutva hegemony, and the (imperial) assertion of power into/across ocean space. Imagining India standing astride "the crossroads of Indian Ocean," Modi himself declares, "We must also assume our responsibility to shape its future." Put another way, when India's leaders dream of an expanding, purportedly sustainable Blue Economy, they simultaneously dream of empire.

Acknowledgments This paper has accrued many debts. Without any obligation, Ramchandra Bhatta provided years of instruction in the geography of coastal Karnataka. Theoretical claims, whenever sound, are the product of my privilege to work with advisors such as Michael Watts and Gill Hart. Aarthi Sridhar and Rosanna Carver, especially, have helped me think through the Blue Economy, oceanic materiality and the meanings of development for coastal peoples. Mangal Shetty, Panchabhuta Conservation Foundation, and WWF-India provided necessary institutional and financial support. Without V. Rajalingam and a team of research assistants and students, the broader 18-month research campaign would have been impossible. Aghanashini residents offered insights during hundreds of hours of interviews and surveys. And that's to say nothing of all the readily shared chai, kaju, imli, and paan.

References

Adams WM (2001) Green development: environment and sustainability in the Third World, 2nd edn. Routledge, London

Ali A, Sowman M, McDaid L, Philipose P (2021) Jury verdict, independent people's public tribunal on the implications of Blue Economy in India (West Coast). http://blueeconomytribunal.org/wp-content/uploads/BE-India-West-Coast-Verdict.pdf

Allen T (ed) (2021) Poverty and development in the 21st century, 3rd edn. Oxford University Press, Oxford

Alvares C (2010) Science. In: Sachs W (ed) The development dictionary: a guide to knowledge as power, 2nd edn. Zed Books, London, pp 243–259

Amrith SS (2018) Unruly waters: how rains, rivers, coasts and seas have shaped Asia's history. Basic Books, London

Anand N, Gupta A, Appel H (eds) (2018) The promise of infrastructure. Duke University Press, Durham

ANI, see Asian News International

Arnimesh S (2021) Modi tells BJP CMs to go for data-driven governance model based on tech, end license raj. ThePrint, December 15. https://theprint.in/india/governance/modi-tells-bjp-cms-to-go-for-data-driven-governance-model-based-on-tech-end-licence-raj/781854/

Arun, Habbu RS (2010) The cursed land of nusikote. The New Indian Express, May 31. http://www.newindianexpress.com/states/karnataka/2010/may/31/the-cursed-land-of-nusikote-157634.html

Asian News International (2017) Blue Economy can improve India's progress: PM Modi. Business Standard, October 7. https://www.business-standard.com/article/news-ani/blue-economy-can-improve-india-s-progress-pm-modi-117100700274_1.html

Asian News International (2021a) India is giving topmost importance to develop its Blue Economy: PM Modi. Livemint, February 14. https://www.livemint.com/news/india/india-is-giving-topmost-importance-to-develop-its-blue-economy-pm-modi-11613306246436.html

Asian News International (2021b) Goa has everything needed for development of 'Aatmanirbhar Bharat': PM Modi, October 23. https://www.aninews.in/news/national/general-news/goa-has-everything-

needed-for-development-of-aatmanirbhar-bharat-pm-modi20211023135405/
Barbesgaard M (2017) Blue growth: savior or ocean grabbing? J Peasant Stud 45(1):130–149. https://doi.org/10.1080/03066150.2017.1377186
Bardhan P (2022) The 'new' India: a political-economic diagnosis. New Left Rev 136:5–27
Baviskar A (1995) In the belly of the river: tribal conflicts over development in the Narmada Valley. Oxford University Press, Oxford
Benjamin W (2002) The arcades project. Belknap Press, Cambridge
Bennett NJ (2019) In political seas: engaging with political ecology in the ocean and coastal environment. Coast Manag 47(1):67–87. https://doi.org/10.1080/08920753.2019.1540905
Bhat M, Bhatta R (2004) Considering aquacultural externality in coastal land allocation decisions in India. Environ Resour Econ 29:1–20. https://doi.org/10.1023/B:EARE.0000035424.64335.6d
Bhat M, Nayak VN, Chandran S, Ramachandra TV (2014) Fish distribution dynamics in the Aghanashini estuary of Uttara Kannada, west coast of India. Curr Sci 106(12):1739–1744
Bhathal B, Pauly D (2008) 'Fishing down marine food webs' and spatial expansion of coastal fisheries in India, 1950–2000. Fish Res 91(1):26–34
Bhatta R (2019) Sustainable Development Goals, Goal 14: conserve and sustainably use the oceans, seas and marine resources for sustainable development, Life below Water. Department of Planning, Program Monitoring and Statistics, Government of Karnataka, Bangalore.
Bhatta R, Bhat M (2002) Impacts of aquaculture on the management of estuaries in India. Environ Conserv 25(2):109–121. https://doi.org/10.1017/S0376892998000162
Bhattacharya N (2019) The Great Agrarian conquest: the colonial reshaping of a rural world. SUNY Press, Albany
Blythe J, Silver J, Evans L et al (2018) The dark side of transformation: latent risks in contemporary sustainability discourse. Antipode 50(5):1206–1223
Bratman EZ (2015) Passive revolution in the green economy: activism and the Belo Monte dam. Int Environ Agreem Polit Law Econ 15:61–77. https://doi.org/10.1007/s10784-014-9268-z
Bratman EZ (2019) Governing the rainforest: sustainable development politics in the Brazilian Amazon. Oxford University Press, Oxford
Broome A, Quirk J (2015) The politics of numbers: the normative agendas of global benchmarking. Rev Int Stud 41(5):813–818
Brown T (2016) Sustainability as empty signifier: its rise, fall, and radical potential. Antipode 48(1):115–133. https://doi.org/10.1111/anti.12164
Buck Morss S (1995) The city as dreamworld and catastrophe. October 73(Summer):3–26
Campling L, Colás A (2021) Capitalism and the sea: the maritime factor in the making of the modern world. Verso Books, London
Carant JB (2017) Unheard voices: a critical discourse analysis of the Millennium Development Goals' evolution into the Sustainable Development Goals. Third World Q 38(1):16–41. https://doi.org/10.1080/01436597.2016.1166944
Carver R (2019) Resource sovereignty and accumulation in the Blue Economy: the case of seabed mining in Namibia. J Polit Ecol 26(1):381–402
Cavanagh CJ, Benjaminsen TA (2017) Political ecology, variegated green economies, and the foreclosure of alternative sustainabilities. J Political Ecology. 24(1):200–216
Cederlöf G, Sivaramakrishnan K (eds) (2014) Ecological nationalisms: Nature, livelihoods, and identities in South Asia. University of Washington Press, Washington, DC
Chandran S, Mesta P, Boominathan M, Rao GR, Vishnu DM, Ramachandra TV (2012) Aghanashini estuary in Kumta taluk, Uttara Kannada—biological heritage site. Environmental Information System, Centre for Ecological Sciences, Indian Institute of Science, Bangalore.
Chari S, Corbridge S (2008) The development reader. Routledge, London
Chatterji S (2021) The importance of Amitabh Kant. Hindustan Times, July 1. https://www.hindustantimes.com/india-news/the-importance-of-amitabh-kant-101625134286809.html
Childs J, Hicks C (2019) Securing the blue: political ecologies of the Blue Economy in Africa. J Polit Ecol 26(1):323–340
Clancy M (2001) Exporting paradise: tourism and development in Mexico, Emerald, Bingley
Conca K, Dabelko G (eds) (2010) Green planet blues: four decades of environmental politics, 4th edn. Westview, Boulder
Cooper F (1997) Modernizing bureaucrats, backward Africans, and the development concept. In: Cooper F, Packard R (eds) International development and the social sciences: essays on the history and politics of knowledge. University of California Press, Berkeley, pp 64–92
Costanza R, d'Arge R, de Groot R et al (1997) The value of the world's ecosystem services and natural capital. Nature 387:253–260. https://doi.org/10.1038/387253a0
Costanza R, de Groot R, Sutton P et al (2014) Changes in the global value of ecosystem services. Glob Environ Chang 26:152–158. https://doi-org.libproxy.berkeley.edu/10.1016/j.gloenvcha.2014.04.002
Cross J (2014) Dream zones: anticipating capitalism and development in India. Pluto Press, London
D'Souza R (2006) Drowned and dammed: colonial capitalism and flood control in eastern India. Oxford University University, New Delhi

Damodaran V, Winterbottom A, Lester A (eds) (2015) The East India Company and the natural world. Palgrave Macmillan, Houndsmills

Davis M (2002) Late Victorian Holocausts: El Niño famines and the making of the Third World. Verso, London

Davis D (2007) Resurrecting the granary of Rome: environmental history and French colonial expansion in North Africa. University of Ohio Press, Athens

Day F (1873) Report on the sea-fishes of India and Burma. Office of Superintendent of (British Indian) Government Printing, Calcutta

Dean M (2015) The Malthus effect: population and the liberal government of life. Econ Soc 44(1):18–39. https://doi.org/10.1080/03085147.2014.983832

Death C, Gabay C (2015) Doing biopolitics differently? Radical potential in the post-2015 MDG and SDG debates. Globalizations 12(4):597–612. https://doi.org/10.1080/14747731.2015.1033172

Desai BN, Bhargava RMS, Sarupria JS (1990) Estimates of fishery potentials of the EEZ of India. Estuar Coast Shelf Sci 30(6):635–639

Drayton RH (2000) Nature's government: science, imperial britain and the "improvement" of the world. Yale University Press, New Haven

Ehrlich PR (1968) The population bomb. Ballantine, New York

Equations (2019) The growth of tourism and the Blue Economy in India. https://equitabletourism.org/initiative/growth-tourism-and-blue-economy-india

Esquivel V (2016) Power and the Sustainable Development Goals: a feminist analysis. Gend Dev 24(1):9–23. https://doi.org/10.1080/13552074.2016.1147872

Ferguson J (1994) The anti-politics machine: development, depoliticization, and bureaucratic power in Lesotho. University of Minnesota Press, Minneapolis

Fischer AM (2019) Poverty as ideology: rescuing social justice from global development agendas. Zed, London

Fletcher R (2014) Romancing the wild: cultural dimensions of ecotourism. Duke, Durham

Foucault M (1980) Power/knowledge: selected interviews and other writings 1972–1977. Pantheon, New York

Foucault M (1991) Governmentality. In: Burchell G, Gordon C, Miller P (eds) The Foucault effect: studies in governmentality. University of Chicago Press, Chicago, pp 87–104

Fukuda-Parr S, McNeil D (2019) Knowledge and politics in setting and measuring the SDGs: introduction to special issue. Global Pol 10(S1):5–15. https://doi.org/10.1111/1758-5899.12604

Fukuda-Parr S, Yamin E (2013) The power of numbers: a critical review of MDG targets for human development and human rights. Development 56(1):58–65

Furnivall JS (1948) Colonial policy and practice: a comparative study of Burma and Netherlands India. Cambridge University Press, Cambridge

Germond-Duret C (2022) Framing the Blue Economy: placelessness, development and sustainability. Dev Chang 53(2):308–334. https://doi.org/10.1111/dech.12703

Ghosh S, Bakshi M, Bhattacharya S et al (2015) A review of threats and vulnerabilities to mangrove habitats: With special emphasis on east coast of India. J Earth Sci Clim Change 6(4). https://doi.org/10.4172/2157-7617.1000270

Griggs D, Stafford-Smith M, Gaffney O et al (2013) Sustainable development goals for people and planet. Nature 495:305–307. https://doi.org/10.1038/495305a

Grove R (1997) Ecology, climate, and empire: colonialism and global environmental history, 1400–1940. White Horse Press, Cambridge

Grove R (2003) Green imperialism: colonial expansion, tropical island Edens and the origins of environmentalism, 1600–1860. Cambridge University Press, Cambridge

Grove R, Damodaran V, Sangwan S (eds) (2000) Nature and the orient: the environmental history of South and Southeast Asia. Oxford University Press, Oxford

Guha R, Martinez-Alier J (1997) Varieties of environmentalism. Earthscan, London

Gunakar S, Jadhav A, Bhatta R (2017) Protections for small-scale fisheries in India: A study of India's monsoon fishing ban. In: Jentoft S, Chuenpagdee R, Barragán-Paladines M, Franz N (eds) The small-scale fisheries guidelines. Springer, pp 291–311. https://doi.org/10.1007/978-3-319-55074-9_14

Hall S (1996) The west and the rest: discourse and power. In: Hall S, Held D, Hubert D, Thompson K (eds) Modernity: an introduction to modern societies. Blackwell, Malden, pp 184–227

Hampton MP (2013) Backpacker tourism and economic development perspectives from the less developed world. Routledge, London

Hart G (2001) Development critiques in the 1990s: cubs de sac and promising paths. Prog Hum Geogr 25(3):649–658

Hart G (2010) D/developments after the meltdown. Antipode 41(s1):117–141

Harvey D (1974) Population, resources and the ideology of science. Econ Geogr 50(3):256–277

Harvey D (2003) The new imperialism. Oxford University Press, Oxford

Hébert K, Brock S (2017) *Counting and counter-mapping*: Contests over the making of a mining district in Bristol Bay, Alaska. Sci Cult 26(1):56–87. https://doi.org/10.1080/09505431.2016.1223112

Hegde B (2014) Proposal for declaration of Aghanashini Estuary as critically vulnerable coastal area (CVCA) under CRZ Notification, 2011. Bivalve Collector's Group and Snehakunja Trust, Honnavar

Höhler S, Ziegler R (2010) Nature's accountability: stocks and stories. Sci Cult 19(4):417–430. https://doi.org/10.1080/09505431.2010.519609

Hulme D (2009) The Millennium Development Goals (MDGs): a short history of the world's biggest promise. BWPI Working Paper 100

Huxley TH (1883) Inaugural Meeting of the Fishery Congress: address. William Clowes and Sons for International Fisheries Exhibition, London

Jadhav A (2015) Coastal development in Karnataka: seeing neoliberalism through legal plural eyes. Conference paper. International conference of the commission on legal pluralism, 14 December 2015

Jadhav A (2018) Undefining small-scale fisheries in India: challenging simplifications and highlighting diversity and value. In: Johnson D, Acott T, Stacey N, Urquhart J (eds) Social wellbeing and the values of small-scale fisheries. Springer, pp 147–173. https://doi.org/10.1007/978-3-319-60750-4_7

Jadhav A (2020) The shore scene: PM-MSY's fisheries development promises anything but *sabka vikas*. The Bastion, June 7. https://thebastion.co.in/covid-19/the-shore-scene-pm-msys-fisheries-development-promises-anything-but-sabka-vikas/

Jadhav A, Anderson S, Dyer MJB, Sutton P (2017) Revisiting ecosystem services: assessment and valuation as starting points for environmental politics. Sustainability 9(10). https://doi.org/10.3390/su9101755

Jaffrelot C (2021) Modi's India: Hindu nationalism and the rise of ethnic democracy. Princeton University Press, Princeton

Jayanthi M, Thirumurthy S, Ravichandran P (2018) Impact of shrimp aquaculture development on important ecosystems in India. Glob Environ Chang 52:10–21. https://doi.org/10.1016/j.gloenvcha.2018.05.005

Kanan K, McDaid L, Satizábal P, Huq S (2020) Jury verdict, independent people's public tribunal on the implications of Blue Economy in India (east coast). http://blueeconomytribunal.org/wp-content/uploads/BE-India-East-Coast-Verdict.pdf

Kant A (2020) PLI scheme for manufacturing to bring investments worth $520 billion: NITI Aayog CEO Amitabh Kant. Swarajya. https://www.youtube.com/watch?v=vxx2Rx9oWXI&t=1979s

Kareiva P, Tallis H, Ricketts TH et al (eds) (2011) Natural capital: theory and practice of mapping ecosystem services. Oxford University Press, Oxford

Kelley J, Simmons B (eds) (2020) The power of global performance indicators. Cambridge University Press. https://doi.org/10.1017/9781108763493

Kohli K, Menon M (2019, January 17) The coast is unclear: on the 2018 CRZ notification. The Hindu. https://www.thehindu.com/opinion/op-ed/the-coast-is-unclear/article26006723.ece

Kohli K, Sekhsaria P (2000) Designs on a river system. The Hindu, October 8. https://www.thehindu.com/todays-paper/tp-miscellaneous/tp-others/designs-on-a-river-system/article28047554.ece

Krishnan M, Birthal PS (2002) Aquaculture development in India: An economic overview with special reference to coastal aquaculture. Aquacult Econ Manage 6(1–2):81–96

Kumar A, Singh A (2022) Land use/land cover changes in coastal districts of Karnataka. In: Singh RB, Kumar M, Tripathi DK (eds) Remote sensing and geographic information systems for policy decision support. Springer, pp 467–481. https://doi.org/10.1007/978-981-16-7731-1_24

Kurien J (1985) Technical assistance projects and socio-economic change: Norwegian intervention in Kerala's fisheries development. Econ Polit Wkly 20(25/26):A70–A88. http://www.jstor.org/stable/4374540

Lakra WS, Gopalakrishnan A (2021) Blue revolution in India: Status and future perspectives. Indian J Fish 68(1):137–150. http://eprints.cmfri.org.in/15085/

Lakshmi R (2013) India's popular beach destination Goa debates price of tourism. The Washington Post, November 27. https://www.washingtonpost.com/world/asia_pacific/indias-popular-beach-destination-goa-debates-price-of-tourism/2013/11/26/144a907c-4c59-11e3-bf60-c1ca136ae14a_story.html

Landell-Mills P, Agarwala R, Please S (1989) Sub-saharan Africa: from crisis to sustainable growth. The International Bank for Reconstruction and Development, Washington, DC

Larkin B (2013) The politics and poetics of infrastructure. Annu Rev Anthropol 42(1):327–343

Le Mons Walker K (2008) Neoliberalism on the ground in rural India: Predatory growth, agrarian crisis, internal Colonization, and the intensification of class struggle. J Peasant Stud 35(4):557–620. https://doi.org/10.1080/03066150802681963

Levien M (2018) Dispossession without development: land grabs in neoliberal India. Oxford University Press, Oxford

Li T (2007) The will to improve: governmentality, development, and the practice of politics. Duke University Press, Durham

Liverman DM (2018) Geographic perspectives on development goals: Constructive engagements and critical perspectives on the MDGs and the SDGs. Dialogues Hum Geogr 8(2):168–185. https://doi.org/10.1177/2F2043820618780787

Ludden D (2005) Development regimes in South Asia: history and the governance conundrum. Econ Polit Wkly 40(37):4042–4051

MacArthur JW (2014) The origins of the Millennium Development Goals. SAIS Rev Int Aff 34(3):5–24. https://doi.org/10.1353/sais.2014.0033

Mallin F, Barbesgaard M (2020) Awash with contradiction: capital, ocean space and the logics of the Blue Economy paradigm. Geoforum 113(July):121–132. https://doi.org/10.1016/j.geoforum.2020.04.021

Malthus T (1798) An essay on the principle of population, as it affects the future improvement of society with remarks on the speculations of Mr. Godwin, M. Condorcet, and other writers. St. Paul's Church Yard

Mawdsley E, Taggart J (2022) Rethinking d/Development. Prog Hum Geogr 46:3

Meadows D, Meadows D, Randers J, Behrens WW III (1972) The limits to growth. Potomac Associates, Falls Church

Menezes V (2014) The paradox of plenty: Goa tourism destroys itself. Times of India, January 12. http://timesofindia.indiatimes.com/articleshow/28693941.cms

Menga F, Goodman MK (2022) The high priests of global development: capitalism, religion and the political economy of sacrifice in a celebrity-led water charity. Dev Chang 53:705. https://doi.org/10.1111/dech.12723

Menon M (2004) Special economic zones—a takeover of coastal areas. The Hindu Business Line, February 20. https://www.thehindubusinessline.com/2004/02/20/stories/2004022000100900.htm

Menon M, Kohli K (2015) A strange proposal to build Karnataka's largest port raises serious (and awkward) questions. Scroll.in, March 22. https://scroll.in/article/715364/a-strange-proposal-to-build-karnatakas-largest-port-raises-serious-and-awkward-questions

Mesta PN, Setturu B, Chandran S, Rajan KS, Ramachandra TV (2014) Inventorying, mapping and monitoring of mangroves towards sustainable management of west coast, India. J Geophys Remote Sens 3(3):130. http://wgbis.ces.iisc.ernet.in/energy/water/paper/Inventorying-Mapping-and-Monitoring/index.html

Ministry of Environment, Forest and Climate Change (2017) Minutes for 12th meeting of Expert Appraisal Committee (Infra-2) for projects related to all ship breaking yard including ship breaking unit, airport, common hazardous waste treatment, storage and disposal Facilities, ports and harbours, aerial ropeways, CETPs, common municipal solid waste management facility, building/construction project, townships and area Development projects. 26–28 December 2016. MoEFCC, Government of India, New Delhi.

Ministry of Ports, Shipping and Waterways (2022) Projects under Sagarmala. Government of India. https://sagarmala.gov.in/projects/projects-under-sagarmala

Ministry of Shipping (2011) Maritime agenda 2010–2020 launched. Press Information Bureau, Government of India. http://pib.nic.in/newsite/PrintRelease.aspx?relid=69044

Ministry of Statistics and Programme Implementation (2019) Sustainable Development Goals national indicator framework baseline report 2015–16. Government of India. https://mospi.gov.in/documents/213904/0/SDG+National+Indicator+Framework+Baseline+Report%2C+2015-16.pdf/290cae2d-900b-a3ee-5091-ea7969d5cde5?t=1594032410420

Modi N (2015) Text of the PM's remarks on the commissioning of Coast Ship *Barracuda*. Press Information Bureau. Government of India. https://pib.gov.in/newsite/printrelease.aspx?relid=116881

Modi N (2016) India has had a glorious maritime history. We are on the path of shaping an even better maritime future: PM. https://www.narendramodi.in/pm-modi-at-the-inauguration-of-maritime-india-summit-2016-in-mumbai-440341

Modi N (2021) Blue Economy is going to be an important source of Aatmanirbhar Bharat: PM. https://www.narendramodi.in/blue-economy-is-going-to-be-an-important-source-of-aatamnirbharbharat-pm-553218

Murphy M (2017) The economization of life. Duke, Durham

Naidu, M. K. (2019). National Institute of Ocean Technology Silver Jubilee address. https://twitter.com/drharshvardhan/status/1191031137415516160 and https://twitter.com/drharshvardhan/status/1191033463102566400

Nandy A (1999) Science, hegemony and violence: a requiem for modernity. Oxford University Press, Oxford

Narrain A (2021) India's undeclared emergency: constitutionalism and the politics of resistance. Context, Chennai

National Environmental Engineering Research Institute (2015) Environmental impact assessment studies for development of sea port at Tadadi, Karwar, Karnataka. Council of Scientific and Industrial Research, New Delhi

National Institution for Transforming India (2017) Voluntary national review report on implementation of Sustainable Development Goals. Government of India. https://www.niti.gov.in/writereaddata/files/India%20VNR_Final.pdf

National Institution for Transforming India (2018) SDG India index baseline report, 2018. Government of India. http://www.niti.gov.in/sites/default/files/2020-07/SDX_Index_India_Baseline_Report_21-12-2018.pdf

National Institution for Transforming India (2019) SDG India index & dashboard, 2019–2020. Government of India. http://www.niti.gov.in/sites/default/files/2020-07/SDG-India-Index-2.0.pdf

National Institution for Transforming India (2020) India VNR 2020: decade of action taking SDGs from global to local. Government of India. https://www.niti.gov.in/writereaddata/files/India%20VNR_Final.pdf

National Institution for Transforming India (2021) SDG India index & dashboard, 2020–2021: partnerships in the decade of action. Government of India. https://www.niti.gov.in/writereaddata/files/SDG_3.0_Final_04.03.2021_Web_Spreads.pdf

NEERI, see National Environmental Engineering Research Institute

Newman RS (1981) Green revolution-blue revolution: The predicament of India's traditional fishermen. J South Asian Stud 4(1):35–46. https://doi.org/10.1080/00856408108723014

NITI Aayog, see National Institution for Transforming India

Parenti C (2015) The environment making state: Territory, nature and value. Antipode 47(4):829–848

Patel R (2013) The long Green Revolution. J Peasant Stud 40(1):1–63

Peck J (2010) Zombie neoliberalism and the ambidextrous state. Theor Criminol 14(1):104–110. https://doi.org/10.11772/1362480609352784

Peet R, Watts MJ (eds) (1996) Liberation ecologies: environment, development and social movements. Routledge, London

Press Trust of India (2015) India should grab share in ocean-based industry: Harsh Vardhan. The Economic Times, January 1. https://economictimes.indiatimes.com/news/politics-and-nation/india-should-grab-share-in-ocean-based-industry-harsh-vardhan/articleshow/45718906.cms
PTI, see Press Trust of India
Ramesh M, Namboothri N (2018) A myth and its manifold effects: maximum sustainable yield. Econ Polit Wkly 53(41):58–63
Rangan H (1996) From Chipko to Uttaranchal: development, environment and social protest in the Garhwal Himalayas, India. In: Peet R, Watts M (eds) Liberation ecologies: environment, development and social movements. Routledge, London, pp 205–226
Rangarajan, M., Sivaramakrishnan K (eds) (2011) India's environmental history, vols 1 & 2. Orient Blackswan, Hyderabad
Ranjan R, Mukhta M, Ghosh S et al (2017) Prioritized species for mariculture in India. Central Marine Fisheries Research Institute, Kochi
Research Collective (2018) Occupation of the coast: Blue Economy in India. Programme for Social Action, New Delhi
Robbins M (ed) (1972) Only one earth: the Stockholm conference. United Nations Television. https://www.youtube.com/watch?v=jKaYPk5YnsU
Roberts (2021) Abolishing India's Planning Commission: the results after five years. In Mehrotra S, Guichard S (eds) Planning in the 20th century and beyond: India's planning commission and the NITI Aayog (New Delhi: Cambridge University Press, 2020) Saksena KD, NITI Aayog and planning commission: some reflections (Delhi: Shipra Publications, 2019) Montek Singh Ahluwalia, Backstage (Delhi: Rupa 2019). Pub Administr Rev 81(4):799–805. https://doi.org/10.1111/puar.13404
Robertson T (2012) The Malthusian moment: global population growth and the birth of American environmentalism. Rutgers University Press, New Brunswick
Rodriguez S, Sridhar A (2010) Harbouring trouble: the social and environmental upshot of port growth in India. Dakshin Foundation, Bengaluru
Rose N (1999) Powers of freedom: reframing political thought. Cambridge University Press, Cambridge
Rostow WW (1960) The stages of economic growth: a non-communist manifesto. Cambridge University Press, Cambridge
Routledge P (2000) Consuming Goa: tourist site as dispensable space. Econ Polit Wkly 35(20):2647–2656. https://www.jstor.org/stable/4409533
Roy B (2019) Shrimp farming at Chilika Lake, Odisha, India. Environmental Justice Atlas. https://ejatlas.org/conflict/fisherfolks-of-chilika-lake-odisha-india
Roy I (2021) Under Modi, India is an 'ethnocracy', not a democracy. The Quint, September 7. https://www.thequint.com/voices/opinion/under-modi-india-is-an-ethnocracy-not-a-democracy
Sachs W (2010a) Introduction. In: Sachs W (ed) The development dictionary: a guide to knowledge as power, 2nd edn. Zed Books, London, pp xv–xx
Sachs W (2010b) Environment. In: Sachs W (ed) The development dictionary: a guide to knowledge as power, 2nd edn. Zed Books, London, pp 24–37
Sathianandan TV, Mohamed KS, Jayasankar J et al (2021) Status of Indian marine fish stocks: modelling stock biomass dynamics in multigear fisheries. ICES J Mar Sci 78(5):1744–1757
Scoones I (2007) Sustainability. Dev Pract 17(4–5):589–596. https://doi.org/10.1080/09614520701469609
Scott JC (1998) Seeing like a state: how certain schemes to improve the human condition have failed. Yale University Press, New Haven
Sénit C (2020) Leaving no one behind? The influence of civil society participation on the Sustainable Development Goals. Environ Plann Polit Space 38(4):693–712. https://doi.org/10.1177/2F2399654419884330
Sexsmith K, McMichael P (2015) Formulating the SDGs: reproducing or reimagining state-centered development? Globalizations 12(4):581–596. https://doi.org/10.1080/14747731.2015.1038096
Shroff K (2022) Two new papers on poverty levels shine a light on need for reliable official data in India. The Wire, April 25. https://thewire.in/economy/imf-world-bank-poverty-papers-missing-data
Silver JJ, Campbell LM (2018) Conservation, development and the blue frontier: the Republic of Seychelles' Debt Restructuring for Marine Conservation and Climate Adaptation Program. Int Soc Sci J 68(229–230):241–256. https://doi.org/10.1111/issj.12156
Silver JJ, Gray NJ, Campbell LM, Fairbanks LW, Gruby RL (2015) Blue economy and competing discourses in international oceans governance. J Environ Dev 24(2):135–160. https://doi.org/10.1177/1070496515580797
Spalding MD, Leal M. (eds) (2021) The state of the world's mangroves 2021. global mangrove alliance. https://www.mangrovealliance.org/wp-content/uploads/2021/07/The-State-of-the-Worlds-Mangroves-2021-FINAL.pdf
Srinath M, Balan K (2003) Potential yield from Indian EEZ. In: Joseph MM, Jayaprakash AA (eds) Status of exploited marine fishery resources of India. Central Marine Fisheries Research Institute, Kochi, pp 286–290
Stone K, Bhat M, Bhatta R, Mathews A (2008) Factors influencing community participation in mangroves restoration: a contingent valuation analysis. Ocean Coastal Manage 51(6):476–484. https://doi.org/10.1016/j.ocecoaman.2008.02.001
Subramanian A (2009) Shorelines: space and rights in South India. Stanford University Press, Stanford
Sud N (2021) The making of land and the making of India. Oxford University Press, Oxford

Sultana F (2018) An(Other) geographical critique of development and SDGs. Dialogues Hum Geogr 8(2):186–190

Thakur AK (2022) How blue economy can help rewrite India's growth story post pandemic. Firstpost, March 27. https://www.firstpost.com/opinion/how-blue-economy-can-help-rewrite-indias-growth-story-post-pandemic-10492831.html

Tilley H (2011) Africa as a living laboratory: empire, development, and the problem of scientific knowledge, 1870–1950. University of Chicago, Chicago

Truman H (1949) Inaugural address. Harry S. Truman Library. https://www.trumanlibrary.gov/library/public-papers/19/inaugural-address

United Nations General Assembly (2017) Work of the Statistical Commission pertaining to the 2030 Agenda for Sustainable Development. A/RES/71/313. https://ggim.un.org/documents/a_res_71_313.pdf

United Nations Statistics Division (2022) Global indicator framework for the Sustainable Development Goals and targets of the 2030 Agenda for Sustainable Development. https://unstats.un.org/sdgs/indicators/Global%20Indicator%20Framework%20after%202022%20refinement_Eng.pdf

Vanaik A (1997) The furies of Indian communalism: religion, modernity and secularization. Verso, London

Vanaik A (2001) The new Indian right. New Left Rev 9:43–67

Vanaik A (2018) India's two hegemonies. New Left Rev 112:29–59

Vase VK, Rajan K, Jayasankar J, Sreenath KR et al (2019) Potential yield estimates for marine fisheries of Gujarat. J Mar Biol Assoc India 61(2):63–67. http://eprints.cmfri.org.in/14245/1/JMBAI_2019_Vase%20V%20K_Potential%20Yield%20Estimates%20for%20marine%20fisheries%20of%20Gujarat.pdf

Vijay V, Biradar RS, Inamdar AB et al (2005) Mangrove mapping and change detection around Mumbai (Bombay) using remotely sensed data. Indian J Mar Sci 34(3):310–315. http://nopr.niscair.res.in/bitstream/123456789/1566/1/IJMS%2034%283%29%20310-315.pdf

Watts M (1983) Silent violence: food, famine and peasantry in northern Nigeria. University of California Press, Oakland.

Watts MJ (1993) Development I: power, knowledge, discursive practice. Prog Hum Geogr 17(2):257–272. https://doi.org/10.1177/030913259301700210

Watts MJ (1994) Development II: the privatization of everything? Prog Hum Geogr 18(3):371–384. https://doi.org/10.1177/030913259401800307

Watts MJ (1996) Development III: the global agrofood system and late twentieth-century development (or Kautsky redux). Prog Hum Geogr 20(2):230–245. https://doi.org/10.1177/030913259602000208

Watts MJ (2003) Development and governmentality. Singap J Trop Geogr 24(1):6–34

WCED, see World Commission on Environment and Development

Wilson D (1997) Paradoxes of tourism in Goa. Ann Tour Res 24(1):52–75. https://doi-org.libproxy.berkeley.edu/10.1016/S0160-7383(96)00051-5

World Commission on Environment and Development (1987) Our common future. Oxford University Press. https://sustainabledevelopment.un.org/content/documents/5987our-common-future.pdf

Adam Jadhav is a PhD candidate at the University of California at Berkeley and a former research associate at Panchabhuta Conservation Foundation. He studies oceanic and agrarian geographies through the lenses of critical political ecology, development studies, and postcolonial theory.

Reinventing Methods of Preserving and Protecting Forests of India with Particular Reference to Odisha

15

Swetaparna Ankita

Abstract

Environmental issues took centre stage in discussions at the global level in 1972, marking a turning point in the development of international environmental politics. One of the major concerns that came to bother civilized people everywhere was the rapid depletion of forests around the world. This chapter seeks to point out the importance of trees and forests in religious books, in the language of poets and creative writers. It makes a brief survey of how forests have borne the brunt of rapid industrial growth worldwide, particularly in India. It also attempts to present a constantly growing concern of the global community about the hazards human beings around the world are going to face if concrete steps are not taken to prevent deforestation and also talks about the various measures taken by the United Nations in this regard. With respect to the Sustainable Development Goal 15 (SDG-15), this chapter examines the various schemes implemented by the Government of India to develop new forest areas and also halt the loss of existing forest cover in different parts of the country. The main thrust of this paper is to suggest how definite steps can be taken by the authorities concerned in trying to connect people to nature and turn it into a movement which can stop the processes of deforestation significantly.

Keywords

Afforestation · Depletion of forests · Environmental politics · Government schemes · SDG-15 · United Nations

15.1 Introduction

Of all the living beings on earth, humans are the greatest moulder of their environment which provides them physical sustenance and offers them the opportunity for intellectual, moral, social and spiritual growth. Nature, which forms a part of our environment, has often been taken for granted. As a result, there has been rampant exploitation of nature and natural resources without any significant conservation efforts.

Forests have always been one of the greatest natural resources for humans. From time immemorial forests have been the lifeline for human development – physical, mental, economic and spiritual. Religious books mention about it; poets and writers celebrate its glory. The Christian religious scripture *The Bible*, narrating the story of creation, talks about God creating plants and trees on the third

S. Ankita (✉)
VIT-AP School of Social Sciences & Humanities (VISH), Vellore Institute of Technology, Amaravati, Andhra Pradesh, India
e-mail: ankita.swetaparna@vitap.ac.in

V. Dutta, P. Ghosh (eds.), *Sustainability: Science, Policy, and Practice in India*, Sustainable Development Goals Series, https://doi.org/10.1007/978-3-031-50132-6_15

day. "And God saw that it was good" (*English Standard Version Bible* 2001, Genesis1:12). The Ramayana, the scripture of the Hindus, talks about the hermits living in dense forests and totally depending on it for their sustenance. Poets around the globe have narrated the mesmerizing beauty of nature, particularly that of the forests, in their poems emphasizing their contribution to the well-being of human mind and soul. Be it the famous British poet William Wordsworth, or the American poet Robert Frost or the Odia poet Radhanath Ray, all of them have described the captivating beauty of nature and the forests.

15.2 The Significance of Forests

In the background study prepared for the 13th session of the United Nation's Forum on Forests, Duncan Brack writes, "More than 1.6 billion people world-wide – almost a quarter of the global population – depend on forests for food, medicine and fuel, as well as their jobs and livelihoods, and 200 million people, including many indigenous people, directly depend on forests for their survival. Forests contain over 80 percent of the world's terrestrial biodiversity." He also says "As the largest storehouse of carbon after the oceans, forests have the potential to absorb and store about one tenth of the global anthropogenic carbon emissions projected for the forest half of this century into their biomass, soils and products – though at present about 10 percent of global greenhouse gas emissions are estimated to derive from deforestation" (p. 24).

The Food and Agricultural Organization (FAO) of the United Nations, after making a survey in 2020 reports that in both low- and high-income countries and in all climate zones, communities living within the forests rely the most directly on forest biodiversity for their lives and livelihoods (Wilkie and Buszko-Briggs 2021). In both low- and high-income countries throughout the world, communities that live within forests are dependent on the forest and its products most directly for their lives and livelihoods.

15.3 India: A Look at the Past

Having an ancient culture, recorded history of India dates back to the Indus Valley Civilization which flourished as early as 2500 B.C. (Karan 2004). In the past, hermits and sages used to live in the forest. Different Vedas make it clear that often people saw the image of God in nature and treated them as divine objects with great devotion and love. "Plants were regarded as possessing divine qualities with reference to their healing powers" (Srivastava and Barman 2019, p. 153).

Many kings and emperors ruled over the land in different centuries till the East India Company came to India and gradually took control over the land and its administration. Some of these kings and emperors like Ashoka took interest in planting trees and the protection of forest areas. The edicts issued by Emperor Ashoka include 'Forest must not be burned', 'Trees shall be planted on both the sides of the roads' (Srivastava and Barman 2019).

Till the inception of the British colonial period, the forests were protected by the tribal people. However, during the British period, forests were no more looked upon as natural resources but came to be treated as a matter of revenue generation. Initially, forests were cleared to make room for cultivation and agriculture. Later there was extensive damage to forest areas as trees were cut down in large numbers and at regular intervals to meet the timber needs for shipbuilding, iron smelting and tanning. Also, oak forests were cut and shipped to England for the use of the English Royal Navy. To serve the imperial cause, a Forest Act was enacted in 1865, the main purpose of which was to facilitate the acquisition of the Indian forest areas to supply timber for the railways and establish claim of the state on the forest land. Many more Acts were passed during the British rule and even after the independence of India, but cutting down of trees continued unabated, and areas under forest cover went on shrinking.

15.4 The Global Scenario vis-à-vis India

According to reports, forests cover 31% of the global land area (FAO and UNEP 2020). Approximately half the forest area is relatively intact out of which more than one third is primary forest. More than half of the world's forests are found in only five countries such as the USSR, Brazil, Canada, the United States and China. Nevertheless, deforestation and forest degradation continue to occur at alarming rates, contributing significantly to the ongoing loss of biodiversity. According to the FAO report 2020, 420 million hectares of forest have been already lost through conversion to other land uses. However, the rest of the forest degradation has decreased over the past three decades (FAO and UNEP 2020). The same report says, "Between 2015 and 2020, the rate of deforestation was estimated at 10 million hectares per year, down from 16 million hectares per year in the 1990s. The area of primary forest worldwide has decreased by over 80 million hectares since 1990s. Agricultural expansion continues to be the main driver of deforestation and forest degradation and the associated loss of forest diversity. Large-scale commercial agriculture (primarily cattle ranching and cultivation of soya beans and oil palm) accounted for 40 percent of tropical deforestation between 2000 and 2010, and local subsistence agriculture for another 33 percent" (FAO and UNEP 2020, p. 16).

Rapid depletion of forest areas has always been a matter of concern for everyone interested in maintaining the overall status of the forest areas in different parts of the globe. According to reports of the State of the World's Forests (SOFO) released on May 2, 2022, "The world has lost 420 million hectares (mha), approximately 10.34 percent of its total forest area in the last 30 years" (Jha 2022). If this trend continues and no significant steps are taken to halt this pace of deforestation, it would not be surprising to conclude that approximately the same area of forests will be lost in the coming two decades.

When we consider India's deforestation scenario, it is no better. According to India State of Forest Report (ISFR) 2021, the country's total forest cover is 809 million hectares which is only 24.62% of the country's geographical area. Compared to the world average of 31%, it is much less. The state of Odisha has a geographical area of about 155, 842 km^2 and forest covers about 57.184 km^2 in 1993–1994 (Economic Survey 1995–1996, p. 48). At present, the forest cover is estimated at 52,155.95 km^2, comprising 33.50% of the total geographical area of Odisha (Forest Survey of India 2020). Although there has been a gradual decrease in forest area before India's independence, it is encouraging to note that the forest cover of the state has increased by 537.44 km^2 between 2019 and 2020 (Press Trust of India, Bhubaneswar, Jan 14, 2022). This positive change is the result of various initiatives taken under the Joint Forest Management Programme of the government to expand the state forest cover (Press Trust of India 2022).

15.5 Causes of Deforestation

Environmental scientists and others have been making conscious efforts to discover the reasons for deforestation that have brought about the ecological imbalance on earth today. Although the reasons can be a combination of physical, political, and economic demands of life, different scholars have tried to trace the actual causes of it at different points of time in specific regions of the world. Jacques Pouchepadass observes that the deforestation in Mesopotamia, North Africa, Lebanon and Japan during the past centuries seems to have taken place to fulfill the needs of agricultural expansion and urban development (Pouchepadass 1995). He further observes that deforestation was a part of military operations, particularly in India. But there are many other more prominent reasons too. Forests in India are under immense pressure today and are reducing at an alarming rate due to rapid population explosion, exploitation of forest resources, conversion of forest land for non-forestry purposes, expansion of agriculture, poaching and unauthorized occupation (Jha et al. 2000).

The distribution of forest within the State of Odisha is uneven. The Central River Basin as well as the Coastal Area are thickly populated, and hence, forested areas have given place to the growth of human settlements. On the other hand, some areas like the Eastern Ghat region have lost their forests mainly due to 'shifting cultivation' (Podu), a practice prevalent among the tribals since the earliest times (Karthikeyan 2020). For conducting this cultivation, forests are cleared and burnt before sowing different varieties of crops. But due to the shortening of the shifting cycle, the ecological balance has started deteriorating which has led to soil erosion, heavy silting of streams, extinguishing perennial water resources and depleting the biodiversity for both human and animal consumption along with lowering the soil fertility (Lal 2005).

Another major cause of deforestation in Odisha is mining. Mining has a huge impact on the environment which is explicitly visible in the form of soil erosion, formation of sinkholes, loss of biodiversity, etc. The contamination of soil, groundwater and surface water takes place because of the chemicals released during the process of mining (Mining in Deforestation 2017). Mostly carried out in districts like Keonjhar, Sundergarh, Angul, Jharsuguda, Koraput and Mayurbhanj, mining is undertaken to extract precious metals, damaging the ecosystem and causing problems for the people living nearby. It has also a huge impact on those who are exposed to the toxic waste (Dash 2007).

15.6 Combating Deforestation

The major conference to discuss environmental issues was held from June 5 to 16, 1972, in Stockholm, Sweden. It was convened under the auspices of the United Nations which felt the immediacy of putting the thrust on the environment, and hence, it can be considered as a turning point in the development of international environmental politics. Known as the conference on the Human Environment, its goal was "to forge a basic common outlook on how to address the challenge of preserving and enhancing the human environment" (United Nations n.d.). In the same year on 15th December, a resolution was adopted in the general assembly to observe the fifth of June every year as World Environment Day. All governments and organizations coming under the United Nations systems were urged to undertake worldwide activities on that day which will reaffirm their concern for the preservation and enhancement of the environment (Lesniewska 2005).

International Day of Forests is also celebrated on 21st of March after its proclamation by the UN General Assembly on November 28, 2012. This celebration is done to raise awareness among the people regarding the protection of our forests which are an integral part of our lives and are pivotal for our existence. A further boost to this kind of endeavour at the international level could be seen when the US General assembly approved and accepted the Sustainable Development Goals (SDG) in 2015 as the core of the 2030 Development Agenda. SDG 15, on ecosystem, deals with forest ecosystems as a whole and sets goals for promoting sustainable forest management, halting deforestation by 2020, restoring degraded forests and increasing afforestation and reforestation (Brack 2018).

In Odisha, under the Forest Conservation Act 1980, about 59383.59 hectares of forest land have been diverted for 538 different developmental projects as of 31st December, 2021 (Sharma 2022). The schedule Tribes and other Traditional Forest Dwellers (Recognition of Forest Rights) Act, 2006, secures individual and/or community tenure. The Ministry of Environment, Forests and Climate Change, Government of India, grants permission for the diversion of forest land up to 1 hectare. This involves the felling of trees not exceeding 75 per hectare for small projects in 13 categories as envisaged under section – 3 (2) of Forest Rights Act, 2006 (Sharma 2022).

15.7 Forest Legislation in India

In 1865, the first Indian Forest Act was passed, which empowered the government to declare any land covered with trees as government forests.

However, the first effort to regulate the Indian forests began in South India in 1880 when a commission was appointed to enquire into the availability of teak in the Malabar forests (Srivastava and Barman 2019). As per the commission's report, felling of teak below 21 inches in girth was considered illegal.

After Independence of India, Forest Conservation Act 1980 was passed when it was felt that 'deforestation caused ecological imbalance and led to environmental deterioration', and so, it needed prohibition. Under the provisions of this Act, one needs an approval of the central government before converting forest lands for non-forestry purposes (Srivastava and Barman 2019).

The Scheduled Tribes and Other Traditional Forest Dwellers Act, 2006 was passed both in the Lok Sabha and the Rajya Sabha on December 18, 2006. This legislation primarily aimed at giving ownership rights over forest land to all the traditional forest dwellers of India. The most significant aspect of this Act is that, for the first time, it included the right to protect regenerate or conserve or manage any community forest resource (Srivastava and Barman 2019).

15.8 Forest Policies of India

The first official government initiative towards protecting forests can be seen in the National Forest Policy 1894 which focused on conserving forests for maintaining environmental stability and meeting the basic requirements of fringe forest user groups. But ironically, the real focus was maximizing revenue generation. This forest policy advocated that claim of cultivation is stronger than the claims of forest preservation in relation to all types of forests. This was a serious drawback as agriculture received prominence over forestry and encouraged the administration to put hold on forest conservation which led to large-scale clearing up of forest lands for agriculture and other purposes.

After India's independence, a new forest policy was announced in 1952 which emphasized the need of a sustainable supply of forest products such as timber for defence, communication and industries. But this ultimately had an adverse effect on the ecology, since large forest areas with great ecological importance were destroyed to make room for the establishment of different industries related to power, mining, irrigation, and so on. In order to raise revenue for the country, infrastructures like roads and railways were also laid causing gradual depletion of forest areas.

The National Forest Policy of 1988 suddenly changed the whole scenario because of its new emphasis on maintaining environmental equilibrium. It not only signalled the changed attitude of the policy-makers but also paved the way for better forest management. No doubt, it was a turning point in forest management in India as it focused on restoration of ecological balance and preservation of the forest areas of the country. The positive aspect of this policy was the emphasis on conserving the natural heritage of the country by preserving the remaining natural forests.

Currently, to increase the forest cover in India, the Ministry of Environment, Forest and Climate Change is implementing a number of schemes which provide financial assistance to the governments of state and union territories. There are basically three strategies which are involved to conserve and develop the forests of India. They are afforestation through natural/artificial regeneration, protection and management. The Indian Government has implemented various schemes for the development of forest areas. Some of these schemes are National Afforestation Programme (NAP) and Green India Mission (GIM), Integrated Development of Wildlife Habitats (IDWH), Intensification of Forest Management Scheme (IFMS), Project Tiger and Project Elephant including funds for Compensatory Afforestation Fund Management and Planning Authority (CAMPA).

The National Afforestation Programme (NAP) scheme aims for sustainable development and management of forest resources. The major objectives of this scheme include ecological restoration of degraded forests and improvement of livelihoods of the forest-fringe communities. Therefore, the overall objective of the National

Afforestation Programme scheme revolves around the ecological restoration of degraded forests and development of forest resources with the help of local people (Ministry of Environment and Forests, Government of India 2009).

Green India Mission (GIM) aims at protecting, restoring and enhancing India's forest cover and also in responding to climate change. GIM was launched in February 2014 with a clear objective to provide a protective cover to the biological resources of our country in the face of the dangers climate change can bring about. It recognizes the vital impact of forests on ecological sustainability and biodiversity conservation. The GIM aims at protecting, restoring, and enhancing India's diminishing forest cover and responding to climate change through adaptation and mitigation measures – a holistic view beyond tree planting. Multiple ecosystems such as biodiversity, water biomass, mangroves, wetlands and critical habitats are focused in the scheme along with carbon sequestration.

Integrated Development of Wildlife Habitats (IDWH) is a centrally sponsored scheme which aims to add more components and activities to the development of National Parks, sanctuaries, conservation reserves and community reserves for the protection and conservation of wildlife and its habitats in protected areas as well as outside these areas for the recovery programmes for saving critically endangered species and their habitats.

Intensification of Forest Management Scheme (IFMS) not only aims to increase the area of forest cover but also aims to improve the existing forests and to protect them from various threats which lead to its degradation. The components of this scheme include controlling forest fires and its management, conducting surveys, demarcating areas, preparing a working plan and strengthening of infrastructure such as roads, camp offices, watch towers, improving mobility and using modern information and communication technology for the overall management.

Compensatory Afforestation Fund Management and Planning Authority (CAMPA) was started on tenth July, 2009 after the Honorable Supreme Court issued orders for this scheme. It aims to monitor, provide technical assistance and evaluate compensatory afforestation activities. The main objective of CAMPA is to promote afforestation and conduct various regeneration programmes as a way of compensating for forest land diverted towards non-forest uses (Instapedia 2019).

The National Forest Policy of India, 1988 envisions a goal of bringing 33% of the country's land under forest and tree cover. Forest cover mapping is carried out in biennial intervals through remote sensing which in turn helps serving as a monitoring mechanism towards achievement of this goal (Forest Survey of India 2021). Carrying out periodic forest cover assessment at intervals has proved to be a very useful method to be sure about status of forests in the country.

Rise in urbanization in India has led to issues like loss of forest cover, pollution and depletion of wildlife. In order to fight these issues, the Government of India has taken some initiatives like creating Nagar Van Udyan Scheme, the aim of which is to develop at least one City Forest in each city having municipal corporation to accommodate a wholesome health environment and contribute to the growth of clean, green and sustainable India. The Forest Conservation Act, 1980 has provisions not only to put restrictions on deforestation but to promote afforestation in non-forest areas. The wildlife (Protection) Act, 1972 has been implemented primarily to protect wild animals and birds, but it has also led to preservation of biodiversity by setting up sanctuaries and parks.

Forest cover of the country is assessed biannually and published as India State of Forest Report (ISFR). According to ISFR-2021, the total forest and tree cover of the country is 809 million hectares which is definitely a matter of concern, since it comprises only 24.62% of the total geographical area of the country. When we compare it with the situation prevalent in 2019, it can be seen that there is an increase of 2261 km^2 of forests and tree cover in the country. Increase in forest cover has occurred in both open and dense forest areas. Top three states showing increase in forest cover are Andhra Pradesh (647 km^2), followed by Telangana (632 km^2) and Odisha (537 km^2) (Forest Survey of India 2022).

Odisha's forest cover is showing a rising trend and has improved over the last few years. Currently, it has gained 893 square kilometres of forest and tree cover (Sharma 2022). This could be possible because of the numerous plantation and conservation initiatives undertaken by the State of Odisha in a participatory mode (Mohanty 2014). Various afforestation activities were taken up under different schemes in the state. During the year 2021–2022, programmes like plantation of medicinal plants at Patrapada, Ekamravana, near the famous pond Bindusagar and maintenance of other major components of increasing green cover were taken up (Sharma 2022). Some of these components such as 'Green Mahanadi Mission' (GMM), where the river Mahanadi, the lifeline of Odisha, needs to be protected and conserved to ensure the perennial flow of the river for the benefit of the people. In view of this, a new scheme known as the 'Green Mahanadi Mission' was launched during 2018–2019 with the aim of creating a green belt of 1 km width on both sides of the river Mahanadi. Various plantations under Mahatma Gandhi National Rural Employment Guarantee Scheme (MGNREGS) were made through which over 92.08 lakh seedlings of different species were planted (Sharma 2022). Apart from this, several other afforestation activities in the state such as plantations through 'Odisha Mining Corporation Ltd. (OMC), District Mineral Fund (DMF) and Ama Jungle Yojana (AJY), under state CAMPA (Compensatory Afforestation Fund Management and Planning Authority), have been implemented to increase forest and tree cover for rehabilitation of degraded forests (Sarap et al. 2013).

15.9 Conservation and Sustainable Use of Forests: The Roads Ahead

Conservation and sustainable use of forests are two most important concerns at present. Around the globe in every country, steps are being taken to conserve forests because of the ecological hazards which loom large before the human society with more and more intensity (Brandon 2014). There are ways to manage the world's biodiversity in order to ensure the conservation of forests. Creation of protected areas has been one of the methods adopted everywhere to fulfill biodiversity objectives. This approach has brought about positive results in terms of conserving species and minimizing deforestation. But natural reserves alone are not sufficient to conserve biodiversity. In this context, FAO records its thoughts in its report with regard to a balanced solution of the world's food insecurity and deforestation: "The conservation and sustainable management of forests within an integrated landscape approach is key to the conservation of world's biodiversity and to food security and well-being of the world's people" (State of Forests 2020, p. 62). But forests are still being burnt in many parts of India for different reasons. An Odia Daily reports that fire rages in 211 forest ranges of the country where Odisha tops the list with forest fire in 43 Forest Ranges (The Samaj, 18th April, 2022). Since deforestation is a country-wide phenomenon, the reasons are also diversified: extension of human settlement, collection of wood, illegal use for agricultural purposes, poaching, etc. Most of the time, it is observed that the local people are responsible for it as there is a glaring lack of awareness among them about how loss of forest cover will ultimately be hazardous for their own lives and others.

15.10 Connecting People to Nature

Apart from all the schemes and programmes 'connecting people to nature' can be a significant motivational factor for people around the world to save trees and forests. Each World Environment Day is organized around a theme that focuses on a particularly pressing environmental concern. Realizing this basic need of man based on human psychology, 'connecting people to nature' was chosen as the theme for World Environment Day, 2017 which 'implores us to get outdoors and into nature, to appreciate its beauty and its importance, and to take forward the call to protect the earth that we share' (Human Touch 2017, p. 6). Research shows that the more connected people are with nature, the happier they are in their lives as nature can generate a multitude of positive emotions such as calmness, joy, creativity and concentration.

Another concept known as deep ecology is relatively modern, but it has drastically transformed the perception about nature. The phrase 'deep ecology' originated in 1972 with Norwegian philosopher Arne Naess who along with an American environmentalist George Sessions developed a platform of eight organizing principles for the deep ecology social movement (Britannica 2022). This is essentially an environmental philosophy but is also a social movement which has taken its roots on the belief that nature must be preserved for its own sake and not because it is useful to the human society. Nature has an intrinsic value which must be respected, and we should not look upon it from a materialistic point of view. It essentially highlights the fact that when a man loves nature, he will protect it (Spanne 2021). This has a philosophical overtone, but human practices in India and elsewhere have proved it time and again that when the initiative comes from the people themselves, it can lead to greater success.

In Bisnoi villages, in India, for example, herds of blackbucks are seen as an integral part of the community and nobody harms them. Bisnoi community is known for protecting the blackbucks as they consider the antelope to be the reincarnation of their religious Guru Bhagwan Jambeshwar. A cluster of villages also conserve blackbucks in Odisha's Ganjam district. As many as 70 villages in Ganjam district which is situated in the northern Eastern Ghats have been conserving blackbucks for a century. The shy antelopes share open fields with the villagers here. Local communities regard the animal as harbingers of wealth and prosperity. The community protection has led to an increase in the blackbuck population over time. Their population was not significant in the 1990s, but lately, due to consistent protection of the animal through community participation from the area, the blackbucks have thrived in this pocket. According to the blackbuck census conducted by the local forest department, the population stood at 1533 in 2011 which has increased to 2809 in 2018 and in 2020 went up to 6875 (Kumar 2021).

This could serve as an example for protecting forests as well. The key ideas are 'connect to nature' and 'community participation'. At one level, this is going to be vital for the upkeep of our environment so that no further damage to its existing status takes place. What we have as a gift of nature must be protected first with all sincerity and devotion. In India, for example, where many of the trees are considered to be embodiments of deities, particularly by the tribal people who live their lives close to nature, efforts may be made to convince people that it is a sacred job to protect trees, and for the protection of forests, local communities can be involved who would develop the idea of 'owning' it. Another expression of love for nature can be seen in the 'Chipko movement'. It was a nonviolent, socio-ecological movement by the rural villagers particularly by the women in India in the 1970s, which aimed at protecting the trees and forests meant for government – backed logging (Karan 1994). Though the movement originated in the Himalayan region of Uttarakhand in 1973, it quickly spreads throughout the Indian Himalayas. As the movement continued, protests became more project-oriented and expanded to include the entire ecology of the region.

15.11 Conclusion: Fulfilling Sustainable Development Goals

"The term Sustainable Development first came to light in a popular report of 1987, titled 'Our Common Future,'published by the United Nations World Commission on Environment and Development. It highlighted the need for economic development with environmental protection" (World Commission on Environment and Development, 1987). It was taken up more seriously in 2015 when the Sustainable Development Goals (SDG) framework adopted by 193 countries redefined developmental policies and government priorities across the world. It would not be out of place to reiterate that SDG 15 aims to protect, restore and promote sustainable use of terrestrial ecosystems, sustainable management of forests, combating desertification, halting and reversing land degradation and halting biodiversity loss. These are not easy objectives to be achieved, but they have certainly gripped the thinking minds of all communities everywhere. Today, there is a growing awareness about preserving biodiversity, which is visible in every aspect of human life. Currently, efforts are made towards sustainable consumption and production of forest products. Measures are being taken to improve wood products' utilization to extend their service life, reuse and recycle these products. At the same time, efforts are being made to encourage national policies to support the sustainable production and consumption of non-wood forest products. Additionally, in order to promote sustainable forest management, measures are to be taken to reform forest and land use legislation, to give support for community forest management and to provide more excellent and more effective protection of critical areas of forest such as those of high conservation value and high carbon stock. There is a greater need to focus on holistic approaches to the restoration of natural forest landscapes. This means, there is a need to be sensitive to local conditions and local communities and expand plantations of different varieties of trees that are suitable for the local ecosystems and climates (Sabatini et al. 2020).

In fact, sustainability is not merely about policy-making; it needs to percolate down to the grassroots levels with every citizen cultivating it as part of their lifestyle. It requires a concerted effort from all quarters – the government, the business community, the scientists and intellectuals and the people in general. What is needed is to stop further deforestation and take active steps for reforestation and afforestation. If a strong motivation can be created in the minds of the people to love nature not only for the sake of its beauty but also for our own survival and well-being, the goals will not be too difficult to achieve, particularly in India, since 'sustainability has always been at the heart of Indian culture, embedded in its core philosophies and value systems' (The Times of India, July 29, 2022).

References

Brack D (2018) Sustainable consumption and production of forest products. United Nations Forum on Forests

Brandon K (2014) Forests and food. In: Ecosystem services from tropical forests: review of current science, pp 44–52. Center for Global Development. http://www.jstor.org/stable/resrep29725.11

Britannica (2022) Deep ecology. https://www.britannica.com/topic/deep-ecology

Dash LN (2007) Economics of mining in Orissa. Orissa Review pp 69–76

English Standard Version Bible (2001) ESV Online. https://esv.literalword.com/

Forest Survey Report (2020) National Forest Policy, *Govt. of Rajasthan, Forest Department*, NFP-1988

Forest Survey Report (2021) Press Trust of India, PTI. Forest Survey of India, *Business Standard*, BS-537.44

Forest Survey Report (2022) Odisha's forest cover, *Newspaper*, Odia Newspaper-2022

Forest Survey Report (2022, January 13) Press Information Bureau. Retrieved July 25, 2022, from https://pib.gov.in/PressReleseDetail.aspx?PRID=1789635

Human Touch (2017) World Environment Day 14(6):5–6

Instapedia (2019, August 30) CAMPA. https://www.insightsonindia.com/2019/08/30/campa/

Jha S (2022, May 02) State of the world's forests 2022: 10% of total forest area on earth lost in 30 years. Retrieved September 07, 2022, from https://www.downtoearth.org.in/news/forests/state-of-the-world-s-forests-2022-10-of-total-forest-area-on-earth-lost-in-30-years-82658

Jha CS, Dutt CBS, Bawa KS (2000) Deforestation and land use changes in Western Ghats, India. Curr Sci 79(2):231–238. http://www.jstor.org/stable/24103455

Karan PPK (1994) Environmental movements in India. Geogr Rev 84(1):32–41

Karan PPK (2004) The non-Western world: environment, development, and human rights. Routledge, New York

Karthikeyan P (2020, May 16) Why jhum cultivation of north-east & podu in Odisha the last nail in the coffin. Retrieved September 1, 2022 from https://www.utkaltoday.com/jhum-cultivation-mizoram/#:~:text=The%20word%20Jhum%20or%20Podu,and%20some%20southern%20Indian%20states

Kumar M (2021, January 26) A cluster of villages conserve shy blackbucks in Odisha's Ganjam. Retrieved July 19, 2022, from https://india.mongabay.com/2021/01/a-cluster-of-villages-conserve-shy-blackbucks-in-odishas-ganjam/

Lal R (2005) Shifting cultivation. Encyclopedia of Soils in the Environment, 488–497. https://doi.org/10.1016/B0-12-348530-4/00485-9

Lesniewska F (2005) International environmental laws and forests. In laws for forests: an introductory guide to international forest and forest related legal materials that shape forest ethics and practice. pp 13–21. International Institute for Environment

and Development http://www.jstor.org/stable/resrep18067.8
Mining in Deforestation (2017, March 17) Deforestation. Mining in deforestation blog. https://miningindeforestation.wordpress.com/author/miningindeforestation/
Ministry of Environemnt and Forests, Government of India (2009) National Afforestation Programme. Retrieved December 28, 2023, from https://pib.gov.in/Pressreleaseshare.aspx?PRID=1596332
Mohanty PK (2014) Plants, trees and animals – an ecological background of ancient Odisha. Proceedings of the Indian History Congress 75:62–70. http://www.jstor.org/stable/44158362
Pouchepadass J (1995) Colonialism and environment in India: comparative perspective. Econ Polit Wkly 30(33):2059–2067
Press Trust of India (2022, January 14) Odisha's forest cover increases by 537.44 sq km in two years. The Business Standard https://www.business-standard.com/article/current-affairs/odisha-s-forest-cover-increases-by-537-44-sq-km-in-two-years-122011400300_1.html
Sharma (2022) Highlights of Odisha forestry sector. Aranya Bhawan
Sabatini FM, Keeton WS, Lindner M, Svoboda M, Verkerk PJ, Bauhus J, Bruelheide H, Burrascano S, Debaive N, Duarte I, Garbarino M, Grigoriadis N, Lombardi F, Mikoláš M, Meyer P, Motta R, Mozgeris G, Nunes L, Ódor P, Kuemmerle T (2020) Protection gaps and restoration opportunities for primary forests in Europe. Divers Distrib 26(12):1646–1662. https://www.jstor.org/stable/26946062
Sarap K, Sarangi TK, Naik J (2013) Implementation of Forest rights act 2006 in Odisha: process, constraints and outcome. Econ Polit Wkly 48(36):61–67. http://www.jstor.org/stable/23528373
Spanne A (2021, July 26) What is deep ecology? Philosophy principles, and criticism. Treehugger. https://www.treehugger.com/what-is-deep-ecology-philosophy-principles-and-criticism-5191550
Srivastava AR, Barman N (2019) Forest laws in India – policy and assessment. Int J Legal Dev Allied Issues 5(2):151–182
State of Forests (2020) The State of the World's Forests 2020. Forests, biodiversity and people. Rome. Retrieved August 01, 2022 from https://doi.org/10.4060/ca8642en
United Nations (n.d.) A movement that arose in the 70's. Retrieved December 28, 2023, from https://www.un.org/en/observances/environment-day/background
Wilkie ML, Buszko-Briggs M (2021) To conserve biodiversity, we must protect our forests. Retrieved August 04, 2022, from https://sdg.iisd.orgcommentary/guest-articles/to-conserve-biodiversity-we-must-protect-our-forests/
World Commission on Environment and Developmentns (1987) Our Common Future. United Nations. Retrieved December 28, 2023, from https://www.are.admin.ch/are/en/home/media/publications/sustainabledevelopment/brundtland-report.html

Swetaparna Ankita is an Assistant Professor in English at VIT-APUniversity, Amaravati, India. She has also contributed significantly in subjects like Geography and Psychology. She has been a resource person and a regular contributor of research articles in English and Linguistics.

Green Diplomacy: A Way to Achieve SDG-17

16

S. Venkata Krishnan

Abstract

In the history of mankind, the sufferings are reduced or eliminated either with the help of technological innovations or well-timed diplomacy. This is true to man-made disasters as well as natural hazards. Nevertheless, climate change and environmental degradation have emerged as a challenge to check the abilities of human beings, i.e. their innovation capabilities and diplomacy. Countries of the world have accepted this challenge and are constantly improving their technologies along with cooperative negotiations in related matters. In this background, this chapter studies India's diplomatic activities in the environmental domain. India's diplomatic stance on the environment has been characterised by a commitment to addressing global environmental challenges while prioritizing the country's development aspirations. India engages in several international forums and negotiations on a wide range of environmental issues, including climate change, biodiversity conservation, and sustainable development. The article suggests that a change in traditional diplomatic actions may yield higher benefits for the country as well as for other countries of the world.

S. Venkata Krishnan (✉)
Pandit Deendayal Energy University,
Gandhinagar, Gujarat, India
e-mail: vksharma80@gmail.com

16.1 Introduction

The challenges we face today make us either believe that they are insurmountable or able to defuse completely by invoking the moral principles of human life. However, we are in a period that is not much different from history. Conflicts and challenges are always part of human life. The world has faced many conflicts and challenges in the form of wars (in the name of religion, ideology, territory, and imperialism) and numerous natural disasters, including the present COVID-19 pandemic. Nevertheless, the survival of human life has not become extinct. This has become possible not through merely applying moral principles to life. This was achieved by the judicious application of technology and diplomacy. Nevertheless, environmental degradation and the associated climate change have emerged as a challenge to check the abilities of human beings, i.e. their innovation capabilities and diplomacy. Countries of the world have accepted this challenge, and they are constantly improving their various technologies along with cooperative negotiations in related matters. Against this background, this chapter aims to describe India's domestic and diplomatic activities and is interested in suggesting a change in traditional diplomatic communication and institutions. This may yield higher benefits for the country as well as for other countries of the world. The first section of this chapter discusses India's distinctive geo-

V. Dutta, P. Ghosh (eds.), *Sustainability: Science, Policy, and Practice in India*,
Sustainable Development Goals Series, https://doi.org/10.1007/978-3-031-50132-6_16

graphical features and its long tradition of environmental protection and conservation. In addition, domestic institutions, acts, actions and programmes are also listed. The narration of the National Action Plan on Climate Change (NAPCC) and Intended Nationally Determined Contributions (INDC) in this section exemplifies India's initiatives to limit climate change. The second section addresses India's external engagements. It starts with an introduction to the Ministry of External Affairs (MEA) and then itemises India's stance with different international treaties and describes India's climate change negotiations. The third section, as a long conclusion, suggests how India could share the successes or lessons it learned in the domestic as well as external spheres with the rest of the world.

Table 16.1 Natural regions: Comparison[a]

Natural region	World: Select countries/region	India
Equatorial and tropical forests	Congo, Brazil, South East Asia Region	Western part of Western Ghats, North East India
Tropical monsoon and grasslands	Australia and Brazil	Peninsular India
Desert [hot and cold]	African countries (hot) and Argentina (cold)	Rajasthan and Ladakh
Temperate	Europe, USA, Japan	Indus & Ganga plain, Lesser Himalayas
Polar	Russia, Canada, Scandinavia	J&K, Himachal Pradesh, Uttarkhand, Sikkim and Arunachal Pradesh
Coral reefs	Maldives	Lakshadweep, Andaman and Nicobar Islands

[a]Author's scheme

16.2 India and Environment: Domestic Institutions, Programmes and Policies

India faces an enormous challenge in coping with the impacts of environmental degradation and climate change compared to most other countries. There are many reasons for this: population (has become the world's most populated country), geographical location (tropical and peninsular country), biodiversity (one of the recognised mega-diverse countries of the world having different natural regions, see Table 16.1), rapid urbanisation (urban population rising at a much faster rate than its total population) and natural hazards (one of the world's most disaster-prone countries). Apart from this, linking environmental problems as part of international politics made this a challenge in external relations as well. The politicisation of the environment has become a unique feature and one of the major issues in the international politics of the twenty-first century (Barua 2017). Global environmental politics (GEP), an area of both practice and academic research, is the politics of the environment on a global scale and practiced in different places on different issues to alter or protect the environment (Harris 2014). The concern about the environment and giving a political shade to it took place only after the 1972 Stockholm UN Conference on the Human Environment and 1992 Earth Summit, respectively (Cass 2014). Moreover, for some, environmental issues, including climate change, are challenges of economic and technical nature, and for others, they are related to justice and equity (Dubash 2019). This late start and realisation in the international system is different for India, and it has a long history of environmental preservation and protection.

India approached the environment not only like others but from the beginning given a spiritual orientation too. The statement of Union Minister for Environment, Forest and Climate Change at UN Biodiversity Conference, "We in India have respected and revered nature for thousands of years", reflects this (Yadav 2021). More than 3000-year-old Indian scriptures have articulated the need for preservation and sustainable use of natural resources, and these are echoed in the constitution, legislation, programmes, policy frameworks and international commitments of the country (Ministry of Environment, Forest and Climate Change n.d.-a). Even before India's independence, environmentally related institutions and legislation existed. A fresh impetus and a constitutional sanction were also given to environmental concerns after the 1972 Stockholm UN Conference on the Human Environment. In India, the National Council for Environmental Policy and Planning within the Department of Science and Technology was established, which evolved into a full-fledged Ministry of Environment and Forests (MoEF) in 1985 and was renamed the Ministry of Environment, Forest and Climate Change in 2014 (Economic Times Bureau 2014). Today, as an apex administrative body and as a nodal agency in the central government, it is implementing India's climate change, environment and forest policies and programmes relating to conservation of the country's natural resources, including lakes and rivers, its biodiversity, forests and wildlife, ensuring the welfare of animals and prevention and abatement of pollution. The following table reveals India's comprehensive approach towards environmental protection and climate change in the domestic sphere (see Table 16.2).

In the list of above, the illustration of the National Action Plan on Climate Change (NAPCC) and Intended Nationally Determined Contributions (INDC) will disclose India's seriousness towards this challenge.

The NAPCC espouses India's vision of sustainable development and steps to be taken to implement it. It is grounded on the awareness that climate change actions must advance simultaneously in numerous closely interrelated domains, such as agriculture, water, forests, mountain environment, energy, industry and urbanisation. This was the background to the Eight National Missions spelt out in the NAPCC. These missions are on conserving water, making agriculture sustainable, sustaining the fragile Himalayan Ecosystem, creating a Green India through expanded forests, enhancing energy efficiency, solar energy, creating a sustainable urban habitat and creating a strategic knowledge platform for aiding all National Missions. The NAPCC has recognised that climate change and energy security were two sides of the same coin and that India had to make a deliberate change from its current dependence

Table 16.2 India's domestic institution, acts, actions and programmes

Botanical Survey of India – 1890.
Zoological Survey of India – 1916.
Indira Gandhi National Forest Academy – 1987 (1938).
Forest Survey of India – 1981 (1965).
Wildlife Institute of India (WII) was established in 1986.
Indian Forest Act, 1927.
Eco-Development Forces (EDF) Scheme was established in 1980s.
National Forest Policy, 1988.
National Afforestation and Eco-Development Board (NAEB) was set up in 1992.
Wildlife (Protection) Act, 1972, as amended in 2006.
Integrated Development of Wildlife Habitats.
Project Tiger 1973.
Project Elephant (PE) 1991–1992.
Asiatic Lion Conservation Project 2019.
18 designated Biosphere Reserves (BRs)
Scheme on Biodiversity Conservation.
Biological Diversity (BD) Act in 2002.
National Biodiversity Action Plan (NBAP) in 2008, 2014.
Air (Prevention and Control of Pollution) Act, 1981.
Environment (Protection) Act (EPA), 1986.
Public Liability Insurance Act, 1991.
Environmental Impact Assessment 1994.
Chemicals Accidents (Emergency Planning, Preparedness and Response) Rules, 1996.
Notification on fly ash utilisation in 1999, amended in 2003, 2009 and in 2016.
Energy Conservation Act, 2001.
Hazardous Waste (Management, Handling and Transboundary Movement) Rules, 2008 and 2016.
Plastic Waste (Management and Handling) Rules, 2011.
National Air Quality Index 2015.
Plastic Waste Management Rules, 2016.
Solid Waste Management Rules, 2016.
Construction and Demolition Waste Management, 2016.
E-waste (Management) Rules, 2016.
BS-VI fuel standards 2020.
Ganga Action Plan (GAP) in 1985.
National Wetland Conservation Programme (NWCP) was initiated in 1987.
National River Conservation Plan (NRCP) in 1995.
Wetlands (Conservation and Management) Rules in 2010.
The National Plan for Conservation of Aquatic Eco-systems (NPCA).

(continued)

Table 16.2 (continued)

National Environmental Policy (NEP) 2006.
National Water Policy 2012.
Common Effluent Treatment Plants.
National Natural Resources Management System (NNRMS) 1985.
National Environmental Awareness Campaign 1986.
The National Green Tribunal (NGT) was set up in 2010 under the NGT Act, 2010.
National Action Plan on Climate Change (NAPCC), 2008.
Climate Change Action Programme since 2014.

on fossil fuels to a pattern of economic activity founded progressively on cleaner sources such as nuclear energy and renewable sources of energy such as solar. Such a paradigm shift would augment India's energy security and contribute to dealing with the risks of climate change. Thus, a cobenefit approach is the basis of India's climate change strategy. The NAPCC not only establishes that India's reaction to climate change is based on its own resources but also recognises that it is closely related to the corresponding multilateral efforts, grounded in the principles and provisions of the UNFCCC, to institute a global climate change regime (Saran 2019).

India's INDC connects commitment to sustainable economic development with its ancient civilisational values of concerning nature, integrating a sense of intergenerational equity and common humanity. The goals that India has willingly committed itself to are unparalleled for a large developing country. Compared to the 2005 base year, the energy intensity of India's growth will drop by 33–35% by 2030. As one of the world's largest emerging economies, India, which at present has a huge global energy footprint, establishes the most important contribution to tackling worldwide climate change. On the strength of the exceptional success of the National Solar Mission, the INDC has set a goal of 175 GW of renewable energy by 2030. India is also expeditiously decreasing the component of coal-based thermal power in its energy mix. It is not commonly known that the country has a very high cess on coal, of the order of Rs.400 per tonne, earnings from which go into a Clean Energy Fund. India has also committed to not building any new thermal plants that are not of the most efficient ultrasupercritical category (Saran 2019).

16.3 India and Environment: External Engagement

In India, the nodal agency for the setup, such as the United Nations Conference on Environment and Development (UNCED)/Earth Summit, International Centre for Integrated Mountain Development (ICIMOD) and United Nations Environment Programme (UNEP), is the Ministry of Environment, Forest and Climate Change. The Ministry also coordinates with regional bodies such as the Economic and Social Council for Asia and Pacific (ESCAP) and South Asian Association for Regional Cooperation (SAARC) and international bodies such as the Commission on Sustainable Development (CSD) and Global Environment Facility (GEF) on matters pertaining to the environment (Ministry of Environment, Forest and Climate Change n.d.-b). However, the Ministry of External Affairs and diplomatic missions abroad play a primary role in the external relations of India, including issues related to environmental protection and climate change.

A new division was set up by the Ministry of External Affairs (MEA) – States Division – in October 2014 to harmonise with States and Union Territories and to promote their exports and tourism and invite further overseas investments and knowledge (Ministry of External Affairs, Government of India 2016). Due to the increasing salience of the Indo-Pacific concept in global discourse, the Ministry of External Affairs established a new division for the Indo-Pacific in April 2019 (Ministry of External Affairs 2019–2020). In January 2020, the Ministry of External Affairs (MEA) took the progressive step to establish India's first, New and Emerging Strategic Technologies (NEST) division (Ministry of External Affairs, GoI 2020). In 2021, there was an induction of two ministers of state to the MEA.[1]

[1]Meenakshi Lekhi and Rajkumar Ranjan Singh are two ministers joined the Ministry of External Affairs (MEA) in the 2021 cabinet expansion.

With the entry of these two to the MEA, a record total of four ministers made available to conduct India's foreign relations (Chaulia 2021). Thus, this unceasing dynamism makes the Ministry of External Affairs of India a vibrant one in the overall administration. Moreover, these are the instances that clearly demonstrate the high priority accorded for foreign policy and international engagements by India. In 2021, MEA has 202 missions and posts operational globally and has approximately 70 administrative divisions (Ministry of External Affairs 2021). Normally, India's diplomatic establishments abroad consist of officers such as Ambassador/High Commissioner (A/HC), Deputy High Commissioner (DCH), Consular General (CG) and Secretary for political, commercial or cultural affairs and educational, military or scientific attaché with supporting staff. The lead officials (A/HC, DCH and CG) in all the diplomatic missions and in the MEA divisions are Indian Foreign Service (IFS) officers. IFS officers will be supported by other experts, such as defence officials (Military Attaché), subject experts and administrative personnel in diplomatic activities.

MEA and diplomatic missions have been involved in different negotiations, such as security and defence, trade, science and technology and environmental protection and climate change. Even in the theme of environment protection and climate change, these institutions have been part of many negotiations, as listed in Table 16.4.

Along with domestic initiatives (see Table 16.2), India's external engagement is also comprehensive. Table 16.4 illustrates India's commitment to and involvement in the different aspects of environmental protection and climate change.

16.4 India and Climate Negotiations

When compared to many of the world's larger nations, India has continuously participated actively in international environmental negotiations (Short 1999). India expressed its desire to work with the major nations to address all pressing global issues, including how to effectively manage the world's natural resources, as soon as it had gained independence (Ali 1948, pp. 22–26). India's negotiating strategy was typically characterised by its civilisational history and its appeal to moral, Third Worldist imperatives, or more generally, by its rigid distributive nature. However, this has been comprehended as a traditional 'nay-sayer', accepting neither compromise nor side payments (Cohen 2001; see also Narlikar 2006, p. 76). Irrespective of such observations, India's support, as listed in Table 16.4, for various international conventions, treaties, protocols and programmes explains the country's commitment to preserving the planet Earth. Moreover, this is a dynamic issue and an intricate subject that are still in progress. In addition, every country's economic development aspirations have made this a common challenge to the international community. Recognising this complexity,

Table 16.3 Attaché level officers in select Indian diplomatic establishments (as of 11 Sep. 22)[a]

Congo (DRC) [Equatorial]	Australia [Tropical]	Belgium, Luxembourg and EU [Temperate]	Finland and Estonia [Polar, Sub-Polar]	Maldives [Island]	Fiji [Island]
Attaché (Admin and Cons)	Attaché (Consular) Attaché (Admin) Attaché (Establishment) Attaché (DHC Office)	Attaché (Ambassador's Office) Attaché O/o FS(EU) and FS(HOC) Attaché (Political) Attaché (PI&C) Military Attaché Air Attaché Attaché (Estt.) Attaché (Consular)	Attaché (Consular and Administration)	Nil	Attaché (Admin) Attaché (PS) Attaché (ITEC)

[a]Author's scheme

Table 16.4 India and international treaties/programmes[a]

India was a signatory of the Man and Biosphere Programme, launched by the UNESCO in 1970.
India is a member of UNEP (inception in 1972) and provides annual financial contribution of USD 100,000 to UNEP Environment Fund.
India became a signatory to the Ramsar Convention in 1982.
Vienna Convention for the Protection of the Ozone Layer in 1985 – India is a party.
The Montreal Protocol on Substances that Deplete the Ozone Layer in 1987 – India is a party.
Basel Convention (1989) on the Control of Transboundary Movements of Hazardous wastes and their Disposal – India is party to the Convention.
Global Environment Facility (GEF) was set up in 1991, and India is one of the founding members.
In 1992, at the Earth Summit in Rio de Janeiro, Brazil, 178 countries adopted Agenda 21 – India is a signatory.
In 1994, India has ratified the Convention on Biological Diversity (CBD).
Cartagena Protocol on Biosafety (CPB) adopted in 2000 – India is a party to the Protocol.
United Nations Convention to Combat Desertification (UNCCD): India became a signatory to it in 1994.
MoEFCC and US Government signed an agreement in 2000 to implement the Globe (Global Learning and Observations to Benefit the Environment) programme in India.
The Rotterdam Convention on the prior informed consent procedure for certain hazardous chemicals and pesticides in International Trade came into force in 2004 – India acceded in 2005.
The Stockholm Convention on persistent organic pollutants (POPs). The Convention came into force in 2004. India ratified the Convention in 2006.
In 2006, over 190 countries including India acceded to the Strategic Approach to International Chemicals Management (SAICM).
Minamata Convention on Mercury, 2013 – India ratified it in 2018.
2030 Agenda for Sustainable Development, which India adopted along with 193 countries in the year 2015
Paris Agreement 2015 to the UNFCCC – India ratified it in 2016.
International Solar Alliance (ISA) conceived as a joint effort by India and France on 21st Conference of Parties (COP21) to the United Nations Framework Convention on Climate Change (UNFCCC) held in Paris in 2015.
India's submitted Intended Nationally Determined Commitments (INDC) to the UN Framework Convention on Climate Change (UNFCCC) in October 2, 2015.
COP26 to the UNFCCC – 2021: India has promised to cut its emissions to net zero by 2070; India to get 50% of its energy from renewable resources by 2030.

[a]Author's scheme

countries of the world are engaging and negotiating with one another in different bilateral and multilateral forms. In particular, climate diplomacy has become a major feature of international relations (Desai 2019). India often emphasizes the principle of "common but differentiated responsibilities" in the context of environmental diplomacy. This principle acknowledges historical differences in contributions to environmental issues and emphasises that developed countries should take greater responsibility for addressing global environmental challenges.

An overview of India's role in international climate negotiations since their start in the late 1980s will help to understand India's climate diplomacy. In climate negotiations, as in other international negotiations, India is of special interest for at least three reasons (Michaelowa and Michaelowa 2011). First, it has recently emerged as a very significant actor in international negotiations pertaining to climate diplomacy. Second, it combines characteristics of extreme climate change susceptibility with a sizable population and rising greenhouse gas emissions. Third, as India's economy grows, it has already exceeded several of the developed nations in Europe. Hence, India's situation is as important as a sufferer of past and present emissions from industrialised nations and one of the polluters that could, on its own, contribute considerably to global emissions (Michaelowa and Michaelowa 2011). Initially, the Indian position was determined by just a small number of officials, without much general interest being taken in the issue (Rajan 1997). At one of the first international conferences, India disagreed with emission reduction goals for developing nations but asked for the inclusion of developing nations in global climate negotiations. The World Resources Institute (WRI) published country-specific greenhouse gas (GHG) emission data in 1990, which sparked the first in-depth discussion regarding the Indian position on global climate policy at both the national and international levels. The WRI report placed more emphasis on the current level of emissions than on the total cumulative amount of emissions. The WRI report pointed at India as a major perpetrator due to its large methane emissions from rice farming. The Centre for Science and Environment (CSE), a New Delhi-based NGO, responded angrily refuted the report of WRI (Agarwal and Narain 1991). CSE not only argued against the questionable assumptions underlying the WRI's report but also claimed for a differentiation between 'survival emissions' in the South and 'luxury emissions' in the North. Furthermore, the CSE has stressed that emissions should be calculated on a per capita basis and that industrialised countries should be given fair regard for their historical responsibility. As a result, India's stance became more strident and pushed for increased responsibility differentiation and technology transfer to developing nations (Michaelowa and Michaelowa 2011).

During subsequent years, India has shown some flexibility by adjusting its initially strong opposition to cautious support and has embraced the Clean Development Mechanism (CDM). Nevertheless, when a new epoch in climate diplomacy was opened by the entry into force of the Kyoto Protocol in early 2005 and the opening of post-2012 climate negotiations at COP 11 in Montreal, India remained steadfast in its defence of 'common but differentiated responsibilities (CBDR)'. At the Paris climate summit, India played a major role and assured its success. The Indian Prime Minister's personal intervention in the adoption of the landmark Paris Agreement was acknowledged by several world leaders. His initiative on the establishment of an International Solar Alliance for promoting solar power worldwide was also welcomed (Saran 2019). However, India is against discussing climate change in the UN Security Council and blocked a discussion in early 2007 with the argument that it was uncertain. This position of India is steady, and it believes that bringing climate security into UNSC discourse has the potential to disrupt the nature of overall discussions (Times of India 2021).

Thus, India's negotiation strategies in climate diplomacy have substantially changed over time, and three distinct periods have been observed (Sengupta 2019). Although the underlying situation had not changed much, there had been considerable changes in the way they were offered and sustained on a global scale. In particular, India's climate diplomacy has moved away from protective and purely distributive policies towards mixed strategies with a variety of value-creating elements. India's shift in strategy appears to be influenced by a number of elements, including changes in the country's political landscape, the minister in charge, the negotiating team's leadership, an overall pattern associated with growing public awareness, tightening domestic energy supplies, financial gains from the Kyoto Protocol's mechanisms, responses to pressure from other developing nations and more reporting by domestic media (Michaelowa and Michaelowa 2012).

India, domestically and externally, is advancing on a broad front to diminish the effects of climate change and has expanded environmental protection, drawing upon its ingrained civilisational attributes and putting in place a wide range of legal frameworks and policy interventions covering all facets of modern life. There is also an active and vibrant civil society that is promoting citizens' awareness of the threats of climate change and environmental degradation and what each of us can do as individuals to meet these threats. It is expected that India's leadership in dealing with these challenges will act as an incentive to many other countries to increase their contributions to meet this existential and global challenge (Saran 2019). There is a similar view from outside that '(m)any look forward to gaining from the Indian experience because India is a big country. India probably has a lot to give and take, especially with the other two biggies in the world, Brazil and China (Clusener-Godt 1996)'. Hence, India should share experiences with other countries.

16.5 Conclusion: Way to Sharing the Success

Numerous whole-hearted initiatives in the domestic sphere along with proportionate external involvement to limit climate change through diplomatic negotiations stimulate to propose that, India should now consider the following suggestions from the perspective of diplomacy.

16.5.1 Perceptional Change

'In today's world, 'Balance Of Payments' and 'Terms of Trade' have replaced 'Balance of Power' in importance (President of India 2016)'. Indian diplomacy should accept that we are in a period of external governance and not in the era of foreign policy (Venkata Krishnan 2013). In the period of the foreign policy, military-centric actions of states had been considered threats. However, in the time of external governance, challenges and threats are not only state-centric, and there are other challenges and threats for survival. This thought, even though it looks like primitive human thinking, the search for scientific solutions to the current challenges makes this rational and valid. Furthermore, today, the 'protection and preservation of environment' is used to promote national interests in a disguised form. Climate negotiations have become less about meeting an elemental challenge to human survival and more about safeguarding the narrowly conceived economic self-interests of nations (Saran 2015). Indian diplomacy should be cautious about pseudo altruism in international negotiations about climate change and biodiversity.

16.5.2 Changes in Diplomatic Communication

Our diplomatic communication, i.e. communication between our diplomats and the foreign governments and Indian government's messages to diplomats stationed in embassies and consulates in India, should highlight matters related to the environment. Our embassies and consulates should observe and report causes for environmental degradation, such as pollution, emission, production of hazardous substances, deforestation, shrinking of wetlands and various threats to wildlife in the host countries. They must observe and report on the availability of renewable resources, pollution-reducing technologies and good environmental management practices in the host countries. Once a Union Minister lamented that 'there is lack of commitment of the developed world in fulfilling their obligations in environmental protection' and asserted that 'India's green energy programmes will be carried out whether we receive support of the western world or not (Goyal, 2016)'. However, the enhanced diplomatic communication, including through Track 1.5, Track 2 and social media platforms such as Facebook, Instagram and Twitter, will be helpful in influencing the detractors, spreading India's sustainable development initiatives and paving the way for sharing the expertise of both the worlds.

16.5.3 Creating an Environmental Division and Appointing Environmental Attaché

However, the above mentioned new practices will increase the workload in the MEA and in our embassies and consulates, which are already overburdened because of manpower shortages in our diplomatic establishments. Therefore, India should create a division within MEA and place an environmental attaché in our embassies and consulates. Even appointing an environmental attaché will appear too ambitious; then, at least we have to do the same for countries that are rich in biodiversity, countries that are more vulnerable to climate change such as the small island developing states (ISDS) and countries with good environmental management practices. Alternatively, an environmental attaché could be appointed for a cluster of countries or for a region based on natural vegetation (Table 16.3). Such appointments are mutually beneficial, i.e. to the host countries and for India.

16.5.4 Benefits of Green Diplomacy

What benefits will accrue to India by these changes in diplomacy? First, with these changes, a better database on the world environment would be possible for India. Second, close observations in different regions, such as polar, temperate, tropical, arid, heavy rainfall and oceanic regions of the world, would be helpful in managing the similar environment prevailing in different parts of India, and vice versa is also possible (Table 16.1). Third, it is feasible to streamline our developmental and industrial projects according to the requirements of local environmental conditions or based on sustainable management practices. Fourth, engaging vulnerable and developing states in this manner would encourage them to be partners in containing the eco-imperialism Alfred (1986) of the North. Finally, this 'green diplomacy' will also demonstrate India's earnestness to the world that, India, which traditionally believes the world as one family, would not hesitate to take a lead in the revitalisation of the global partnership for sustainable development.

References

Agarwal A, Narain S (1991) Carbon emissions in an unequal world. Centre for Science and Environment, New Delhi

Alfred C (1986) Ecological imperialism: the biological expansion of Europe, 900–1900. Retrieved from Oxford reference: https://www.oxfordreference.com/display/10.1093/oi/authority.20110803095741676

Ali A (1948) India's role in one world. Ann Am Acad Pol Soc Sci 258:22–26

Barua M (2017) India's environmental strategy in the global climate negotiations. Int Stud 51:195–211

Cass LR (2014) The discipline of global environmental politics: a short history. In: Harris PG (ed) Routledge handbook of global environmental politics. Routledge, London, pp 17–29

Chaulia S (2021, July 21) How a four-minister MEA could amplify India's diplomatic power. Retrieved from Livemint: https://www.livemint.com/opinion/how-a-four-minister-mea-could-amplify-india-s-diplomatic-power-11626796198986.html

Clusener-Godt M (1996, November 30) Many want to gain from India's experience. Retrieved from Down To Earth: https://www.downtoearth.org.in/interviews/many-want-to-gain-from-indias-experience-27086

Cohen S (2001) India: emerging power. Brookings Institution, Washington

Desai N (2019) Foreword. In: Dubash NK (ed) India in a Warming World. Oxford University Press, New Delhi

Dubash NK (2019) An introduction to India's evolving climate change debate: from diplomatic insulation to policy integration. In: Dubash NK (ed) India in a warming world. Oxford University Press, New Delhi, p 1

Economic Times Bureau (2014, May 28) Ministry of environment and forests undergoes a nomenclature change; government serious to tackle climate change. Retrieved from The Economic Times. https://economictimes.indiatimes.com/news/economy/policy/ministry-of-environment-and-forests-undergoes-a-nomenclature-change-government-serious-to-tackle-climate-change/articleshow/35651292.cms

Goyal P (2016, April 02). Press Information Bureau. Retrieved from PIB: https://pib.gov.in/newsite/PrintRelease.aspx?relid=138511

Harris PG (2014) Delineating global environmental politics. In: Harris PG (ed) Routledge handbook of global environmental politics. Routledge, London, pp 1–14

Michaelowa K, Michaelowa A (2011) India in the international climate negotiations: from traditional nay-sayer to dynamic broker. ETH Zurich and University of Zurich. Center for Comparative and International Studies. Retrieved from https://www.files.ethz.ch/isn/133820/WP70.pdf

Michaelowa K, Michaelowa A (2012) India as an emerging power in international climate negotiations. Clim Pol 12(5):575–590. https://doi.org/10.1080/14693062.2012.691226

Ministry of Environment, Forest and Climate Change (n.d.-a) Chapter 2. Retrieved from Ministry of Environment, Forest and Climate Change. https://moef.gov.in/wp-content/uploads/wssd/doc2/ch2.html

Ministry of Environment, Forest and Climate Change (n.d.-b) About MoEFCC. Retrieved from PARIVESH. https://parivesh.nic.in/About.aspx

Ministry of External Affairs (2019–2020) Annual Report. Policy Planning and Research Division

Ministry of External Affairs (2021, December 13) About us: divisions. Retrieved from Ministry of External Affairs: http://www.mea.gov.in/divisions.htm

Ministry of External Affairs, GoI (2020, February 5) Media Center. Retrieved from Ministry of External Affairs. https://mea.gov.in/lok-sabha.htm?dtl/32359/QUESTION_NO552_NEW_AND_EMERGING_STRATEGIC_TECHNOLOGIES_DIVISION

Ministry of External Affairs, Government of India (2016, May 11) Media Center. Retrieved from Ministry of External Affairs: http://www.mea.gov.in/lok-sabha.htm?dtl/26812/question+no2970+new+division+for+centrestate+relations

Narlikar A (2006) Peculiar chauvinism or strategic calculation? Explaining the negotiation strategy of a rising India. Int Aff 82(1):59–76

President of India (2016, April 06) On the occasion when Officer Trainees of the Indian Foreign Service (2014 and 2015 batches) called on the President of India. Retrieved from: http://pib.nic.in/newsite/PrintRelease.aspx?relid=138659

Rajan MK (1997) Global environmental politics: India and the North–South politics of global environmental issues. Oxford University Press, New Delhi

Saran S (2015, November 23) Paris climate talks: developed countries must do more than reduce emissions. Retrieved from The Guardian: https://www.theguardian.com/environment/2015/nov/23/paris-

climate-talks-developed-countries-must-do-more-than-reduce-emissions
Saran S (2019, November 8) India's climate change policy: toward a better future. Retrieved from Ministry of External Affairs, Government of India: https://www.mea.gov.in/articles-in-indian-media.htm?dtl/32018/Indias_Climate_Change_Policy_Towards_a_Better_Future
Sengupta S (2019) India's engagement in global climate negotiations from Rio to Paris. In Dubash NK (ed) India in a warming world integrating climate change and development. https://doi.org/10.1093/oso/9780199498734.001.0001
Short P (1999) Mao: a life. Owl Books, New York
Times of India (2021, September 23) Bringing climate security into UNSC discourse. Times of India. Retrieved October 31, 2021, from Times of India: https://timesofindia.indiatimes.com/india/bringing-climate-security-into-unsc-discourse-has-potential-to-disrupt-nature-of-overall-discussions-india/articleshow/86458464.cms
Venkata Krishnan S (2013, August 12) Foreign Policy to External Governance: India's changing soft infrastructure. Retrieved from: http://www.academia.edu/4731357
Yadav B (2021, October 12) Ministry of environment, forest and climate change. Retrieved from PIB Delhi: https://pib.gov.in/PressReleaseIframePage.aspx?PRID=1763341

S. Venkata Krishnan is an Assistant Professor in the School of Liberal Studies, PDEU, Gandhi Nagar, Gujarat, teaching political science and international relations. He is also managing head of the CCE, PDEU which gives guidance to UPSC Civil Services examination. He has qualified UGC-NET in three disciplines: political science, international and area studies, and public administration. His areas of interest are diplomacy, regionalism, political, administrative and geographical thoughts, modern Indian history, and Indian economy.

17 Sustainable Development Goals: Challenges, Opportunities, and the Way Forward

Priyanka Ghosh and Tania Chakravarty

Abstract

Realizing the 17 Sustainable Development Goals (SDGs) set by the United Nations is a pressing global issue. India has implemented various initiatives and programs to align with the SDGs. These initiatives address issues such as poverty, health, education, gender equality, water and sanitation, renewable energy, and environmental sustainability. The progress made towards one goal can have a bearing on how things turn out for the other. The National Institution for Transforming India (NITI) Aayog serves as the country's focal point for tracking SDG development, and it has developed the SDG India Index to enable a comparison of state-level success. Using examples from the work of several authors who have contributed to this collection, this chapter examines the obstacles and possibilities associated with SDGs in India. Individual behavioural and lifestyle changes, forest and ecosystem conservation, renewable energy capacity growth, and collaboration between the private sector, civil society, academic institutions, and international development organizations are just a few of the initiatives suggested in this chapter. All of these are necessary if we are to meet the Sustainable Development Goals (SDGs) by the 2030 deadline.

Keywords

SDGs · India · Challenges · Opportunities · Policy-making

17.1 Introduction

The Sustainable Development Goals (SDGs), also called Global Goals, are a set of 17 goals adopted by the United Nations in 2015 to end poverty and hunger, protect the planet, and ensure peace and prosperity of all people by 2030. These 17 Goals are integrated in such a manner that actions taken with respect to one goal will impact the outcome of other. The SDGs are applicable to all 193 member countries of the United Nations, and all member countries need to fulfill their time-bound targets in key development sectors such as education, health, clean water, sanitation, energy, and environment. In India, NITI Aayog is the nodal

P. Ghosh (✉)
VIT-AP School of Social Sciences and Humanities (VISH), VIT-AP University, Amaravati, Andhra Pradesh, India
e-mail: priyanka.ghosh@vitap.ac.in

T. Chakravarty
VIT-AP School of Social Sciences and Humanities (VISH), VIT-AP University, Amaravati, Andhra Pradesh, India

Vivekananda College (Affiliated to University of Calcutta), Kolkata, India
e-mail: taniachakravarty@gmail.com

V. Dutta, P. Ghosh (eds.), *Sustainability: Science, Policy, and Practice in India*, Sustainable Development Goals Series, https://doi.org/10.1007/978-3-031-50132-6_17

institution to implement SDGs, and since 2018, it has introduced SDG India Index & Dashboard—a powerful tool to measure progress made by states and union territories (UTs). SDG India Index has also helped to identify key areas which need improvement and interventions. With respect to 17 SDGs, the index is also a way of monitoring India's achievements at the national and subnational levels. In certain sectors of development, India has made substantial progress such as clean water and sanitation (SDG6), affordable and clean energy (SDG 7), responsible consumption and production (SDG 12), and peace, justice, and strong institutions (SDG 16) (SDG India Index and Dashboard 2020–21, NITI Aayog). In 2014, India had a sanitation coverage of mere 38.70%. After Swachh Bharat Mission was launched, the country has achieved a sanitation coverage of 100% in rural areas. Close to 6 million villages, 636 districts, and 35 states/UTs have been verified to be ODF (open-defecation-free) in December 2019. Despite such gains, India still leads the world in number of people who do not have access to toilets (Rubin and Kapur-Gomes 2020).

Although India was lauded for its sanitation services, much improvement needs to be done in case of groundwater quality (Green Governance Initiative). In India, groundwater is contaminated by arsenic, fluoride, nitrates, and irons. Four states of India including Bihar, Chhattisgarh, West Bengal, and Uttar Pradesh are suffering from high level of arsenic contamination. Fluoride contamination has been found in 19 states of India, and an area of about 2 lakhs sq. km is suffering from the problem of salinity. Additionally, India also faces challenges of fulfilling the demand of drinking water supply in its major cities and towns (Green Governance Initiative). By 2030, India will have 68 cities with more than 1 million people, and a number of mega cities will also increase from 5 to 6 with populations of 10 million or more (Mint 2010). This will further pose challenge to provide clean water, sewage, and sanitation.

Even though in certain SDGs, India has performed well, the country still needs to work on zero hunger (SDG 2); quality education (SDG 4); gender equality and empowerment of all women and girls (SDG 5); decent work and economic growth (SDG 8); industry, innovation, and infrastructure (SDG 9); climate action (SDG 13); and life on land (SDG 15). Progress made in case of a specific SDG is uneven and varies geographically. Goa and Lakshadweep are the top states/UTs in terms of SDG 6 having a SDG India Index Score of 100. On the other hand, the states of Assam, Rajasthan, and Delhi have only moderately performed in case of SDG 6. It should also be noted that a particular state is performing well in one SDG and stagnating in another. For example, Goa has performed exceptionally well in SDG 6 and 7 but has performed poorly in SDGs 12 and 13 (SDG India Index and Dashboard 2020–21, NITI Aayog).

This book addresses India's contemporary sustainability narrative from various parts of the country and addresses the challenges and opportunities in the context of SDGs. The contributors of this book also come from varied background ranging from academicians, research scholars, policy-makers, and practitioners. The book brings in-depth empirical case studies from different parts of India which delve into the constraints and potentials of achieving SDGs in India. The success of achieving 17 sustainable development goals in India will also make a difference at the global scale considering India's staggering 1.4 billion population. Therefore, an India-centric book providing empirical case studies adds timely values into and extends the literature of SDGs. Further, in a diverse country like India, achieving the SDGs can be equally challenging for government, policy-makers, and common people. This chapter attempts to lay out the challenges and opportunities of achieving SDGs in India as examined by the various authors of this volume and explores their future implications.

17.2 SDGs in India: Challenges and Opportunities

One of the major challenges of achieving SDGs in India is poverty. India has made progress in reducing poverty, with initiatives like the Pradhan

Mantri Jan Dhan Yojana, which focuses on financial inclusion, and social welfare programs aimed at providing financial assistance to vulnerable populations. This volume illustrates the imperfections and limitations of poverty measurements in a vast and diverse country like India. Ravi Patni shows that poverty is multidimensional, and the definition of what constitutes "human needs" is rapidly changing in the post-COVID-19 scenario. At present, poverty is not only about the lack of basic need of food and shelter but encompasses the wider attributes of living standards like sanitation, housing, clean water, energy, education, health, and employment. Patni recommends a recalibration of the poverty line that should be done based on changes in income, consumption patterns, and prices. Similarly, Shewli Shabnam's chapter addresses the problem of using inadequate indicators to collect and evaluate data in a changing demographic scenario in India. Her chapter focuses on the importance of developing correct indicators for assessment of people's health using summary measures of population health (SMPH) which combines morbidity and mortality data.

The right to decent working conditions is directly associated with economic development and sustainability. However, it has been a major challenge for India. The idea of decent work is rooted in equity, security, and respect of human rights. A decent work place not only exhibits economic value but also social value. Though migrant workers contribute to economic development of a country, their migrant status makes them vulnerable to systemic exploitation. Female migrant workers in India often work in unorganized sectors such as domestic help, construction work, farming, small businesses, etc. Female migrant workers contribute towards economic growth in both their places of origin and destinations. But they lack the legal protection and community support offered to the workers in organized sectors. Basu and Chattoraj suggest that the only way to rectify the situation is to encourage workers' trade unions to intervene in unorganized work sectors. Trade unions can effectively negotiate with government, policymakers, and employers to help the migrant workers and provide them with better working conditions.

While discussing SDG 4, Bhatnagar recognizes education as the means of fulfilling the objectives of other SDGs. It not only creates awareness about unsustainable practices but also helps in the development of innovative solutions to withstand/rectify such challenges. In addition, she uniquely manifests how the SDGs are related to and impact the outcomes of each other. For instance, SDG 1 (poverty) and SDG 8 (economic growth) are interrelated, and both can be addressed through the attainment of SDG 4 (quality education for all). Phogat's chapter deals with equality and inclusiveness in education in India, where socio-economic discrimination exists in varied forms, especially in the form of caste. She makes the appropriate suggestion that NEPs should not only include education for all, but all learners should feel included in a classroom environment. She recommends that teachers' training programmes should be regularly updated and modified so that teachers from the very onset should learn to be sensitive towards all children and their respective background.

The challenges and prospects of achieving the targets of SDG6 are also discussed in this volume with respect to Ganga River Basin (GRB) and associated *Namami Gange* Programme. GRB plays a critical role in overall improvement of water quality and sanitation in India as it is home to 400 million of people. This particular river basin has always attracted Indian government's attention to fight against water pollution and provide clean and safe water through various pollution abatement initiatives. One such recent initiative is *Namami Gange* Programme, implemented by the National Mission for Clean Ganga (NMCG), that wants to set an ideal model for sustainable water management in India. Even after spending a huge amount of money in various Clean Ganga initiatives, there is not enough change in water quality along the main channel of the river, and at certain sites, the water is not even safe for drinking and bathing (Kaur 2018). The target of providing universal access to safe and affordable drinking water in urban setting is equally challenging in India considering the fact that the urban population has rapidly increased in India in the past few decades.

Affordable and clean energy (SDG 7) is a prerequisite for achieving several other SDGs in India such as education, health, gender equality, and poverty alleviation. Embracing a sustainable energy future, India is the only G-20 country to meet the 2 °C global warming targets. By 2030, India's total primary energy demand is going to increase by 63%. In addition, India's contribution to global energy-related CO_2 emission is going to increase from 6.7% to 10.6 %. Therefore, India needs to focus on expanding her capacity in the renewable energy industry as well as in indigenous technological development. The chapter suggests some measures to increase power supply in the country. These include Faster Adoption and Manufacturing of (Hybrid and) Electric Vehicles in India (FAME) I and II, PLI Scheme for National Program on Advanced Chemistry Cell, Battery Storage, Battery Swapping Policy, etc. There is also an emphasis on ethanol blending programme which will reduce the carbon footprint while boosting the income of our farmers. India has also identified the need for expanding nuclear energy which currently contributes only 3.11% of the total energy generation in India. In sum, government of India is taking initiatives to achieve energy security through policy reforms, international collaborations, public-private partnerships, and research and innovations in the clean energy sector.

India has a total forest cover of 71.38 million hectares which is 22% of the total geographical area of the country (Narain Down to Earth 2022). Although there has been a negligible increase of 0.16 million hectares (2019–2020), it is still below the standard requirement of 33% as envisioned in the National Forest Policy, 1988. According to the India State of Forest Report (ISFR) 2021, the increase in India's forest cover inside recorded forest area (i.e. forest land under control of the state government's forest department) between 2019 and 2021 has been insignificant, while the growth outside is 0.76% driven primarily by open forest (Narain 2022). The United Nations recognizes the importance of sustainable management of natural resources such as forests, mountains, and drylands as these resources play a major role in a country's socio-economic development. Along this line, Ankita Swetaparna's article on forest conservation in Odisha examines various schemes developed by the Indian government to create new forests land and to prevent the loss of existing forest cover in the country. At the national level, such schemes include National Afforestation Program (NAP), Green India Mission (GIM), Integrated Development of Wildlife Habitats (IDWH), and Intensification of forest Management Scheme (IFMS). In spite of these government's efforts, it must be recognized that the loss of forest cover is uneven throughout the country. For example, Northeast India lost 79% of its tree cover in 2020, recording the biggest dip in the country (Roy 2021). Although there are such cases, forest cover at the sub-national level can increase due to plantation and participatory conservation initiatives taken by the state government. An example can be drawn from Odisha where afforestation activities and conservation programmes such as Joint Forest Management, plantation of medicinal plants, and Green Mahanadi Mission have contributed to an increase of forest cover in the recent past. Overall, the chapter demonstrates the significance of policy-making at the national and sub-national levels for sustainable management of forest and ecosystems.

17.3 SDGs: Future Implications and Policy Recommendation

The chapters in this book illustrate that the Sustainable Development Goals (SDGs) are not a mere set of directives that should only be used by policy-makers to formulate policies at government level. India's success in achieving the SDGs is only possible if we, the people, take up the collective duty to live a life that is in harmony with our planet and does not harm it. Pro-planet activities are not a complicated set of practices. Through small changes in lifestyles, using biodegradable materials, sun-drying clothes, conserving electricity and water, banning single use plastics, etc. can help us preserve and protect the environment. Besides small changes at the individual level, conscious government's efforts at national and sub-national scales are required to fulfill the gaps between specific SDG targets and outcomes. For instance, careful planning is needed in case of water resource management to reduce the gap between supply and demand especially in the urban areas. In other words, the proper management of hydrological cycle and its components such as surface water, groundwater, storm water, and wastewater is a necessary step to bridge the gap between demand and supply. In the recent past, several integrated approaches have emerged for sustainable water management such as Integrated Water Resource Management (IRWM), Integrated Urban Water Management (IUWM), and Water Sensitive Urban Design (WSUD). Among these, WSUD is believed to be crucial for future urban water management as it enables regenerative water services and reduces the impact of hydrological climatic events on urban environments (Brown et al. 2008). WSUD can play an important role in sustainable water management in the coming decade as it conceptualizes cities as water supply catchments and important sites for provision of ecosystem services (Niekerk et al. 2018). The Smart Cities Mission and AMRUT (Atal Mission for Rejuvenation and Urban Transformation) aim to address urban challenges and promote sustainable urban development. Existing urban infrastructure in many cities is often outdated and inadequate. Upgrading or retrofitting infrastructure to meet smart city standards can be a complex and costly task. Issues such as inadequate public transport, insufficient waste management, and water supply systems need to be addressed. Many cities face challenges in securing adequate funding for the development and maintenance of smart infrastructure and services. Limited budgets and competing priorities can hinder the timely implementation of projects. Smart city initiatives involve multiple stakeholders and agencies, including municipal bodies, local government departments, and private partners. Ensuring effective coordination and collaboration among these entities can be challenging. Integrated planning and execution are crucial for the success of smart city projects.

As a country of astounding 1.4 billion population, India needs to focus on eradicating all forms of poverty and reduce its poverty rate from 21.92% to 10.96% by 2030 (SDG India Index and Dashboard 2020–21). Although India has made considerable progress in eradicating poverty in the past two decades, a more holistic anti-poverty approach is needed for permanent alleviation of poverty. This permanent poverty eradication requires paying attention to employment, equality, and empowerment (Bhukta 2020). As per NITI Aayog, poverty eradication should focus on two aspects: sustained rapid growth leading to employment generation and successful implementation of anti-poverty schemes (NITI Aayog 2016). Reduction of inequality is crucial for a country where significant gap exists between have and have nots. It should be noted that in the past, Indian government did not pay much attention towards reducing inequality due to several misbeliefs such as inequality cannot be avoided with rapid economic growth as well as inequality in India is relatively lower than other developing countries. The government of India should address this issue and promote inclusive growth. Additionally, empowerment of the poor is a must for the removal of poverty, and Indian government should focus on bringing poor and marginalized groups into policy-making and implementation of various poverty eradication programmes (Bhukta 2020). For instance, sustainable development requires a balanced and collective approach to eliminate all forms of poverty, inequality, and injustice and ensure the rights of indigenous people, local communities, migrants, women, children, persons with disabilities, and people living in vulnerable conditions. The success of the SDGs should be assessed in the context of complex and challenging global geopolitical situations and its impacts on the climate, energy, food, and economic situations. Although India is the fifth largest economy in the world in terms of GDP, the economic growth will not be sustainable if India does not focus on combatting climate change and its impacts (SDG-13). In addition, climate change action must be integrated with other SDGs including reduction of poverty and hunger, education, gender equality, health, clean water, and energy.

Currently, India is the third largest contributor of global CO_2 emissions after China and the United States. Though India is taking a centre stage in climate diplomacy, it is fast emerging as a polluter in terms of greenhouse gas emissions. India submitted its intended nationally determined contribution (NDC) to the United Nations Framework Convention on Climate Change (UNFCCC) on 2nd October, 2015. India's NDC is ambitious and demonstrates her commitment to disengage economic growth from greenhouse gas emissions. The government of India updated NDC in 2022. This updated NDC will be implemented from 2021 to 2030 through various programmes and schemes. These schemes will focus on adaptation and mitigation across many sectors like water, agriculture, forest, energy, housing, and waste management. The concrete goals of updated NDC include the following:

Reduction of Emissions Intensity of its GDP by 45% by 2030, from 2005 level.
Achieving about 50% cumulative electric power installed capacity from non-fossil fuel-based energy resources by 2030.
Creating an additional carbon sink of 2.5–3 billion tonnes of CO_2 through afforestation by 2030.
The updated framework intends to create and promote green jobs and low emission products such as electric vehicles, super-efficient appliances, and renewable fuels like green hydrogen (Govt. of India, NDC 2022).

India's speech in COP27 was a call to foster strong international cooperation through climate action and solution-oriented coalitions. At the heart of India's vision of a safe planet "LIFE" or "Lifestyle for Environment", that was introduced by Prime Minister Modi during COP26 on 1st November, 2021. LIFE was launched as an international mass movement to promote "mindful and deliberate utilization, instead of mindless and destructive consumption" in order to conserve environment. LIFE inspires citizens to contribute towards environmental conservation through simple activities in their everyday life and live in harmony with the natural environment (Govt. of India, NDC 2022).

The success of SDGs in India will depend on harnessing the power of partnerships. At present, NITI Aayog is trying to develop three categories of partnerships:

Leverage/exchange: These types of partnerships are based on mutual benefit. For example, NITI Aayog partnered with Japan International Cooperation Agency (JICA) whereby JICA is forwarding financial support and NITI Aayog is providing an ecosystem in solving development challenges.
Combine/integrate: This category involves multiple parties collaborating to achieve a common goal. In case of India's voluntary national review (VNR) 2022 preparation process, the NITI Aayog has taken the lead and initiated the process; United Nations (India) is providing the financial and technical support; and CSO umbrella organizations are mobilizing and documenting various population groups across the country.
Transform: Multiple actors collaborate in this type of partnership to address complex challenges. The SDG localization efforts undertaken by NITI Aayog are an example of this type of partnership (SDG India Index and Dashboard 2020–21).

One of the reasons of building collaborative partnership is that government alone cannot achieve all the SDGs. The government of India has recognized the role of private sector and civil society to achieve these 17 SDGs. It has also identified that building partnerships with civil society organizations (CSOs) are crucial in the localization of SDGs as it would provide required contextualization of SDGs at the sub-national and local scales (SDG India Index and Dashboard 2020–21). The role of CSOs is also going to be vital in providing a voice to the poor and marginalized communities in India as well as in monitoring the progress of SDGs on the ground. The future success of achieving SDGs by 2030 would therefore largely depend on building and nurturing strong partnerships with multiple stakeholders including private sector, CSOs, academia, and international development organizations.

References

Bhukta A (2020) No poverty: how much, how far. In: Hazra S, Bhukta A (eds) Sustainable development goals: an Indian perspective. Springer, Cham, pp 1–11

Brown R, Keath N, Wong T (2008) Transitioning to water sensitive cities: historical, current and future transition states. In: 11th international conference on urban drainage, Edinburgh, Scotland

Govt. of India, NDC. 2022. India's Updated First Nationally Determined Contribution Under Paris Agreement (2021-2030). Retrieved December 05, 2023, from https://unfccc.int/sites/default/files/NDC/2022-08/India%20Updated%20First%20Nationally%20Determined%20Contrib.pdf

Green Government Initiatives (GGI). Goal 6: clean water and sanitation. Retrieved January 25, 2023, from https://ggiindia.in/goal-6-clean-water-and-sanitation/

Kaur B (2018, October 15) Namami Gange: 5 reasons why Ganga will not be clean by 2020. Down to Earth. Retrieved January 25, 2023, from https://www.downtoearth.org.in/coverage/pollution/namami-gange-5-reasons-why-ganga-will-not-be-clean-by-2020-61891

Ministry of Environment, Forest and Climate Change, Govt. of India. India delivers National Statement at COP 27. Retrieved January 26, 2023, from https://pib.gov.in/PressReleasePage.aspx?PRID=1876119

Mint (2010) India city population to double by 2030: report. Retrieved January 26, 2023, from https://www.livemint.com/Politics/RbumlfgLl93Swg9oB3RRcO/India-city-population-to-double-by-2030-report.html

Narain S (2022, January 13) Missing: One-third of India's recorded forests. Down to Earth. https://www.downtoearth.org.in/news/environment/missing-one-third-of-india-s-recorded-forests-81592

Niekerk VW, Duncker L, Maditse K, Davis C, Pieterse A (2018) Water sensitive urban planning as adaptation strategy. Water sensitive urban planning 54th ISOCARP Congress 2018. Retrieved January 25, 2023, from https://isocarp.org/app/uploads/2021/08/ISOCARP_2018_VanNiekerk_42.pdf

NITI Aayog (2016) Eliminating poverty: creating jobs and strengthening social programs. Government of India. Retrieved January 26, 2023, from http://www.indiaenvironmentportal.org.in/content/427070/eliminating-poverty-creating-jobs-and-strengthening-social-programs/

Roy P (2021, October 2021) North East India saw largest share of the country's tree cover loss: study. Retrieved January 26, 2023, from https://www.downtoearth.org.in/factsheet/north-east-india-saw-largest-share-of-the-country-s-tree-cover-loss-study-79653

Rubin BF, Kapur-Gomes S (2020) India spent $30 billion to fix its broken sanitation. It ended up with more problems. CNET. Retrieved January 26, 2023, from https://www.cnet.com/culture/india-spent-30-billion-to-fix-its-broken-sanitation-it-ended-up-with-more-problems/

SDG India Index and Dashboard 2020–21. NITI Aayog. Retrieved January 26, 2023, from https://sdgindiaindex.niti.gov.in/#/

Priyanka Ghosh is a Sr. Assistant Professor of Geography at VIT-AP University, Amaravati, Andhra Pradesh, India. Her areas of interest are political ecology, biodiversity conservation, protected area management, human-animal conflicts, traditional ecological knowledge, tourism, and sustainable development. Her research on the Indian Sundarbans has been published in leading international journals such as *GeoJournal*, *Geographical Review*, and *Environmental Management*. She has also published several chapters on environmental conservation, water resources, and tourism governance in peer-reviewed international books.

Tania Chakravarty completed her PhD in History from the School of Oriental and African Studies (SOAS), University of London. She did her BA, MA, and MPhil from Department of History, Jadavpur University, Kolkata. She is an Assistant Professor at Vivekananda College (Affiliated to University of Calcutta), Kolkata. Her areas of expertise include social and environmental history of modern India, urban history of late nineteenth and early twentieth century India, and history of constitutional reforms in early twentieth century India.

MIX
Papier aus verantwortungsvollen Quellen
Paper from responsible sources
FSC® C105338

If you have any concerns about our products,
you can contact us on
ProductSafety@springernature.com

In case Publisher is established outside the EU,
the EU authorized representative is:
Springer Nature Customer Service Center GmbH
Europaplatz 3, 69115 Heidelberg, Germany

Printed by Libri Plureos GmbH
in Hamburg, Germany